D1788882

1979-90

LOWELL W. FOSTER
President and Director
Technology Concepts and Engineering, International
Minneapolis, Minnesota

GEO-METRICS II
The Application of Geometric Tolerancing Techniques
(using customary inch system)

ADDISON-WESLEY PUBLISHING COMPANY

Reading, Massachusetts

Menlo Park, California · London · Amsterdam · Don Mills, Ontario · Sydney

This book is based on the author's earlier volumes, *A Treatise on Geometric and Positional Dimensioning and Tolerancing*, originally published in 1963 by Honeywell Inc., *A Treatise on Geometric Dimensioning and Tolerancing*, originally published in 1966 by Honeywell Inc., and *Geometric Dimensioning and Tolerancing—A Working Guide*, published in 1970 by Addison-Wesley Publishing Company, Inc., and *Geo-Metrics: The Metric Application of Geometric Tolerancing*, published in 1974 by Addison-Wesley Publishing Company, Inc. These publications have been widely used by many companies, the military, and schools. The original works have been expanded, updated, and revised to present the subject in terms of the latest state of the art.

The figures on pages 23, 42, 50–51, 52 (top), 54–57, 63–65, 69, 73, 77, 83, 85, 87, 91, 231, and 269 were extracted from the American national standard on drafting practices, "Dimensioning and Tolerancing for Engineering Drawings," American National Standard ANSI Y14.5-1973 with the permission of the publisher, The American Society of Mechanical Engineers, United Engineering Center, 345 East 47th Street, New York, N. Y. 10017.

The drawings on pages 275 and 276 appear through the courtesy of Sandia Laboratories and the U. S. Atomic Energy Commission.

Library of Congress Cataloging in Publication Data

Foster, Lowell W
    Geo-metrics II.

    Includes index.
    1. Drawing-room practice. 2. Tolerance (Engineering)
I. Title.
T352.F68   604'.2   78-67959
ISBN 0-201-01936-1

ISBN 0-201-01936-1
CDEFGHIJKL-AL-89876543210

PREFACE

This text has been prepared to promote worldwide use of geometric dimensioning and tolerancing and to up-grade knowledge on the subject. Its title, "GEO–METRICS II," attempts to represent and encourage the giant stride being taken by United States industry in its transition toward greater use of national and international standards on this subject. This text also aspires to be of value to all users worldwide as a catalyst for better understanding and to improve the state of the art.

Another major objective of this text is to promote and encourage use of the principles involved as an effective engineering, production, and quality control language or tool which can provide economic and technical advantages.

To further the foregoing objectives, this text is, therefore, presented in a format suitable for use as a teaching mechanism, both for the classroom and in-industry situation, as well as for providing a daily ready reference for engineering or production work.

This text is dedicated to the promoting of standardization of engineering drawing techniques in support of the American National Standard ANSI Y14.5–1973 "Dimensioning and Tolerancing." It is also dedicated to furthering worldwide efforts through the International Standards Organization and its standards and programs (e.g., ISO/TCIO/ SC5 "Dimensioning and Tolerancing.")

The data contained in this book must be considered advisory and are to be used at the discretion of the reader.

The author wishes to express his gratitude to his many colleagues and professional friends in this field of endeavor around the world who also have made contributions to the state-of-the-art. He also wishes to give due credit to the reference documents listed and to those responsible for them.

Finally, the author wishes to express particular gratitude to his wife, Marion and number two son John for their patience and understanding in enduring the often untimely sacrifices of family together time. Without their encouragement and support, this work could not have been completed.

Minneapolis, Minnesota   L. W. F.
February 1979

ACKNOWLEDGMENTS

Credit is gratefully given to the following persons and to the documents used as reference material:

1. American National Standard ANSI Y14.5-1973. "Dimensioning and Tolerancing for Engineering Drawings," American Society of Mechanical Engineers, New York City, and American National Standards Institute, New York City.

2. American National Standard ANSI Y14.5-1966 "Dimensioning and Tolerancing for Engineering Drawings," American Society of Mechanical Engineers, New York City, and American National Standards Institute, New York City.

3. International Standards Organization Standards, ISO/R1101−1969, *Tolerances of Form and of Position, Part I, Generalities, Symbols, Indications on Drawings;* ISO/DIS2691, 1972, *Technical Drawings−Tolerances of Form and Position, Part II,* ISO/R1660−1971, *Technical Drawings−Tolerances of Form and Position, Part III, Dimensioning and Tolerances of Profiles.*

4. SAE Aerospace−Automotive Drawing Standards, Section A 6, 7, and 8, September 1963.

5. Military Standard 8C, 16 October 1963.

6. "True Position Dimensioning"−Rev. 1962," Bibeau & Sweet, Scintilla Division, Bendix Aviation Corporation, Sidney, New York.

7. "Process Standard 9900011" Sandia Corporation, Albuquerque, New Mexico.

8. "Concepts of the True Position Dimensioning System" Sandia Corporation, Albuquerque, New Mexico.

9. Ordnance Corps, "Standard for Dimensioning and Tolerancing ORD 30-1-7."

10. "Synopsis of MIL-STD-8B" J. V. LaPointe, Honeywell Inc., Minneapolis, Minnesota.

11. "Fundamentals of Position Tolerance" John V. Liggett, Massey Ferguson Inc., Detroit, Michigan.

12. Military Standard 8B, 16 November 1959.

13. "Clarification of Problem Areas of MIL-STD-8B Dimensioning and Tolerancing−code ident 10001 OD15292, Department of the Navy, Bureau of Naval Weapons, 19 June 1961."

14. "A Treatise on Geometric and Positional Dimensioning and Tolerancing," Lowell W. Foster, Honeywell Inc., Minneapolis, Minnesota. First printing December 1963, revised edition November 1964.

15. "A Treatise on Geometric Dimensioning and Tolerancing," Lowell W. Foster, Honeywell Inc., Minneapolis, Minnesota. First printing January 1966; revised edition first printing September 1966, second printing March 1967, third printing July 1967, fourth printing February 1968, fifth printing July 1968.

16. "Geometric Dimensioning and Tolerancing," Lowell W. Foster, Honeywell Inc., Minneapolis, Minnesota. Society of Automotive Engineers, New York City. SAE Paper 680488.

17. "True Position Tolerancing−Before and After the Fact," Lowell W. Foster, Honeywell Inc., Minneapolis, Minnesota. American Society of Tool and Manufacturing Engineers, Dearborn, Michigan. ASTME Paper I068-409.

18. *Geo-Metrics: The Metric Application of Geometric Tolerancing,* Lowell W. Foster, Minneapolis, Minnesota. Published by Addison-Wesley Publishing Company, Inc., 1974.

CONTENTS

Introduction 1
Glossary 3
Why use geometric dimensioning and tolerancing? 8
What is geometric dimensioning and tolerancing? 8
When should geometric dimensioning and tolerancing be used? . . . 8
Geometric characteristics and symbols 9
Using symbols 10
Maximum material condition principle 11
Regardless of feature size 12
Basic and datum 13
Symbolic method of stating a basic theoretical exact value—recommended 14
Placement of the datum identification symbol 15
Feature and feature control symbol 16
Combined feature control symbol and datum identifying symbol . . . 17
Geometric characteristics—form, runout, and locational
    tolerance—other symbols and terms 20
Geometric characteristics, symbols and terms 21
General rules 22
Virtual condition 28

TOLERANCES OF FORM AND RUNOUT 29

Tolerance of form 30
Tolerance of form—individual features—no datum 31
Flatness 32
Straightness 33
Straightness tolerance applied to flat surface 35
Roundness 40
Cylindricity 43
Evaluation of roundness and cylindricity 45
Datums 48
Tolerances of form—related features using datums 49
Perpendicularity 50
Angularity 58
Parallelism 62
Tolerances of form—profile tolerancing 66
Profile of a surface 69
Profile of a line 72
Profile—coplanar surfaces 74
Runout tolerances—related features using datums 76
Circular runout 80
Circular runout and total runout 84
Total runout 92

TOLERANCES OF LOCATION 97

Position theory 98
Mating parts—floating fastener 103

CONTENTS

Mating parts—fixed fastener 106
Relation to implied datum surfaces 110
Relation to specified datum surfaces 112
Combination coordinate and position tolerancing 114
Composite position tolerancing 116
Projected tolerance zone 132

POSITION OF NONCYLINDRICAL FEATURES 139

POSITION OF COAXIAL FEATURES 155

POSITION EXTENDED PRINCIPLES 171

DATUMS 201
Definitions 204
Establishing datums 210
Datum application 220
Extended datum principles 252

CONCENTRICITY 257

SYMMETRY 267
Conclusion 272

APPENDIX 273
Index 295

INTRODUCTION

Geometric dimensioning and tolerancing is a means of specifying engineering design and drawing requirements with respect to actual "function" and "relationship" of part features. Furthermore, it is a technique which, properly applied, ensures the most economical and effective production of these features. Thus geometric dimensioning and tolerancing can be considered both an engineering design drawing language and a functional production and inspection technique. Uniform understanding and interpretation among design, production, and inspection groups are the major objective of the system. This text discusses the subject step by step, focusing on practical application. Before presenting these details, however, we wish to provide the reader with a brief overview of the basis and status of geometric dimensioning and tolerancing.

The authoritative document governing the use of geometric dimensioning and tolerancing in the United States is ANSI Y14.5, "Dimensioning and Tolerancing for Engineering Drawings." ANSI Y14.5 (current issue 1973) evolved out of a consolidation of earlier standards, ASA Y14.5-1957, SAE Automotive Aerospace Drawing Standards (Sections A6, 7, and 8–September 1963), and MIL–STD–8C, October 1963. This consolidation was accomplished over several years by committee action representing military, industrial, and educational interests. The work of the committee had three prime objectives:

1) to provide a single standard for practices in the United States,

2) to update existing practices in keeping with technological advances, and extend the principles into new areas of application, and

3) to establish a single basis and "voice" for the United States in the interest of international trade, in keeping with the United States' desire to be more active, gain greater influence, and pursue a more extensive exchange of ideas with other nations in the area of international standards development.

The historical evolution of geometric dimensioning and tolerancing in the United States is an interesting story which, however, is not discussed in this text. It suffices to say that the early introduction of functional gaging, giving rise to the possibility of new techniques, along with the growing need for more specifically and economically stated engineering design requirements, has caused its growth. Advancing product sophistication and complexity, rapid industrial expansion, diversification, etc., have created an environment in which more exacting engineering drawing communication is not only desirable but mandatory for comptetitive and effective operation.

Updated and expanded practices have been initiated in the present Y14.5 standard. Further expansion will no doubt occur as growth in the area continues. In the process of extending into new areas, this expansion is confronted by the challenge of ensuring progress without upsetting stability. Rapid advances in this subject, although desirable, must be tempered by the ability to make the transition with no loss of continuity or understanding.

United States coordination and compatibility with international dimensioning and tolerancing practices have been extended significantly in the current Y14.5-1973 standard.

The symbology and influence from ISO (International Standards Organization) and ABCA (America, Britain, Canada, Australia) documents, activities, and committees have found their way into the current Y14.5 standard. This influence continues in the on-going development and use of geometric tolerancing in the United States.

INTRODUCTION

Many of the United States industrial concerns and the military have oversea's affiliations. Thus, the increasing need for understanding and for more uniform practices throughout the world is evident.

This text presents the subject of dimensioning and tolerancing in order of complexity of the details, and attempts to clarify and promote the use of Y14.5. It also emphasizes the importance of the on-going effort to expand the principles and to more closely relate to international practices.

GLOSSARY

Actual Size – An actual size is the measured size of the feature.

Angularity – Angularity is the condition of a surface, axis, or center plane which is at a specified angle (other than 90°) from a datum plane or axis. Symbol: $\angle$.

Basic Dimension – A dimension specified on a drawing as BASIC (or abbreviated BSC) is a theoretical value used to describe the exact size, shape, or location of a feature. It is used as the basis from which permissible variations are established by tolerances on other dimensions or notes. A basic dimension is symbolized by boxing it: 1.270 .

Basic Size – The basic size is that size from which limits of size are derived by the application of allowances and tolerances.

Bilateral Tolerancing – A bilateral tolerance is a tolerance in which variation is permitted in both directions from the specified dimension, e.g., 38.1 ± 0.13.

Center Plane – Center plane is the middle or median plane of a feature.

Circular Runout – Circular runout is the composite control of circular elements of a surface independently at any circular measuring position as the part is rotated through 360°. Symbol: $\nearrow$.

Circularity – See Roundness.

Clearance Fit – A clearance fit is one having limits of size so prescribed that a clearance always results when mating parts are assembled.

Concentricity – Concentricity is a condition in which two or more features (cylinders, cones, spheres, hexagons, etc.) in any combination have a common axis. Symbol: $\circledcirc$ (or, formerly, $\odot$).

Contour Tolerancing — See Profile of a Line or Surface.

Coaxiality — Coaxiality of features exists when two or more features have coincident axes, i.e., a feature axis and a datum feature axis.

Cylindricity — Cylindricity is a condition of a surface of revolution in which all points of the surface are equidistant from a common axis. Symbol: $\cancel{O}$.

Datum — Datums are points, lines, planes, cylinders, axes, etc., assumed to be exact for purposes of computation or reference, as established from actual features, and from which the location or geometric relationship of other features of a part may be established.

Datum Axis — The datum axis is the theoretically exact center line of the datum cylinder as established by the extremities or contacting points of the actual datum feature cylindrical surface, or the axis formed at the intersection of two datum planes.

Datum Surface — A datum surface or feature (hole, slot, diameter, etc.) refers to the actual part surface or feature coincidental with, relative to, and/or used to establish a datum.

Datum Identification Symbol — The datum identification symbol contains the datum reference letter in a drawn rectangular box, e.g. $\boxed{-A-}$.

Datum Feature — A datum feature is a feature used to establish a datum.

3

GLOSSARY

Datum Line — A datum line is that which has length but no breadth or depth such as the intersection line of two planes, center line or axis of holes or cylinders, reference line for functional, tooling, or gaging purposes.

Datum Plane — A datum plane is a theoretically exact plane established by the extremities or contacting points of the actual feature surface with a reference plane (surface plate or other checking device).

Datum Point — A datum point is that which has position but no extent such as the apex of a pyramid or cone, center point of a sphere, or reference point on a surface for functional, tooling, or gaging purposes.

Datum Reference — A datum reference is a datum feature.

Datum Reference Framework (Frame) (System) — A datum reference framework is a system of three mutually perpendicular datum planes or axes established from datum surfaces or features as a basis for dimensions for manufacture and measurement. It provides complete orientation for the features involved.

Datum Target — A datum target is a specific datum point, line, or area (identified on the drawing with a datum target symbol ⊕) used to establish datum points, lines, planes, or areas for special function, or manufacturing and inspection repeatability.

Dimension — A dimension is a numerical value expressed in appropriate units of measure and indicated on a drawing.

Feature — Features are specific component portions of a part and may include one or more surfaces such as holes, screw threads, profiles, faces, or slots. Features may be "individual" or "interrelated."

Feature Control Symbol — The feature control symbol is a rectangular box containing the geometric characteristic symbol and the form, runout, or location tolerance. If necessary, datum references and modifiers applicable to the feature or the datums are also contained in the box, e.g. ⟋ .002 A .

Fit — Fit is the general term used to signify the range of tightness or looseness which may result from the application of a specific combination of allowances and tolerance in the design of mating part features. Fits are of four general types: clearance, interference, transition, and line.

Flatness — Flatness is the condition of a surface having all elements in one plane. Symbol: ⟋⟍.

Form Tolerance — A form tolerance states how far an actual surface or feature is permitted to vary from the desired form implied by the drawing. Expressions of these tolerances refer to flatness, straightness, parallelism, perpendicularity, angularity, roundness, cylindricity, profile of a surface, and profile of a line.

Full Indicator Movement (FIM) (FIR) (TIR) — Full indicator movement is the total movement observed with the dial indicator (sometimes referred to as a "clock" in international references) in contact with the part feature surface during one full revolution of the part about its datum axis. Full indicator movement (FIM) is the term used internationally. United States terms FIR and TIR, used in the past, have the same meaning as FIM.

Full indicator movement also refers to the total indicator movement observed while in traverse over a fixed noncircular shape.

Full Indicator Reading (FIR) (TIR) (FIM) — full indicator reading is the total indicator reading observed with the dial indicator in contact with the part feature surface during one full revolution of the part about its datum axis. (Use of the international term FIM is recommended.)

Full indicator reading also refers to the full indicator reading observed while in traverse over a fixed noncircular shape.

Geometric Characteristics — Geometric characteristics refer to the basic elements or building blocks which form the language of geometric dimensioning and tolerancing. Generally, the term refers to all the symbols used in form, runout, and location tolerancing.

Implied Datum — An implied datum is an unspecified datum whose influence on the application is implied by the dimensional arrangement on the drawing; e.g., the primary dimensions are tied to an edge surface; this edge is implied as a datum surface and plane.

Interference Fit — An interference fit is one having limits of size so prescribed that an interference always results when mating parts are assembled.

Interrelated Datum System — An interrelated datum system is one which has one or more common datums with another datum system.

Least Material Condition (LMC) — This term implies that condition of a part feature wherein it contains the least (minimum) amount of material, e.g. largest hole size and smallest shaft size. It is opposite to maximum material condition (MMC). Symbol sometimes used: Ⓛ (symbol not per ANSI Y14.5-1973)

Limits of Size — The limits of size are the applicable maximum and minimum sizes of a feature.

Limit Dimensions (Tolerancing) — In limit dimensioning only the maximum and minimum dimensions are specified. When used with dimension lines, the maximum value is placed above the minimum value, e.g., $.\frac{300}{295}$. When used with leader or note on a single line, the minimum limit is placed first, e.g., .295-.300.

Line Fit — A line fit is one having limits of size so prescribed that surface contact or clearance may result when mating parts are assembled.

Location Tolerance — A location tolerance states how far an actual feature may vary from the perfect location implied by the drawing as related to datums or other features. Expressions of these tolerances refer to the category of geometric characteristics containing position, concentricity, and symmetry.

Maximum Material Condition (MMC) — Maximum material condition is that condition of a part feature wherein it contains the maximum amount of material, e. g. minimum hole size and maximum shaft size. Symbol used: Ⓜ.

Maximum Dimension — A maximum dimension represents the acceptable upper limit. The lower limit may be considered any value less than the maximum specified.

Median Plane — Median plane is the middle or center plane of a feature.

GLOSSARY

Minimum Dimension — A minimum dimension represents the acceptable lower limit. The upper limit may be considered any value greater than the minimum specified.

Minimum Material Condition — See Least Material Condition

Modifier — A modifier is the term sometimes used to describe the application of the "maximum material condition" or "regardless of feature size" principles. The modifiers are maximum material condition (MMC), symbol Ⓜ, and regardless of feature size (RFS), symbol Ⓢ.

Nominal Size — The nominal size is the stated designation which is used for the purpose of general identification, e.g., 1.400, .060, etc.

Normality — See Perpendicularity.

Parallelepiped — Shape of tolerance zone. The term is used where TOTAL WIDTH is required and to describe geometrically a square or rectangular prism, or a solid with six faces, each of which is a parallelogram.

Parallelism — Parallelism is the condition of a surface, line, or axis which is equidistant at all points from a datum plane or axis. Symbol: // (or, formerly ||).

Perpendicularity — Perpendicularity is the condition of a surface, axis, or line which is 90° from a datum plane or a datum axis. Symbol: ⊥ .

Position Tolerance — A position tolerance (formerly called true position tolerance) defines a zone within which the axis or center plane of a feature is permitted to vary from true (theoretically exact) position. Symbol: ⊕ .

Profile of Line — Profile of line is the condition permitting a uniform amount of profile variation, either unilaterally or bilaterally, along a *line* element of a feature. Symbol: ⌒ .

Profile of Surface — Profile of a surface is the condition permitting a uniform amount of profile variation, either unilaterally or bilaterally, on a *surface*. Symbol: ⌓ .

Projected Tolerance Zone — A projected tolerance zone is a tolerance zone applied to a hole in which a pin, stud, screw, or bolt, etc., is to be inserted. It controls the perpendicularity of the hole to the extent of the projection from the hole and as it relates to the mating part clearance. The projected tolerance zone extends *above* the surface of the part to the functional length of the pin, screw, etc., relative to its assembly with the mating part. Symbol: Ⓟ.

Regardless of Feature Size (RFS) — This is the condition where the tolerance of form, runout, or location must be met irrespective of where the feature lies within its size tolerance. Symbol: Ⓢ.

Roundness — Roundness is the condition on a surface of revolution (cylinder, cone, sphere) where all points of the surface intersected by any plane (1) perpendicular to A common axis (cylinder, cone) or (2) passing through a common center (sphere) are equidistant from the center. Symbol: ◯.

Runout — Runout is the composite deviation from the desired form of a part surface of revolution during full rotation (360°) of the part on a datum axis. Symbol: ↗ .

Runout Tolerance — Runout tolerance states how far an actual surface or feature is permitted to deviate from the desired form implied by the drawing during full rotation of the part on a datum axis. There are two types of runout: circular runout and total runout.

Size Tolerance — A size tolerance states how far individual features may vary from the desired size. Size tolerances are specified with either unilateral, bilateral, or limit tolerancing methods.

Specified Datum — A specified datum is a surface or feature identified with a datum identification symbol or note.

Squareness — See Perpendicularity

Straightness — Straightness is a condition where an element of a surface or an axis is a straight line. Symbol: —— .

Symmetry — Symmetry is a condition in which a feature (or features) is symmetrically disposed about the center plane of a datum feature. Symbol: ≡ .

Tolerance — A tolerance is the total amount by which a specific dimension may vary; thus, the tolerance is the difference between limits.

Transition Fit — A transition fit is one having limits of size so prescribed that either a clearance or an interference may result when mating parts are assembled.

True Position — True position is a term used to describe the perfect (exact) location of a point, line, or plane of a feature in relationship with a datum reference or other feature.

Total Indicator Reading (TIR) (FIR) (FIM) — Total indicator reading is the full indicator reading observed with the dial indicator in contact with the part feature surface during one full revolution of the part about its datum axis. Total indicator reading also refers to the total indicator reading observed while in traverse over a fixed noncircular shape. (Use of the international term FIM is recommended.)

Total Runout — Total runout is the simultaneous composite control of all elements of a surface at all circular and profile measuring positions as the part is rotated through 360°. Symbol: ⤢ (with word TOTAL added beneath feature control symbol).

Unilateral Tolerance — A unilateral tolerance is a tolerance in which variation is permitted only in one direction from the specified dimension, e.g., $1.400 \, ^{+.000}_{-.005}$.

Virtual Condition (Size) — Virtual condition (size) of a feature is the collective effect of size, form, and location error that must be considered in determining the fit or clearance between mating parts or features. It is a derived size generated from the profile variations permitted by the specified tolerances. It represents the most extreme condition of assembly at MMC. (The term Virtual Condition is now preferred over Virtual Size.)

WHY USE GEOMETRIC DIMENSIONING AND TOLERANCING?

Why is it that we should be so interested in this subject?

FIRST AND FOREMOST ITS USE SAVES MONEY!

It saves money directly by providing for maximum producibility of the part through maximum production tolerances. It provides "bonus" or extra tolerances in many cases.

It ensures that design dimensional and tolerance requirements, as they relate to actual function, are specifically stated and thus carried out.

It ensures interchangeability of mating parts at assembly.

It provides uniformity and convenience in drawing delineation and interpretation, thereby reducing controversy and guesswork.

Aside from these primary reasons there are others of a more general nature:

The intricacies of today's sophistricated engineering design demand new and better ways of accurately and reliably communicating requirements. Old methods simply no longer suffice.

Diversity of product line and manufacture makes considerably more stringent demands of the completeness, uniformity, and clarity of drawings.

It is increasingly becoming the "spoken word" throughout industry, the military, and, internationally, on engineering drawing documentation. Every engineer or technician involved in originating or reading a drawing should have a working knowledge of this new state of the art.

WHAT IS GEOMETRIC DIMENSIONING AND TOLERANCING?

In particular, it is a means of dimensioning and tolerancing a drawing with respect to the actual function or relationship of part features which can be most economically produced. *Function* and *relationship* are the key words.

In general, it is a system of building blocks for good drawing practice which provides the means of stating necessary dimensional or tolerance requirements on the drawing not otherwise covered by implication or standard interpretation.

WHEN SHOULD GEOMETRIC DIMENSIONING AND TOLERANCING BE USED?

When part features are critical to function or interchangeability;

when functional gaging techniques are desirable;

when datum references are desirable to ensure consistency between manufacturing and gaging operations;

when standard interpretation or tolerance is not already implied.

GEOMETRIC CHARACTERISTICS AND SYMBOLS

The geometric characteristics and symbols that are used as the building blocks for geometric dimensioning and tolerancing are:

▱	FLATNESS
—	STRAIGHTNESS
○	ROUNDNESS (CIRCULARITY)
⌀	CYLINDRICITY
⌒	PROFILE OF A LINE
⌓	PROFILE OF A SURFACE
⊥	PERPENDICULARITY (SQUARENESS)
∠	ANGULARITY
//	PARALLELISM
↗	CIRCULAR RUNOUT
↗	TOTAL RUNOUT *
⊕	POSITION
◎	CONCENTRICITY
≡	SYMMETRY

---

* Word TOTAL must be added beneath feature control symbol. Anticipated new symbol in future standards (ANSI, ISO), ⟲.

USING SYMBOLS

The general use of symbols instead of notes on a drawing provides a number of advantages. The illustrations below incorporate the geometric characteristic symbols with datum and feature control symbols. Some of the advantages of symbols over notes are:

1. The symbol has uniform meaning. A note can be stated inconsistently, with a possibility of misunderstanding.

2. Symbols are compact, quickly drawn, and can be placed on the drawing where the control applies.

   Notes require much more time and space, tend to be scattered on the drawing, often appear as footnotes which separate the note from the feature to which it applies.

3. Symbols are the international language and surmount individual language barriers.

   Notes may require translation if the drawing is used in another country.

4. Symbols can be applied with drafting templates and retain better legibility in various forms of copy reproduction.

5. Geometric tolerancing symbols follow the established precedent of other well known symbol systems, e.g., electrical and electronic, welding, surface texture, etc.

USING SYMBOLS

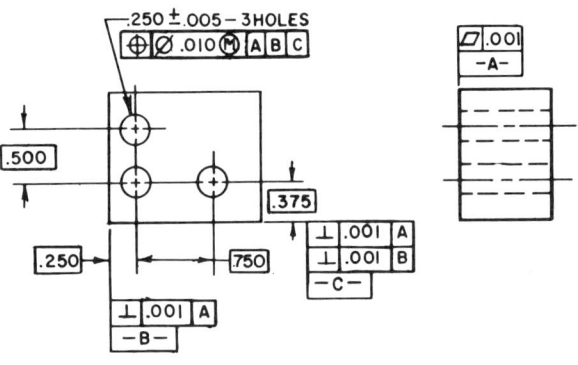

USING NOTES

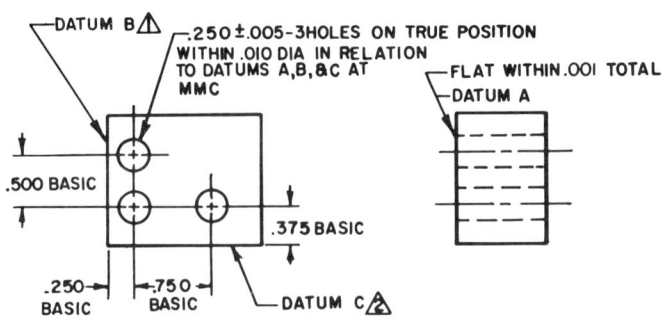

⚠2 SURFACE C PERPENDICULAR TO DATUM A WITHIN .001 & WITH DATUM B WITHIN .001

⚠1 SURFACE B PERPENDICULAR TO DATUM A WITHIN .001

MAXIMUM MATERIAL CONDITION PRINCIPLE
SYMBOL Ⓜ ABBREVIATION (MMC)

One of the fundamental and most important principles of geometric dimensioning and tolerancing is MAXIMUM MATERIAL CONDITION. A thorough understanding of its meaning is therefore essential.

Note in the figure below that the "maximum material condition" size of the .250 ± .005 diameter hole is .245, or its *low* limit size. The hole at its low limit obviously retains more material than if it were at its *high* limit or larger size; thus the term "maximum material condition" defines the *low* limit when it applies to a hole or similar feature.

Note similarly that the .235 diameter pin is at its "maximum material condition" size when it is at its *high* limit of size of .240. In this instance it is more readily seen that more material exists in the pin when it is at its maximum permissible size. However, the same principle exists in both hole and pin MMC situations. Relating mating part features in this manner ensures their functional relationships, and as will be seen later in the text, establishes the criteria for determining necessary form and position tolerances.

The symbol for "maximum material condition," the M enclosed in a circle, and the occasionally used abbreviation MMC are shown above. The symbolic method is to be used with feature control symbols only. The abbreviation MMC may be used with note callouts but not with symbolic representations. We shall discuss later the application of the "maximum material condition" principle and illustrate it with practical examples.

Generally, the use of the "maximum material condition" principle permits greater possible tolerance as part feature sizes vary from their calculated "maximum material condition" limits. It also ensures interchangeability and permits functional gaging

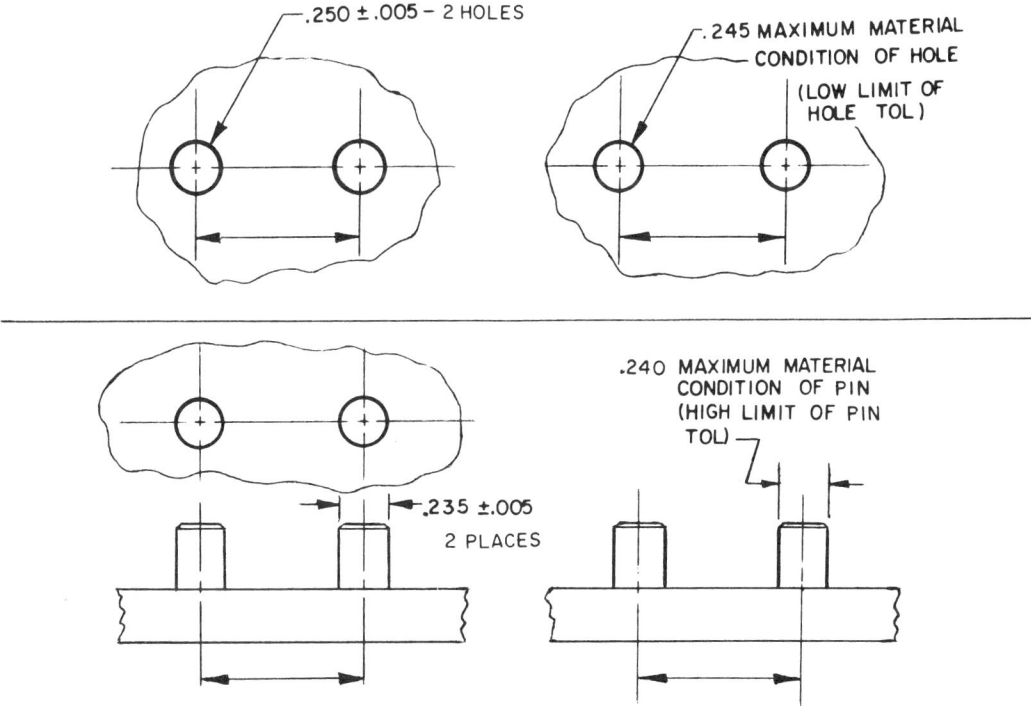

techniques. It is one of the fundamental principles upon which the system of geometric dimensioning and tolerancing is based. Below is the definition of maximum material condition and the usual prerequisites for application. We shall later expand the use of the principle by means of examples.

Definition. That condition of a part feature wherein it contains the maximum amount of material. Example: Minimum hole size and maximum shaft size.

The "maximum material condition" principle is normally valid only when both of the following conditions exist:

1. Two or more features are interrelated with respect to location or form (e.g. a hole and an edge or surface, two holes, etc.). At least *one* of the related features is to be a feature of size.

2. The feature (or features) to which the MMC principle is to apply must be a feature of size (e.g. a hole, slot, pin, etc.) with an axis or centerplane.

"Maximum material condition" might also be considered as a "new" term for an "old" situation, such as the familiar terms "worst condition," "critical size," etc., used in the past for relating mating part features.

Where the maximum material condition principle is not appropriate, the "regardless of feature size" principle may be applied. (See below.)

REGARDLESS OF FEATURE SIZE
SYMBOL Ⓢ ABBREVIATION (RFS)

Definition. The tolerance of form or position of a feature must be met regardless of where the feature lies within its size tolerance.

"Regardless of feature size" is another principle of geometric dimensioning and tolerancing which must be well understood. Unlike maximum material condition, the "regardless of feature size" principle permits *no* additional positional or form tolerance, no matter to which size the related features are produced. It is really the independent form of dimensioning and tolerancing which has always been used prior to the introduction of the MMC principle.

The symbol for "regardless of feature size" and the occasionally used abbreviation RFS is shown above. We shall later clarify the principle by means of examples.

The RFS principle is valid only when applied to features of size (for example, a hole, slot, pin, etc., with an axis or center plane). The size connotation cannot be applied to a feature which does not have "size."

BASIC AND DATUM

The terms BASIC and DATUM are most important. Proper application of the principles implied by these terms greatly contributes to effective geometric dimensioning and tolerancing.

BASIC

Definition. A dimension specified on a drawing as BASIC is a theoretical value used to describe the exact size, shape, or location of a feature. It is used as the basis from which permissible variations are established by tolerances on other dimensions or in notes.

Use of a BASIC dimension, which is a theoretical and exact value, requires also a *tolerance* stating the permissible variation from this exact value (most often relative to a position or form requirement). A BASIC dimension states only *half* the requirement. To complete it, a tolerance must be associated with the features involved in the BASIC dimension.

A BASIC dimension is identified on the drawing by the word BASIC (or the accepted abbreviation BSC) adjacent to, or below, the dimension, or by a general note on the drawing.

EXAMPLE

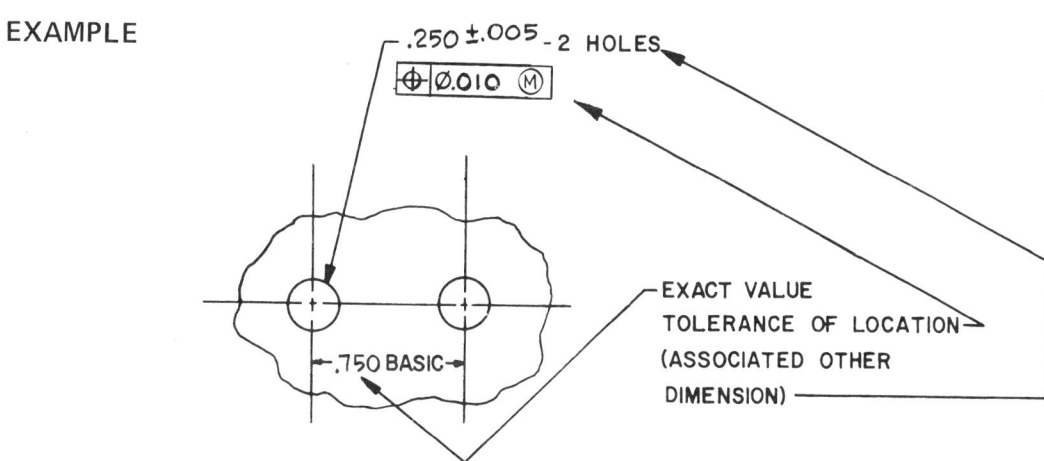

Some companies use a "naked" or untoleranced dimension instead of BASIC. The same meaning as BASIC is invoked by adding a drawing footnote, title block notation, or company standards.

The term TP (for True Position) derived from British standards has also been used in the past. It has the same meaning as BASIC.

BASIC dimensions are also used in applications such as tapers for which tolerances must be derived from associated size dimensions.

The use of BASIC dimensions on datum targets assumes standard tooling or gagemaker's tolerances (see DATUM section for more detail).

13

SYMBOLIC METHOD OF STATING A BASIC
OR THEORETICAL EXACT VALUE — RECOMMENDED

The preferred method of stating an "exact" value replacing BASIC, BSC, TP, etc.,
is the international (ISO) method recommended by ANSI Y14.5. According to this
method, the exact value is enclosed in a frame or box (see example below). The
meaning is the same as that of BASIC.

EXAMPLE

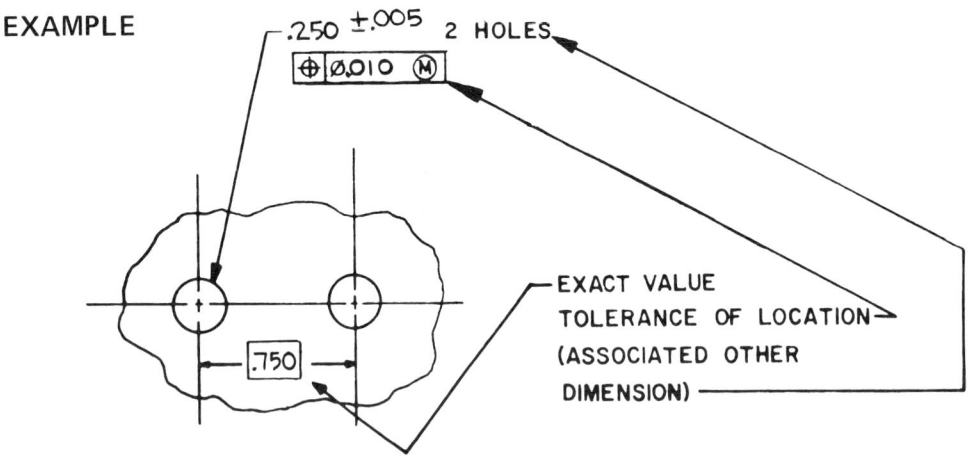

The symbolic method for "exact" values may be used with symbolic *or* notated
geometric tolerancing. Because of the need to standardize U. S. practices and encourage compatibility with international practices, using the symbolic or boxed "exact"
value is strongly recommended.

DATUMS AND DATUM IDENTIFICATION SYMBOL

Definitions. Datums are points, lines, planes, cylinders, etc., which are assumed to be
exact for purposes of computation or reference and from which the location or form
of features of a part may be established. Datums are established by, or relative to,
actual part features or surfaces.

Datum surfaces and datum features are actual part surfaces or features used to
establish datums. They include all the surface or feature inaccuracies.

To identify a feature as a datum, we use the following datum identification symbol:

Each datum requiring identification is assigned a different reference letter.

Note the following conventions:

1. If the datum identification symbol is shown on the *extension line* of a feature, the datum is for that feature only (see Figs. 1 and 2).

2. If the datum identification symbol is shown on a *dimension line* — below or adjacent to a DIMENSION, with a leader, note or feature control symbol — it applies to the entire dimension or feature with which it is associated (see Fig. 3).

PLACEMENT OF THE DATUM IDENTIFICATION SYMBOL

APPLICATION TO PLANE SURFACES (OR CYLINDERS)

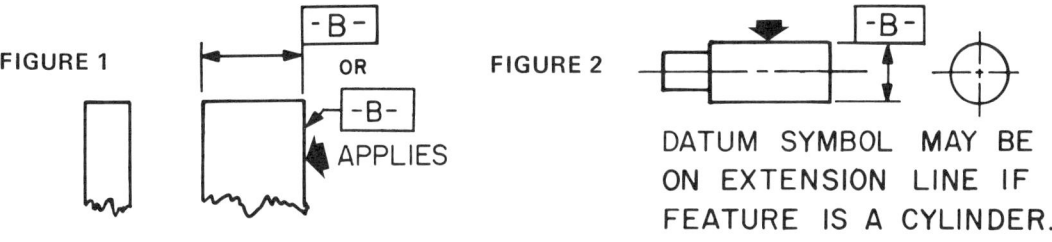

FIGURE 1 · · · OR · · · -B- APPLIES

FIGURE 2 · · · DATUM SYMBOL MAY BE ON EXTENSION LINE IF FEATURE IS A CYLINDER.

APPLICATION TO SIZE FEATURES

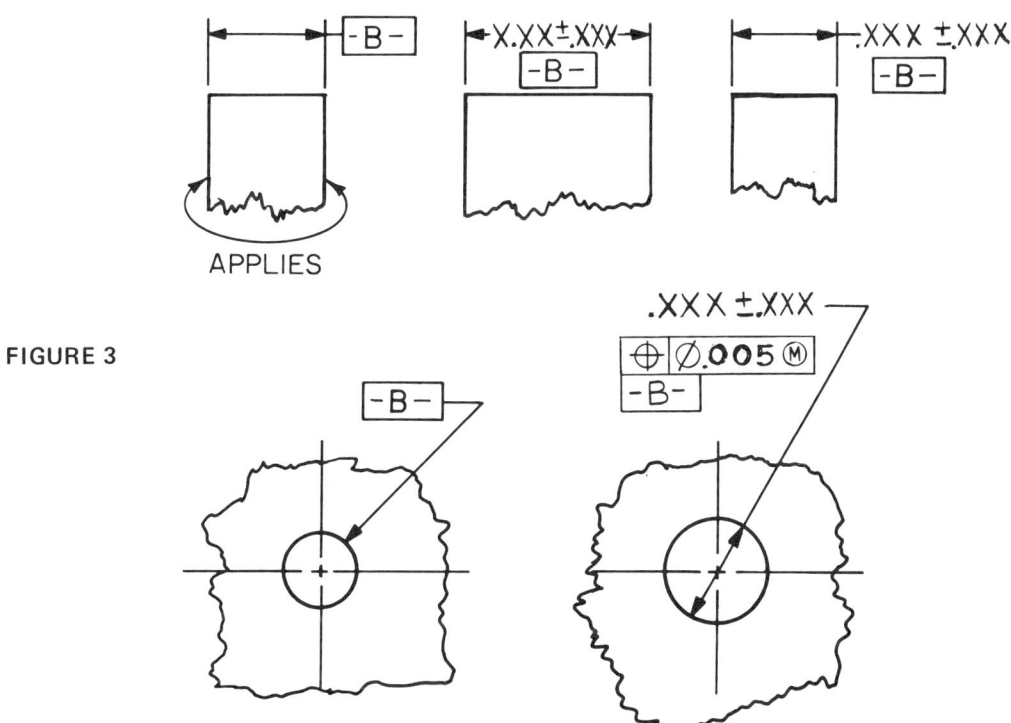

APPLIES

FIGURE 3

FEATURE AND FEATURE CONTROL SYMBOL

FEATURE

Features are specific component portions of a part and may include one or more surfaces such as holes, faces, screw threads, profiles, or slots. Features may be "individual" or "related."

FEATURE CONTROL SYMBOL

The feature control symbol consists of a frame containing the geometric characteristic symbol, datum references, tolerance, and the MMC symbol if applicable.

The following illustration shows a typical feature control symbol.

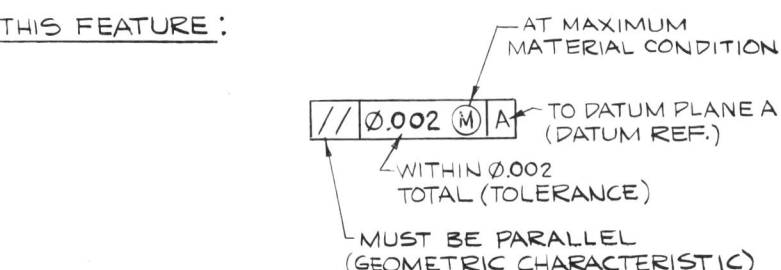

The example below shows this feature control symbol as used on a part drawing.

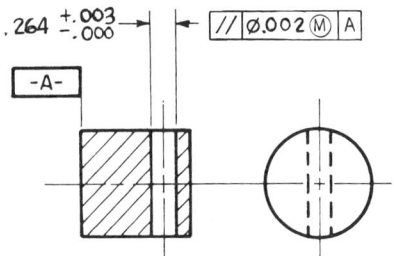

PLACEMENT OF THE FEATURE CONTROL SYMBOL

The feature control symbols are associated with the feature(s) being tolerated by one of the following methods:

1. Attaching a side, end, or corner of the symbol box to an EXTENSION LINE or leader from the feature (used on most form tolerances). See Fig. 1 on the facing page.

2. Attaching a side or end of the symbol box to the DIMENSION LINE or EXTENSION LINE pertaining to the feature when it is cylindrical. See Fig. 2 on the facing page.

3. Placing the symbol box below or closely adjacent to the dimension or note pertaining to that feature. See Fig. 3.

4. Running a leader line from the symbol to the feature. See Fig. 4.

FIGURE 1

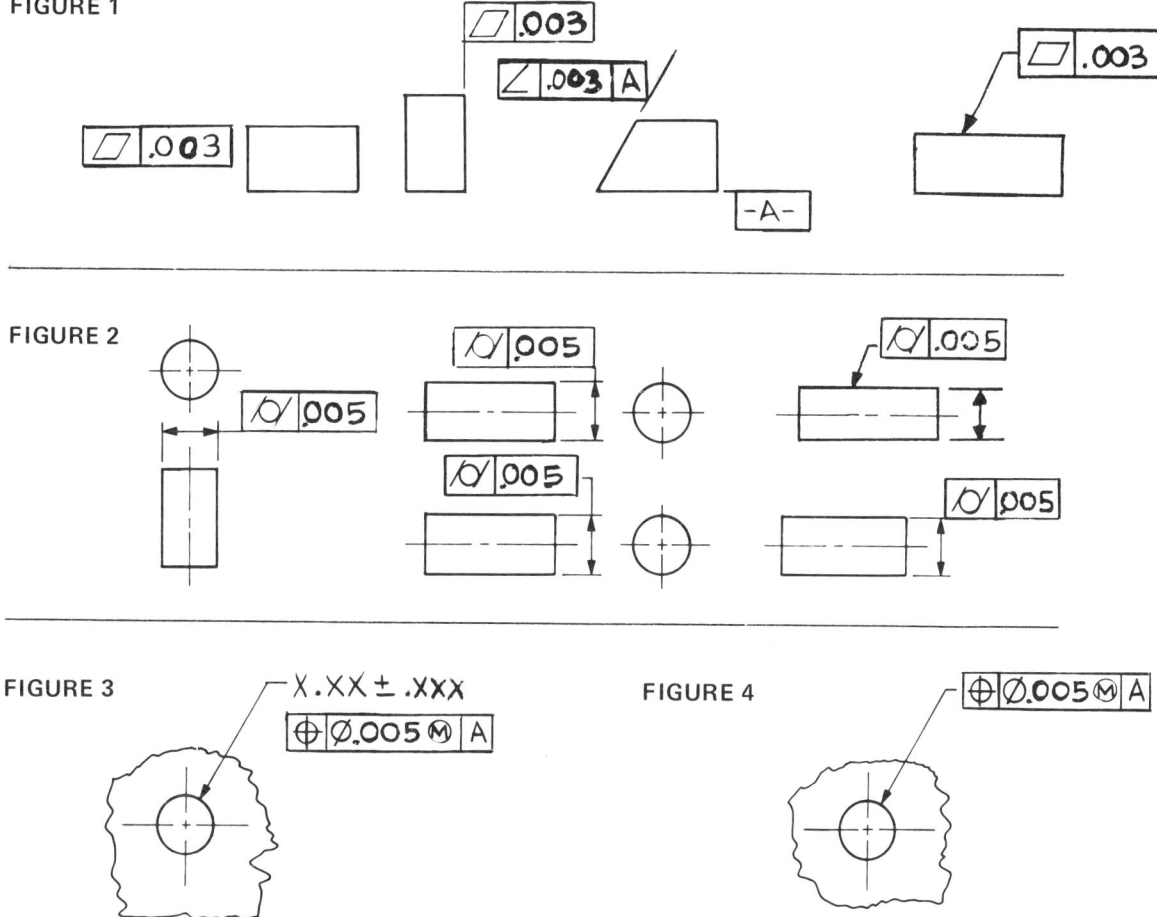

FIGURE 2

FIGURE 3

FIGURE 4

COMBINED FEATURE CONTROL SYMBOL AND DATUM IDENTIFYING SYMBOL

When a feature serves as a datum and is also controlled by a form, locational, or runout tolerance, the feature control symbol and the datum identifying symbol should be combined as shown.

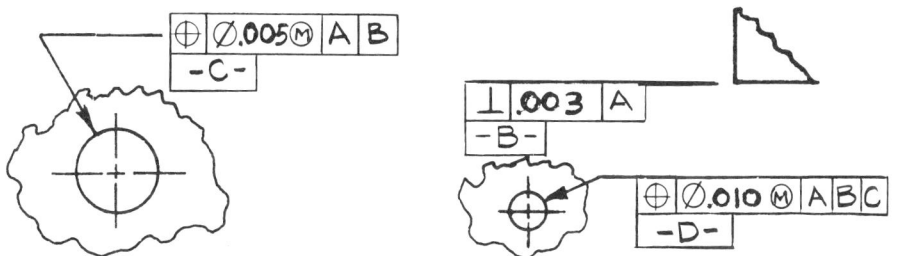

REFERENCE TO DATUM

When a form, runout, or location tolerance must be related to a datum, this relationship is stated by placing the datum reference letter following the geometric characteristic symbol and the tolerance.

The illustrations on the next page show additional examples of the feature control symbols with reference to datums.

Figure 1 is a typical feature control symbol using a single datum reference. The symbol reads "This feature shall be within a .002 tolerance zone perpendicular to datum A."

Figure 2 shows a feature control symbol with *two* datums. The symbol reads "This feature shall be located at true position within .005 diameter at maximum material condition with respect to both datums A and B."

Note that vertical lines are used to separate the characteristic symbol, the feature tolerance, and the datum references. These vertical lines are used on all feature control symbols to ensure clarity. One reason for this is illustrated in Fig. 3, in which the maximum material condition symbol is used. The vertical lines clearly show that the MMC condition symbols apply only to those datums or tolerances with which they appear in the subdivision of the symbol box.

Figure 4 illustrates primary, secondary, and tertiary datums showing the order of precedence. When the order of precedence of datums is significant to function, datum references should be classified as primary, secondary, and tertiary. The datum precedence is shown by placing each datum reference letter in the proper order. The first datum letter (left to right) is considered the primary datum, the second letter secondary, and the third letter tertiary. Thus the datum reference letters will not necessarily be in alphabetical order. See section on DATUMS for further explanation.

Figure 5 illustrates a feature control symbol in which multiple datum features are used simultaneously to establish a single datum reference (equal precedence of datum features) e.g., to establish a common datum axis. See section on DATUMS and RUNOUT for further details.

Figure 6 illustrates a rather unadvisable use of a datum reference. Note that datum A applies at MMC, whereas the feature controlled applies at RFS. This means that the datum reference is subject to variation and cannot serve as a fixed reference for any RFS relationship. Although there may be exceptions under special circumstamces. generally, wherever MMC is used on any datum, the feature controlled should also be controlled at MMC.

DIAMETER SYMBOL

The symbol used to designate a diameter (DIA) or cylindrical feature or tolerance zone is shown above. Where used in a feature control symbol, the symbol ($\emptyset$) precedes the specified tolerance. See Figs. 2, 3, and 4 at right. The symbol may be used elsewhere on the drawing in place of the abbreviation DIA.

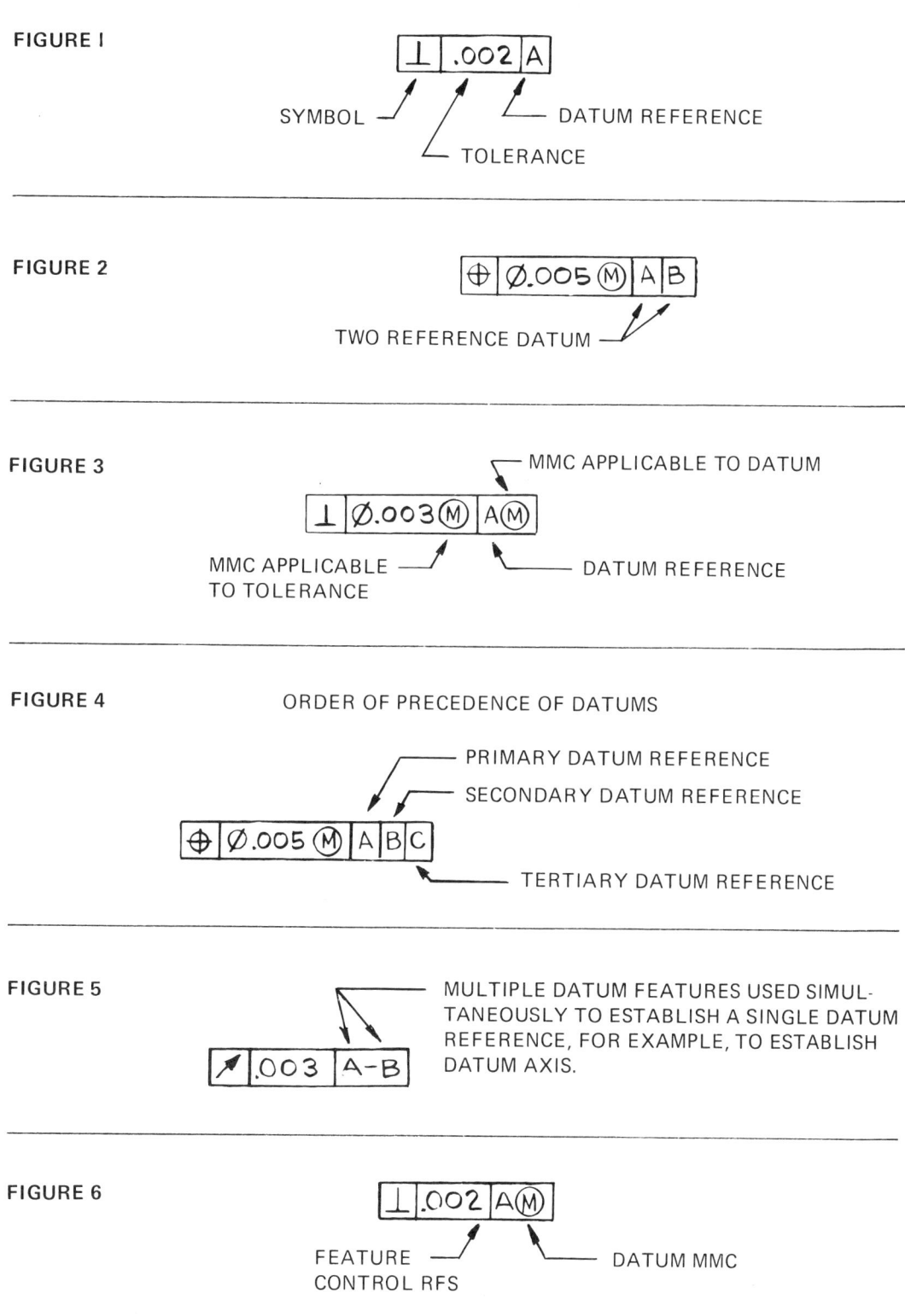

FIGURE 1

SYMBOL — TOLERANCE — DATUM REFERENCE

FIGURE 2

TWO REFERENCE DATUM

FIGURE 3

MMC APPLICABLE TO DATUM

MMC APPLICABLE TO TOLERANCE — DATUM REFERENCE

FIGURE 4

ORDER OF PRECEDENCE OF DATUMS

PRIMARY DATUM REFERENCE
SECONDARY DATUM REFERENCE
TERTIARY DATUM REFERENCE

FIGURE 5

MULTIPLE DATUM FEATURES USED SIMULTANEOUSLY TO ESTABLISH A SINGLE DATUM REFERENCE, FOR EXAMPLE, TO ESTABLISH DATUM AXIS.

FIGURE 6

FEATURE CONTROL RFS — DATUM MMC

NOT NORMALLY PRACTICAL

GEOMETRIC CHARACTERISTICS — FORM, RUNOUT, AND LOCATIONAL TOLERANCE — OTHER SYMBOLS AND TERMS

Illustrated at right are the geometric characteristics and symbols which are the basis for the language of geometric dimensioning and tolerancing.

THREE TYPES OF GEOMETRIC CHARACTERISTICS

Expanding on preceding text explanation, it is seen that the geometric characteristics are of three types (see "3-TYPES" column at right):

1. FORM tolerance — A form tolerance states how far an actual surface or feature is permitted to vary from the desired form implied by the drawing.

2. RUNOUT tolerance — A runout tolerance states how far an actual surface or feature is permitted to vary from the desired form implied by the drawing during full (360°) rotation of the part on a datum axis.

3. LOCATION tolerance — A location tolerance states how far an actual feature is permitted to vary from the perfect location implied by the drawing as related to a datum, or datums, or other features.

KINDS OF FEATURES TO WHICH A GEOMETRIC CHARACTERISTIC IS APPLICABLE

The geometric characteristics are also divisible into three "kinds" of features to which a particular characteristic is applicable. (See "KIND OF FEATURE" column at right.)

1. *INDIVIDUAL* feature — A single surface, element, or size feature which relates to a perfect geometric counterpart of itself as the desired form; no datum is proper nor used. (characteristics ▱, — , ◯, ⌖).

2. RELATED feature — A single surface or element feature which relates to a datum, or datums, in form and attitude (orientation). (characteristics ⊥, ∠, ∥).
     — A size feature (e.g. hole, slot, pin, shaft) which relates to a datum, or datums, in form, attitude (orientation), runout, or location: (characteristics ⊥, ∠, ∥, ⟊, ⊕, ◎, ⩦)

3. *INDIVIDUAL* or *RELATED* feature — A single surface or element feature whose perfect geometric profile is described which may, or may not, relate to a datum, or datums. (characteristics ⌒, ⌓).

OTHER SYMBOLS AND TERMS

For review, other symbols and terms are shown at right. The Projected Tolerance Zone and Datum Target symbols are explained in later text.

GEOMETRIC CHARACTERISTICS, SYMBOLS, AND TERMS

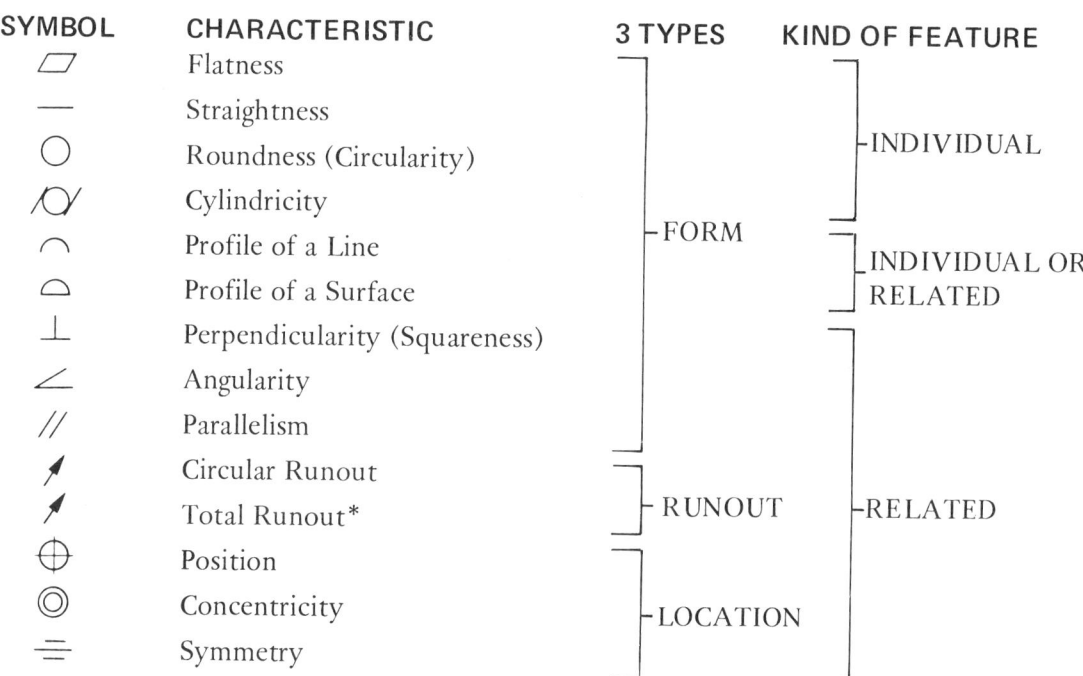

SYMBOL	CHARACTERISTIC	3 TYPES	KIND OF FEATURE
▱	Flatness		
—	Straightness		INDIVIDUAL
○	Roundness (Circularity)		
⌖	Cylindricity	FORM	
⌒	Profile of a Line		INDIVIDUAL OR RELATED
⌓	Profile of a Surface		
⊥	Perpendicularity (Squareness)		
∠	Angularity		
//	Parallelism		
↗	Circular Runout		
↗	Total Runout*	RUNOUT	RELATED
⊕	Position		
◎	Concentricity	LOCATION	
≐	Symmetry		

OTHER SYMBOLS

- Ⓜ Maximum Material Condition MMC
- Ⓢ Regardless of Feature Size RFS
- Ⓟ Projected Tolerance Zone
- Datum Target
- ⌀ Diameter (cylindrical) Tolerance Zone
- .XXX Basic, Exact Dimension
- -A- Datum Identification Symbol
- // ⌀ .002 Ⓜ A   Feature Control Symbol (Frame)

TERMS

BASIC (BSC)	= Theoretically exact dimension.
DATUM	= Reference Points, Lines, Planes, Surfaces
FEATURE	= Component portion of a part, e.g., surface, hole, slot, etc.
LEAST MATERIAL CONDITION (LMC)	= Size opposite from MMC.
VIRTUAL CONDITION	= Collective effect of all tolerance variations on a feature.

---

* Word TOTAL required under feature control symbol. Anticipated new symbol in future standards (ANSI, ISO), ⟋⟍.

GENERAL RULES

Like any discipline, geometric dimensioning and tolerancing is based on certain fundamental rules. Some of these follow from standard interpretation of the various characteristics, some govern specification, and some are General Rules applying across the entire system.

The various rules appropriate to given geometric characteristics and related nomenclature will be discussed later. The General Rules are described below and on succeeding pages.

APPLICABILITY OF GENERAL RULES

ANSI Y14.5 contains five General Rules. ANSI Y14.5 must be referenced whenever these General Rules are to be applied.

INDIVIDUAL FEATURES OF SIZE

RULE 1

Unless otherwise specified, the limits of the dimension of an individual feature of size control the form of the feature as well as the size.

a) No element of the actual feature shall extend beyond a boundary of perfect form at MMC. This boundary is the true form implied by the drawing.

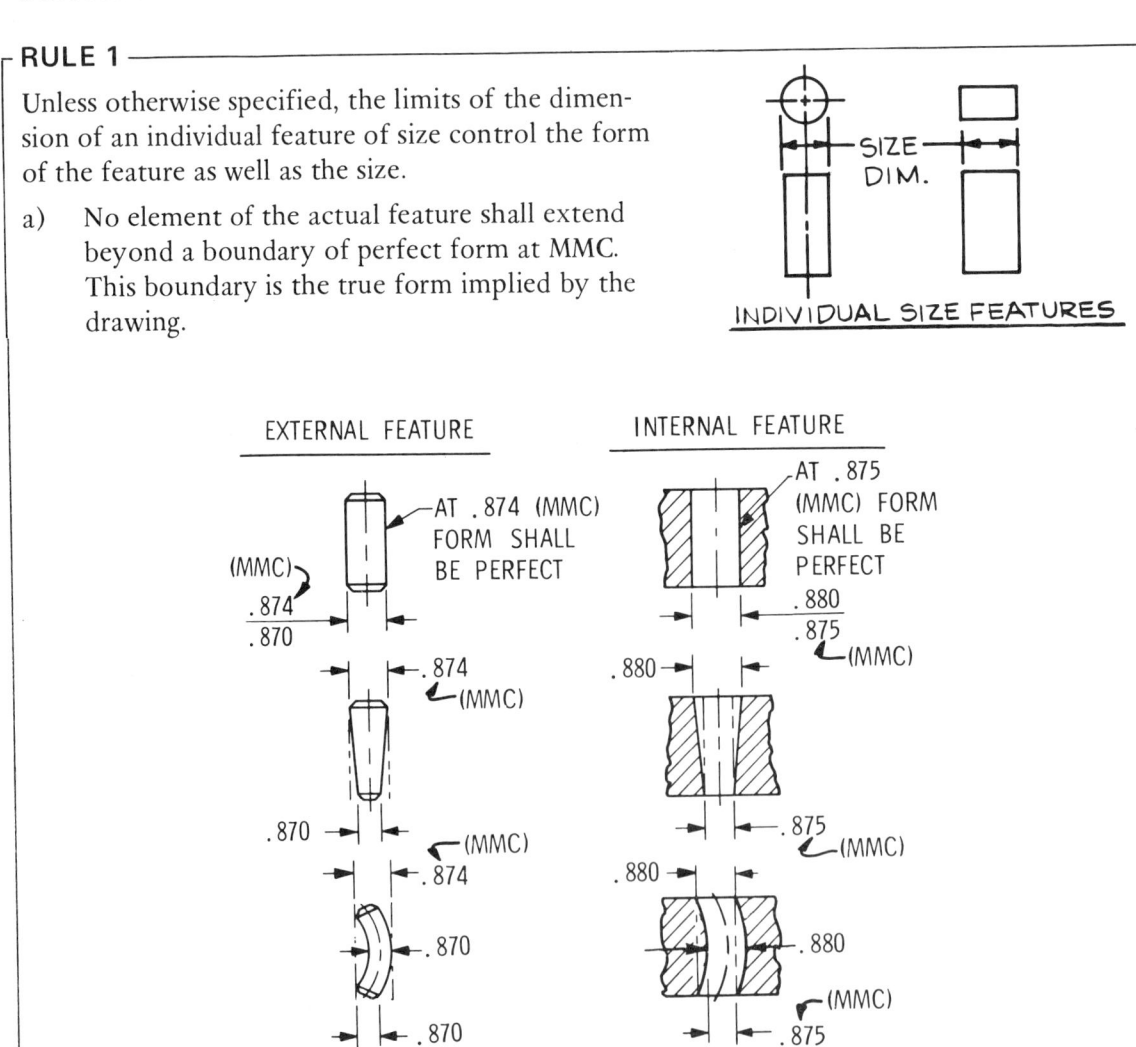

INDIVIDUAL SIZE FEATURES

EXTERNAL FEATURE · INTERNAL FEATURE

b) The actual size of the feature at any cross-section shall be within the LMC limit of size.

c) The form control provision of paragraph (a) does not apply to commercial stock such as bars, sheets, and tubing, where established industry standards prescribe straightness, flatness, and other conditions.

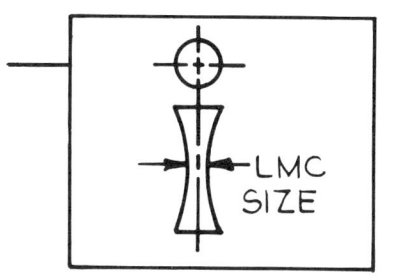

INTERRELATED FEATURES

The form control provision of RULE 1, paragraph (a) applies only to individual features and not to the interrelationship of features.

Where such control of interrelated features is necessary, one of the following methods should be used to the extent dictated by the design requirements:

a) Specify a zero form tolerance at MMC for the features.

b) Indicate this control for the features involved by a note such as, PERFECT FORM AT MMC REQUIRED FOR INTERRELATED FEATURES.

c) Relate the dimensions to a datum reference frame.

PERFECT FORM AT MMC *NOT* REQUIRED

Where it is desired to permit a specified tolerance of form to exceed the boundary of perfect form at MMC, this may be done by adding to the drawing the suitable form tolerance and a note specifically exempting the pertinent size dimensions from the perfect form RULE 1, paragraph (a) requirement. A suitable note might read "PERFECT FORM AT MMC NOT REQUIRED."

GENERAL RULES

RULE 2

(Preferred Practice – Specify whether MMC or RFS).
For a tolerance of POSITION (formerly called true position), Ⓜ or Ⓢ shall be specified on the drawing with respect to the individual tolerance, datum reference(s), or both, as applicable.

(Ⓜ OR Ⓢ SPECIFIED ON SIZE FEATURE AS NECESSARY)

| ⊕ | ⌀.005Ⓜ | AⓂ | BⓂ |

| ⊕ | ⌀.005Ⓢ | AⓈ | BⓈ |

RULE 2a

Alternate Practice* (MMC implied on ⊕ unless specified RFS).
For a tolerance of POSITION (formerly called true position), Ⓜ applies with respect to an individual tolerance, datum reference(s), or both, where no condition is specified. Ⓢ must be specified where it is required.

(IMPLIES Ⓜ ON SIZE FEATURES UNLESS Ⓢ SPECIFIED)

| ⊕ | ⌀.005 | A | B |

| ⊕ | ⌀.005Ⓢ | AⓈ | BⓈ |

RULE 3

For other than a tolerance of position (formerly called true position), RFS applies with respect to an individual tolerance, datum reference(s), or both, where no condition is specified. Ⓜ must be specified on the drawing where it is required.

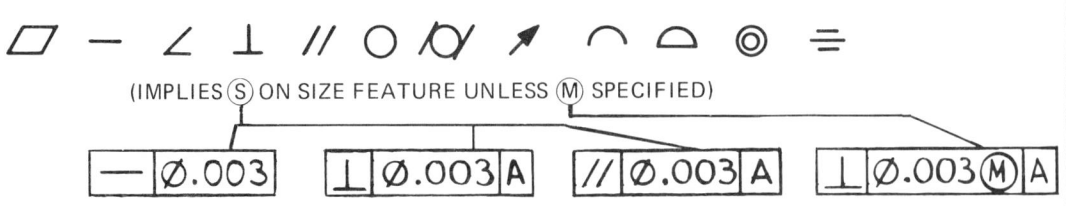

(IMPLIES Ⓢ ON SIZE FEATURE UNLESS Ⓜ SPECIFIED)

| — | ⌀.003 |

| ⊥ | ⌀.003 | A |

| // | ⌀.003 | A |

| ⊥ | ⌀.003Ⓜ | A |

*NOTES: 1. Former recommended practice under ANSI–Y14.5–1966.

2. This rule is contrary to international practice and should be used with caution on drawings subject to international use and interpretation.

RULE 4

Each tolerance of form or location and datum reference for a screw thread applies to the PITCH DIAMETER. Where design requirements necessitate an exception to this rule, a qualifying notation (such as MINOR DIA ($\oslash$), MAJOR DIA ($\oslash$), OD) shall be shown beneath the feature control symbol or datum identification symbol, as applicable.

For gears and splines, each form or location tolerance and datum reference shall designate the specific feature of the gear or spline to which it applies (e.g., MAJOR DIA ($\oslash$), MINOR DIA ($\oslash$), PITCH DIA ($\oslash$), PD).

RULE 5

*Although referenced in a feature control symbol at MMC, a datum feature of size controlled by a separate tolerance of location or form applies at its virtual condition.

Where it is *not* intended for the virtual condition to apply, a zero tolerance at MMC should be specified for the appropriate datum features.

Where no tolerance of location or form is specified for these features, a perfect form at MMC interrelationship is implied relative to each other as datums.

---

* See pages 122–131 for further explanation.

GENERAL RULES

APPLICABILITY OF MMC OR RFS

Characteristic		Applicability to feature	Applicability to datum reference
▱	Flatness	MMC Ⓜ or RFS Ⓢ applicable if tolerance applies to axis or center plane of a feature with size; not applicable if considered feature is a single plane surface	No datum reference
—	Straightness		
○	Roundness	Not applicable	
⌯	Cylindricity		
⌒	Profile of a surface	Not applicable	Not applicable †
⌒	Profile of a line		
⊥	Perpendicularity	MMC Ⓜ or RFS Ⓢ applicable if tolerance applies to axis or center plane of a feature with size; not applicable if considered feature is a single plane surface	MMC Ⓜ or RFS Ⓢ on datum reference applicable if datum feature has size and has an axis or center plane; not applicable if datum feature is a single plane surface
//	Parallelism		
∠	Angularity		
⊕	Position		
═	Symmetry	Not recommended (use position)	Not recommended
⟋	Runout { Circular / Total	Not applicable	Not applicable
◎	Concentricity		

† May have exceptions under special conditions. (See PROFILE section)

SHAPE OF TOLERANCE ZONE FOR LOCATIONAL, FORM, OR RUNOUT TOLERANCES

Where the specified tolerance value represents the diameter of a cylindrical zone, the diameter symbol ∅ shall be included in the feature control symbol.

Where the tolerance zone is other than a diameter, the tolerance value represents the distance between two parallel straight lines or planes or the distance between two uniform boundaries.

EXAMPLES

TOLERANCE ZONE SHAPE IS: Where the diameter (cylindrical) symbol ∅ is specified, the tolerance is a diameter (or cylindrical) shape. Where no ∅ is specified, the tolerance zone is between two parallel lines or planes in the direction of the dimension arrows. The tolerance indicated is the TOTAL tolerance permitted.

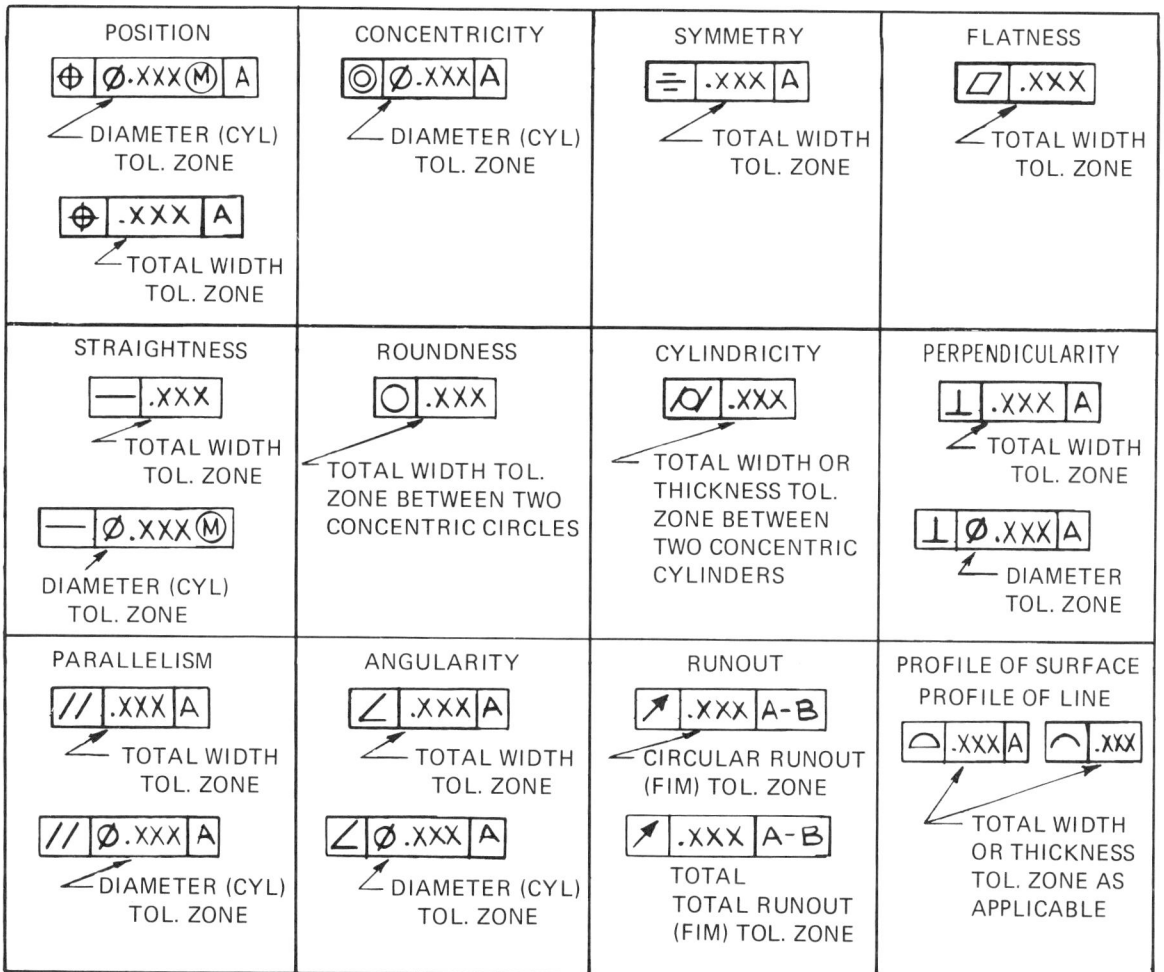

VIRTUAL CONDITION

It is necessary to understand Virtual Condition as it applies to features. The definition below and the examples throughout the text will clarify its meaning.

Definition. The virtual condition of a feature is a derived size generated from the collective effect of all profile variations permitted by the specified tolerances. It represents the most extreme condition of assembly at MMC.

The virtual condition of a feature is the effective size of the profile that must be considered in determining the clearance between mating parts or features.

Size + form or position error = shaft virtual condition
Size − form or position error = hole virtual condition

EXAMPLE

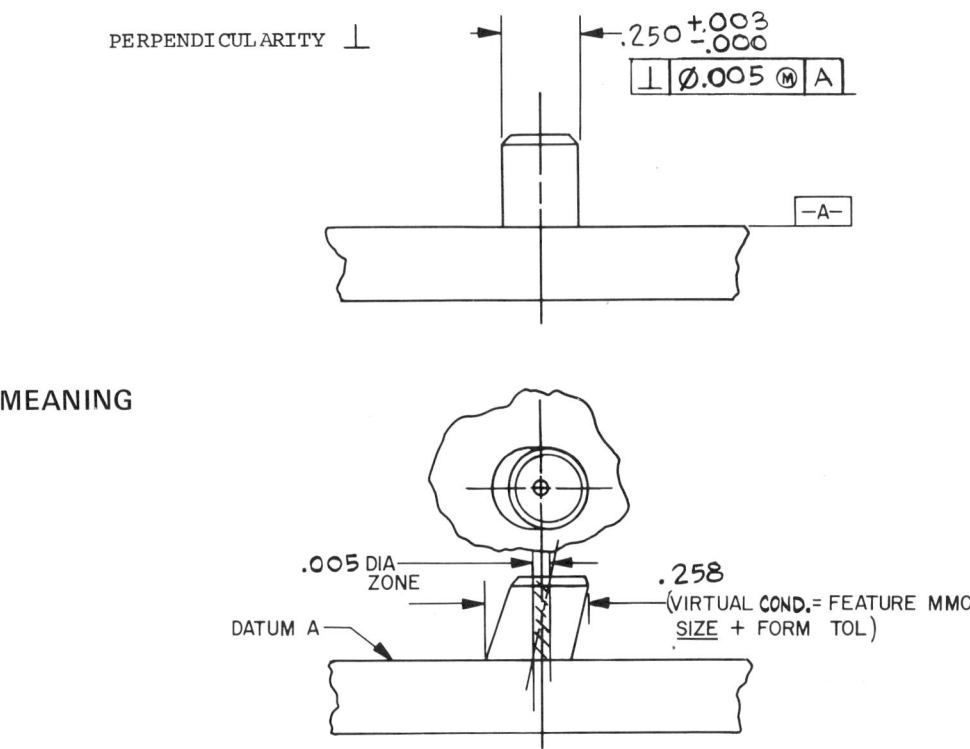

MEANING

VIRTUAL SIZE, a term also used based on earlier U.S. standard practice has the same meaning as VIRTUAL CONDITION.

TOLERANCES
OF FORM
AND
RUNOUT

Tolerances of form and runout state how far actual surfaces or features are permitted to vary from those implied by the drawing. Expressions of form tolerances refer to flatness, straightness, roundness, cylindricity, parallelism, perpendicularity, angularity, profile of a surface, and profile of a line. Runout tolerance is a unique variation of form tolerance and is considered as a separate type of characteristic.

Form or runout tolerances should be specified for all features critical to the design requirements:

a) where established workshop practices cannot be relied upon to provide the required accuracy;

b) where documents establishing suitable standards of workmanship cannot be prescribed;

c) where tolerances of size and location do not provide the necessary control.

The various tolerances of form and runout often have an effect upon one another; that is, parallelism could include flatness or straightness, and runout could include roundness or cylindricity.

The following series of form and runout tolerance examples ignore the effects of combination with other tolerances of form, size, or location for purposes of explanation of the basic principles. Symbolic notation of form and runout tolerances is recommended and emphasized. More detailed examples will be discussed in later portions of the text.

* Anticipated new symbol in future standards (ANSI, ISO), ⟋⟋.

TOLERANCES OF FORM

A FORM TOLERANCE states how far an actual surface or feature is permitted to vary from the desired form implied by the drawing.

KINDS OF FEATURES TO WHICH A FORM TOLERANCE IS APPLICABLE

To correctly apply form tolerances, an understanding of the kind of features, INDIVIDUAL or RELATED, upon which each characteristic can be used is required. Form tolerances can be applied as follows:

INDIVIDUAL feature — A *single surface, element,* or *size* feature which relates to a perfect geometric counterpart of itself as the desired form; no datum is proper or used.

FORM CHARACTERISTICS which can be applied:

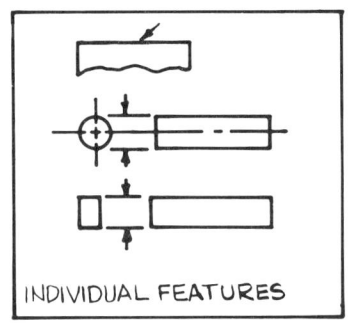

INDIVIDUAL FEATURES

RELATED features — A *single surface or element* feature which relates to a datum, or datums, in form and attitude *(orientation).

FORM CHARACTERISTICS which can be applied:

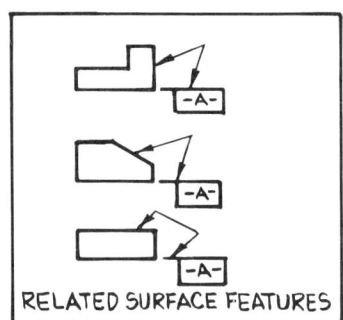

RELATED SURFACE FEATURES

RELATED features — A *size* feature which relates to a datum, or datums, in form and attitude *(orientation).

FORM CHARACTERISTICS which can be applied:

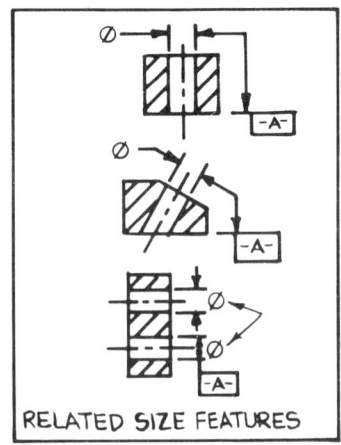

RELATED SIZE FEATURES

---

* Those FORM tolerances involving related features and datums are often referred to as attitude (or orientation) tolerances.

INDIVIDUAL or *RELATED* features — A *single surface or element* feature whose perfect geometric profile is described, which may or may not relate to a datum or datums in form and attitude *(orientation).

FORM CHARACTERISTICS which can be applied:

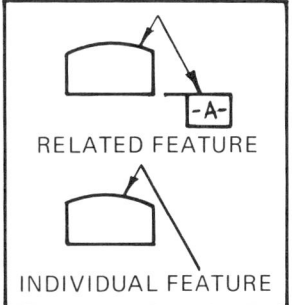

RELATED FEATURE

INDIVIDUAL FEATURE

TOLERANCES OF FORM — INDIVIDUAL FEATURES — NO DATUM

Tolerances of form used on individual features where no datum is proper nor used involve the characteristics below:

(FLATNESS)   (STRAIGHTNESS)   (ROUNDNESS)   (CYLINDRICITY)

These characteristics are used to describe form tolerances of *single surface, element,* or *size* features and relate to a perfect geometric counterpart of itself.

See following pages for details of application.

---

* Those FORM tolerances involving related features and datums are often referred to as attitude (or orientation) tolerances.

FLATNESS ▱

Definition. Flatness is the condition of a surface having all elements in one plane.

FLATNESS TOLERANCE

Flatness tolerance specifies a tolerance zone confined by two parallel planes within which the entire surface must lie.

FLATNESS TOLERANCE APPLICATION

The example below shows how a flatness symbol is applied.

The symbol is interpreted below to read "This surface shall be flat within .002 total tolerance zone over entire surface." Note that the .002 tolerance zone is *total* variation. To be acceptable, the entire actual surface must fall within the parallel plane extremities of the .002 tolerance zone.

A flatness tolerance is a form control of all elements of a surface as it compares to a simulated perfect geometric counterpart of itself. The perfect geometric counterpart of a flat surface is a plane. The tolerance zone is established as a width or thickness zone relative to this plane as established from the actual part surface.

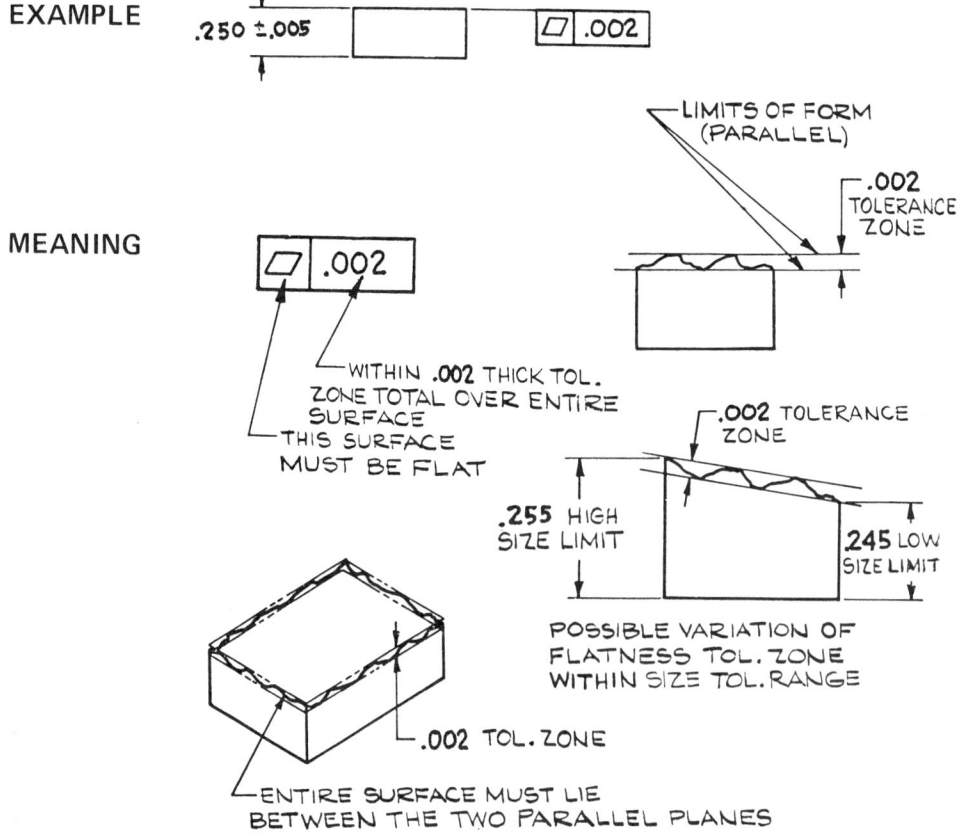

Note that the extremities or high points of the surface determine one limit or plane of the tolerance zone, with the other limit or plane being established 0.05 (the specified tolerance) parallel to it.

Since flatness tolerancing control is essentially a relationship of a feature to itself, no datum references are required or proper.

Also, note that since flatness is a form tolerance controlling surface elements only, it is not applicable to RFS or MMC considerations. *

In the absence of a flatness tolerance specification, the size tolerance and method of manufacture of a part will exercise some control over its flatness. However, when a flatness tolerance is specified, as applicable to a single surface, the flatness tolerance zone must be contained *within* the size tolerance limits. ** It cannot be additive to the size tolerance.

Where necessary the terms "MUST NOT BE CONCAVE" or "MUST NOT BE CONVEX" may be added beneath the feature control symbol.

STRAIGHTNESS —

Definition. Straightness is a condition where an element of a surface or an axis is a straight line.

STRAIGHTNESS TOLERANCE

A straightness tolerance specifies a tolerance zone within which an axis or all points of the considered element must lie.

STRAIGHTNESS TOLERANCE APPLICATION

A straightness tolerance is applied in the view where the elements to be controlled are represented by a straight line.

STRAIGHTNESS TOLERANCE — SURFACE ELEMENT CONTROL

Straightness tolerance is typically used as a form control of individual surface elements such as those on cylindrical or conical surfaces. Since surfaces of this kind are made up of an infinite number of longitudinal elements, a straightness requirement applies to the entire surface as controlled in single line elements in the direction specified.

---

* Under special circumstances, where the flatness control is applicable to the total thickness of sheet metal, square rods, etc., the methods illustrated under Straightness on an RFS or MMC basis may be used (tolerance applicable to size feature and dimension).

** Under special circumstances, a note specifically exempting the pertinent size dimension, i.e., PERFECT FORM AT MMC NOT REQUIRED, may be specified.

STRAIGHTNESS

The example below illustrates straightness control of individual longitudinal surface elements on a cylindrical part. Note that the symbol is directed to the feature surface (or extension line) and not to the dimension lines. The straightness tolerance must be less than the size tolerance.

All circular elements of the surface must be within the specified size tolerance and the boundary of perfect form at MMC. Also, each longitudinal element of the surface must lie in a tolerance zone defined by two parallel lines spaced apart by the amount of the prescribed tolerance where the two lines and the nominal axis of the part share a common plane.

Note: Since surface element control is specified, the tolerance zone applies uniformly whether the part is of a bowed, waisted, or barreled shape.

EXAMPLE

MEANING

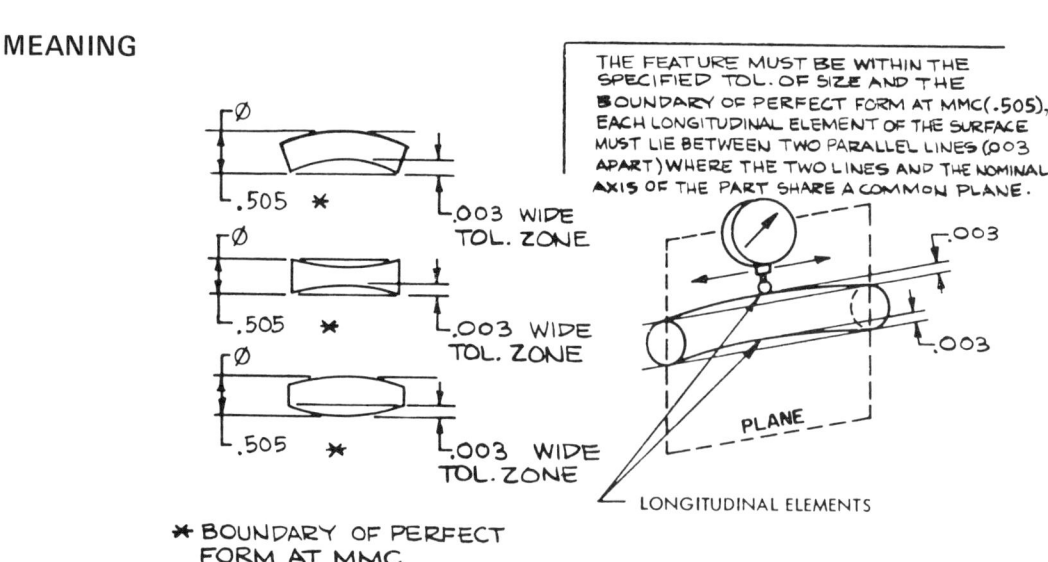

Note the absence of any datum reference when straightness tolerancing is used. Straightness is a form control of a single element, or an axis, as it relates to a perfect geometric counterpart of itself. Therefore, datum references are neither required nor proper.

STRAIGHTNESS

STRAIGHTNESS TOLERANCE APPLIED TO FLAT SURFACE

Straightness tolerancing may be applied to flat surfaces to provide surface element control in a specific direction as a refinement of size tolerance or other form tolerance such as flatness. If so used, the straightness tolerance must be less than the refined tolerance.

The example below illustrates straightness tolerance of surface elements as a refinement of size tolerance. Note that the tolerance is applied in the view in which the elements to be controlled appear as a straight line.

The individual straightness elements must be within both the size tolerance and the straightness tolerance zone of .003, whereas element to element in the other view, variation within the size tolerance may occur.

EXAMPLE

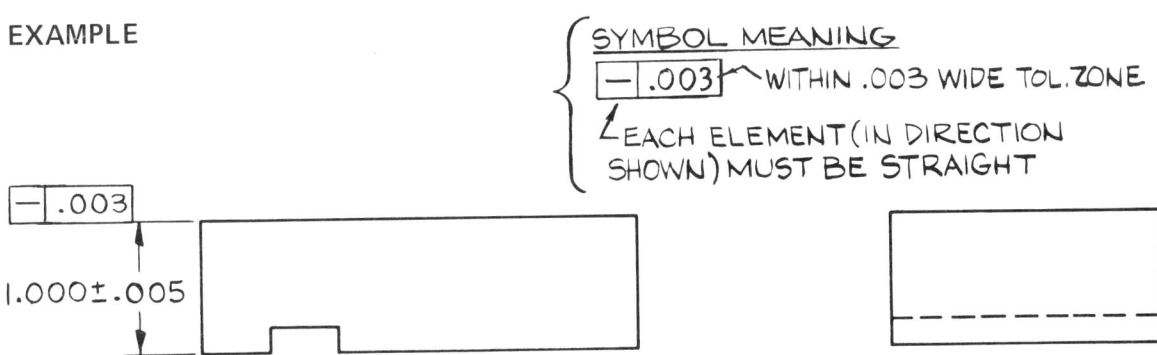

SYMBOL MEANING

─ .003 — WITHIN .003 WIDE TOL. ZONE

EACH ELEMENT (IN DIRECTION SHOWN) MUST BE STRAIGHT

─ .003

1.000 ± .005

MEANING

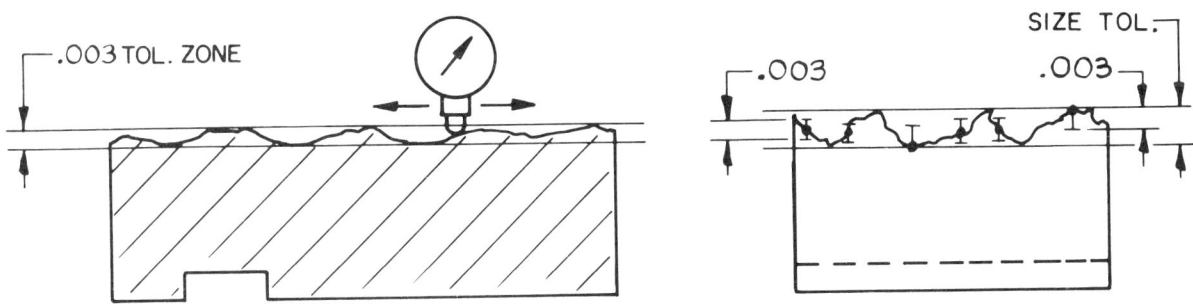

.003 TOL. ZONE

SIZE TOL.

.003   .003

STRAIGHTNESS —

STRAIGHTNESS TOLERANCE RFS AND MMC

Where function of a size feature permits a collective result of *size and form* variation known as the virtual condition,* the RFS or MMC principles may be used.

In this instance, where the appropriate symbology and specifications are used, the part is not confined to the perfect form at MMC boundary. All sectional elements of the surface are to be within the specified size tolerance, but the total part surface may exceed the perfect form at MMC boundary to the extent of the straightness tolerance.

This principle may be applied to individual size features such as pins, shafts, bars, etc. where the longitudinal elements are to be specified with a straightness tolerance independent of, or in addition to, the size tolerance.

STRAIGHTNESS – RFS – VIRTUAL CONDITION

Where a cylindrical feature is to be controlled on an RFS basis as below, the feature control symbol must be located with the size dimension or attached to the dimension line, and the diameter symbol must precede the straightness tolerance.

EXAMPLE

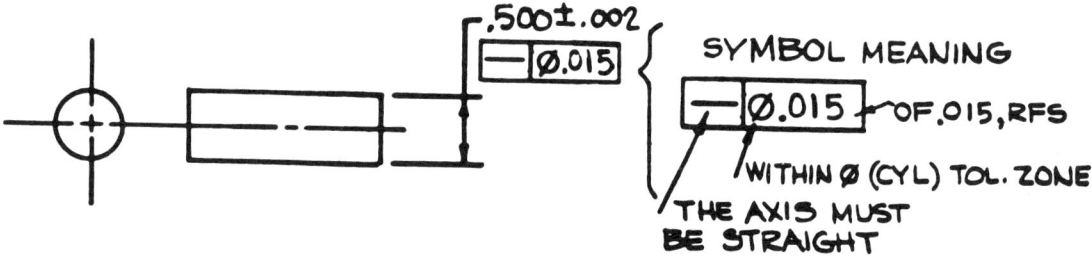

MEANING

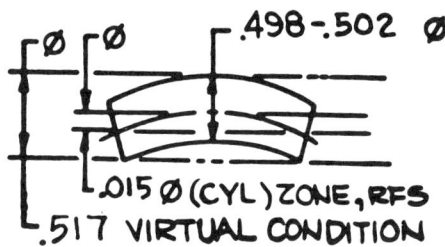

EACH CIRCULAR ELEMENT OF THE FEATURE MUST BE WITHIN THE SPECIFIED TOLERANCE OF SIZE. THE DERIVED AXIS OF THE ACTUAL FEATURE MUST LIE WITHIN A DIA. OR CYLINDRICAL TOLERANCE ZONE .015 REGARDLESS OF FEATURE SIZE.

---

* VIRTUAL CONDITION – Virtual condition of a feature is the collective effect of size and form error that must be considered in determining the fit or clearance between mating parts or features. See above example.

STRAIGHTNESS — MMC — VIRTUAL CONDITION

Where a cylindrical feature has a functional relationship with another feature, such as a pin or shaft and a hole, the control of straightness on an MMC basis may be desirable. If the pin or shaft, for example, is to fit into a hole of a given diameter, the collective effect of the pin size and its straightness error must be considered in relationship to the hole size minimum; i.e., their virtual conditions must be considered relative to one another.

On the part below, the size must be maintained at all sectional elements within stated limits. Likewise its straightness, using the axis as the criteria, must be within tolerance, but only when the part is at MMC. Therefore, the part develops (or is based upon) a virtual condition of .517 which represents the extreme condition the part can have and yet perform its function, or fit the mating part.

By stating the requirements on an MMC basis, the allowable straightness tolerance may increase an amount equal to the size departure from MMC. The feature control symbol must be located with the size dimension, or attached to the dimension line; the diameter symbol must precede the straightness tolerance; and the MMC symbol must be inserted following the tolerance. In this manner maximum tolerance is achieved, part fit is guaranteed, and functional gaging techniques may be used.

EXAMPLE

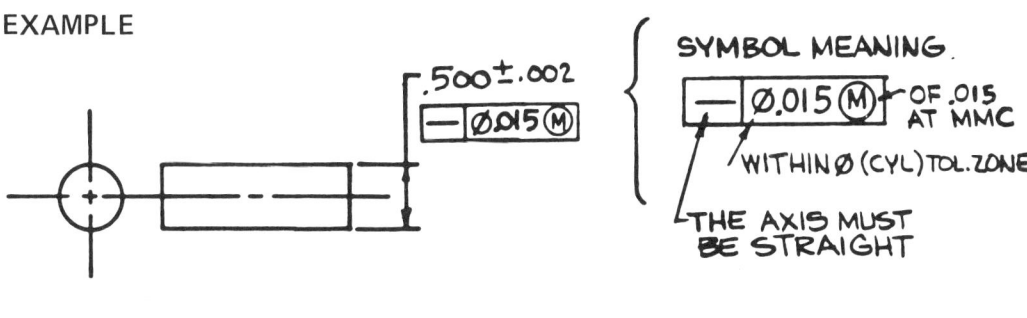

MEANING

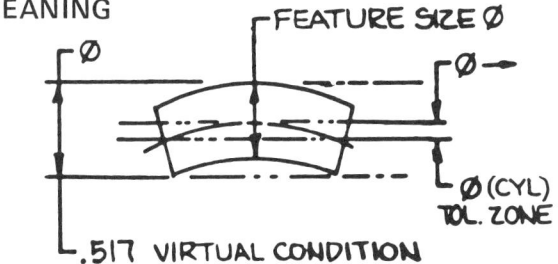

IF FEATURE SIZE Ø	Ø TOL. ZONE IS
.502 MMC	.015
.501	.016
.500	.017
.499	.018
.498 LMC	.019

EACH CIRCULAR ELEMENT OF THE FEATURE MUST BE WITHIN THE SPECIFIED TOLERANCE OF SIZE. THE DERIVED AXIS OF THE ACTUAL FEATURE MUST LIE WITHIN A DIA. OR CYLINDRICAL TOLERANCE ZONE OF .015 AT MMC. AS THE FEATURE DEPARTS FROM MMC, AN INCREASE IN THE STRAIGHTNESS TOLERANCE IS ALLOWED WHICH IS EQUAL TO THE AMOUNT OF SUCH DEPARTURE.

STRAIGHTNESS — MMC — VIRTUAL CONDITION

Where straightness tolerance on an MMC basis is specified, functional gaging techniques may be used. The below gage and conditions demonstrate how these principles can be applied to the preceding part.

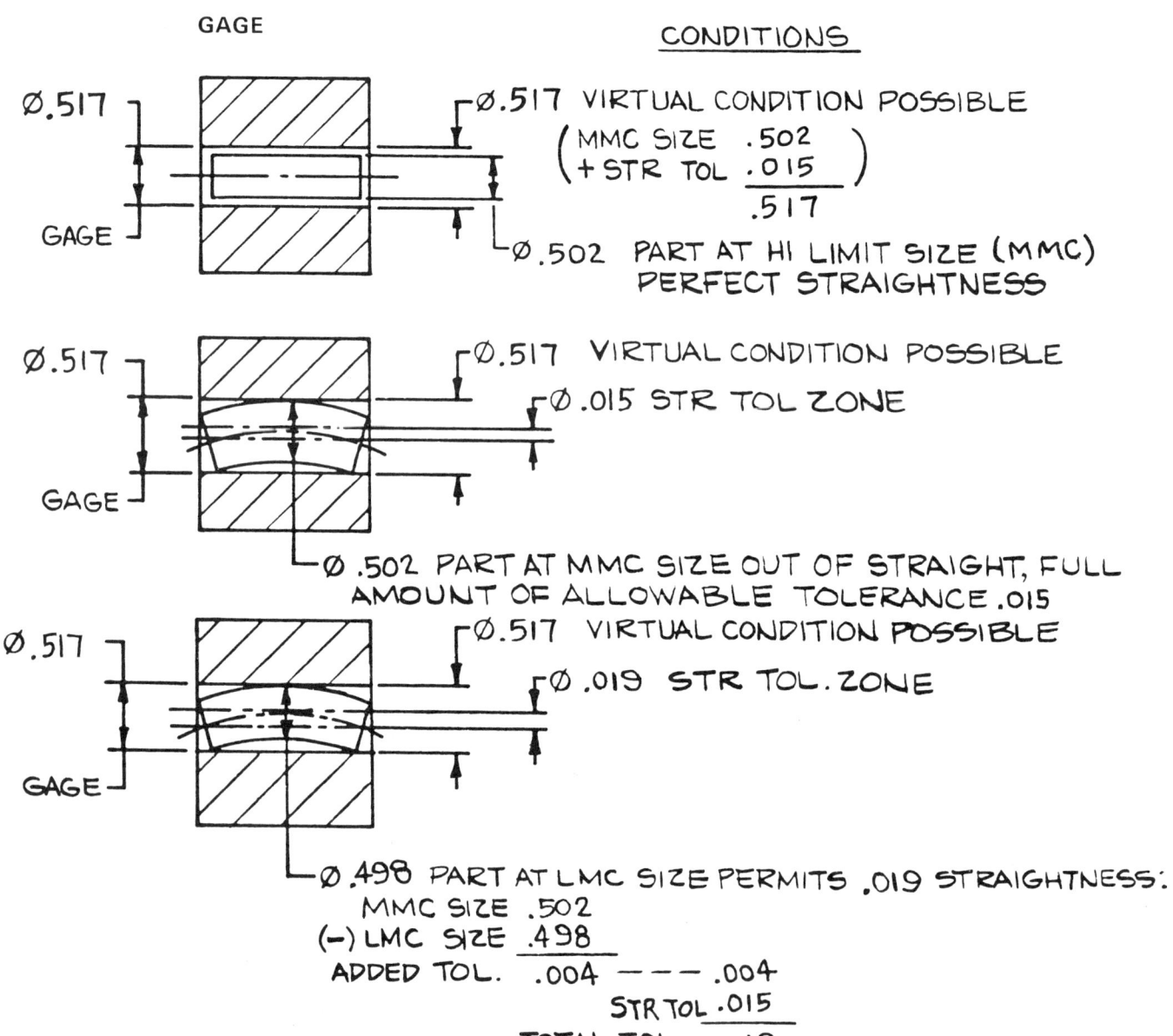

GAGE

<u>CONDITIONS</u>

Ø.517 VIRTUAL CONDITION POSSIBLE

$$\begin{pmatrix} \text{MMC SIZE} & .502 \\ \text{+ STR TOL} & .015 \end{pmatrix}$$
.517

Ø.502 PART AT HI LIMIT SIZE (MMC) PERFECT STRAIGHTNESS

Ø.517 VIRTUAL CONDITION POSSIBLE

Ø.015 STR TOL ZONE

Ø.502 PART AT MMC SIZE OUT OF STRAIGHT, FULL AMOUNT OF ALLOWABLE TOLERANCE .015

Ø.517 VIRTUAL CONDITION POSSIBLE

Ø.019 STR TOL. ZONE

Ø.498 PART AT LMC SIZE PERMITS .019 STRAIGHTNESS:

MMC SIZE .502
(-) LMC SIZE .498
ADDED TOL. .004 --- .004
STR TOL .015
TOTAL TOL .019

From the above it is seen that straightness applied on an MMC basis provides control of mating part conditions to facilitate design and provides maximum production tolerance. Functional gaging techniques are permissible. Note that the above gage simulates both the extreme permissible condition of the part as well as the condition of the mating part.

STRAIGHTNESS ON UNIT LENGTH BASIS

Where required, straightness may be applied on a unit length basis. This method is occasionally used to facilitate special design requirements and to prevent abrupt surface variations within a relatively short length of the feature. To prevent extreme variations of bow over the total length of the part, the amount of permissible straightness variation allowable on the total length of the part should be specified. If the unit variations are permitted to continue along the length of the part with no maximum tolerance indicated an unsatisfactory part could result.

The example below illustrates a part with unit straightness specified on an RFS basis. (MMC principles could also be used if desired)

EXAMPLE

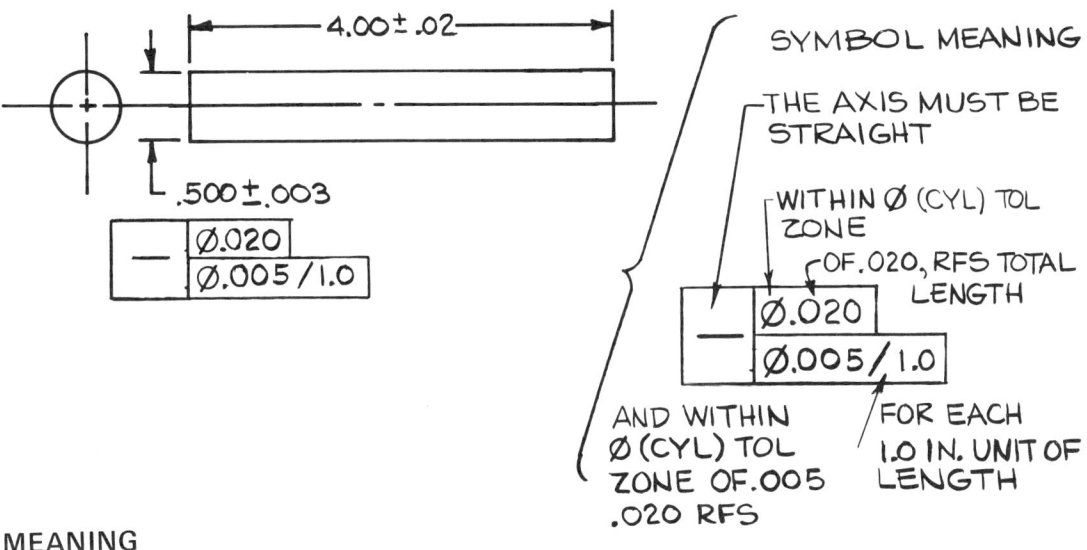

MEANING

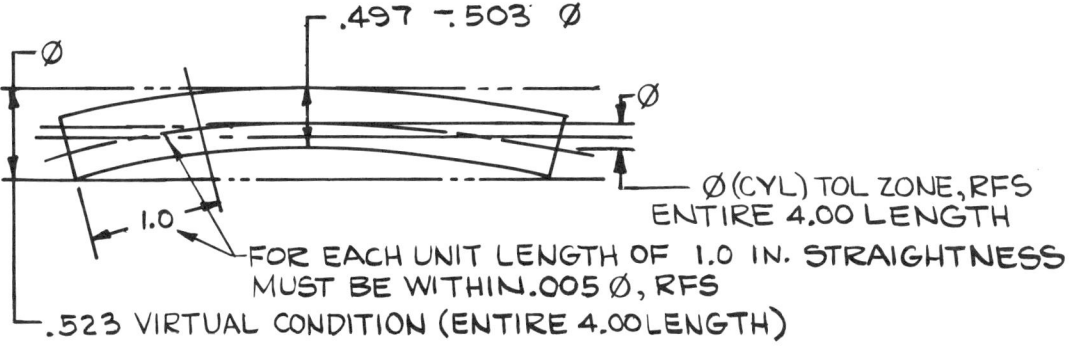

EACH CIRCULAR ELEMENT OF THE FEATURE MUST BE WITHIN THE SPECI-
FIED TOLERANCE OF SIZE. THE DERIVED AXIS OF THE ACTUAL FEATURE
MUST LIE WITHIN A CYLINDRICAL TOLERANCE ZONE OF .020 FOR THE
TOTAL 4.0 LENGTH, AND WITHIN A .005 CYLINDRICAL TOLERANCE ZONE
FOR EACH 1.0 LENGTH, RFS.

ROUNDNESS ○
(CIRCULARITY)

Definition. Roundness is the condition on a surface of revolution where:

1. in the case of a cylinder or cone, all points of the surface intersected by any plane perpendicular to a common axis are equidistant from their axis;

2. in the case of sphere, all points of the surface intersected by any plane passing through a common center are equidistant from that center.

ROUNDNESS TOLERANCE

A roundness tolerance specifies a tolerance zone bounded by two concentric circles within which each circular element of the surface must lie and applies independently at any plane as described above.

ROUNDNESS TOLERANCE APPLICATION

Limits of size exercise control of roundness within the size tolerance. Often this provides adequate control. However, where necessary to further refine form control, roundness tolerancing can be used on any figure of revolution or circular cross section.

The example illustrates a part with a roundness tolerance of .002 specified on a cylindrical part.

The interpretation shows how one establishes the .002 tolerance zone. Note that the tolerance zone is the width of the annular zone between the two concentric circles.

A roundness tolerance zone is established relative to the actual size of the part when measured at the surface periphery at any cross section perpendicular to the part axis. It should be noted that the roundness tolerance applies only at the cross-sectional point of measurement, and is relative to the *size* at that point. Therefore, a cylindrical part with roundness tolerance control could taper or otherwise vary in its surface contour within its size tolerance range, yet still meet roundness requirements if it is within the roundness tolerance at that point.

The part size in this example has been assumed to measure .503 at its largest point at the cross section selected for measurement. The .002 roundness tolerance zone is then established by two theoretically perfect concentric circles, one at the .503 diameter and the other .004 *smaller* at the .499 diameter. This establishes the tolerance zone of *.002 width* between the concentric circles. To be acceptable, the part surface at that cross section must fall within the .002 wide tolerance zone.

As is seen, the tolerance zone is established relative to the part size wherever it may fall in its size tolerance range. That is, the part *size* is first determined and its *roundness* is then defined as a refinement of the part *form* relative to that *size*. Unless otherwise specified, any established size at any point along the surface can be used to determine the roundness tolerance zone. It is, therefore, seen that the roundness tolerance may be based on *different* sizes on the same part. The roundness tolerance zone, however, remains constant.

Note again that the roundness tolerance must always be contained *within* the part *size* tolerance range. The roundness tolerance cannot exceed the size tolerance limits. Furthermore, a roundness tolerance cannot be modified to an MMC application since it controls surface elements only.

A roundness tolerance is a form control of a single part element as it compares to a perfect counterpart of itself. Therefore, no datum references are required nor proper. A roundness tolerance does have a reference center, but this is considered only as a part of the perfect frame of reference (concentric circles and their common center) for measurement, just as a straightness tolerance refers to a perfect line of reference for its measurement. You may compare a roundness tolerance to a straightness tolerance zone curled around a circle.

ROUNDNESS OF A CYLINDER

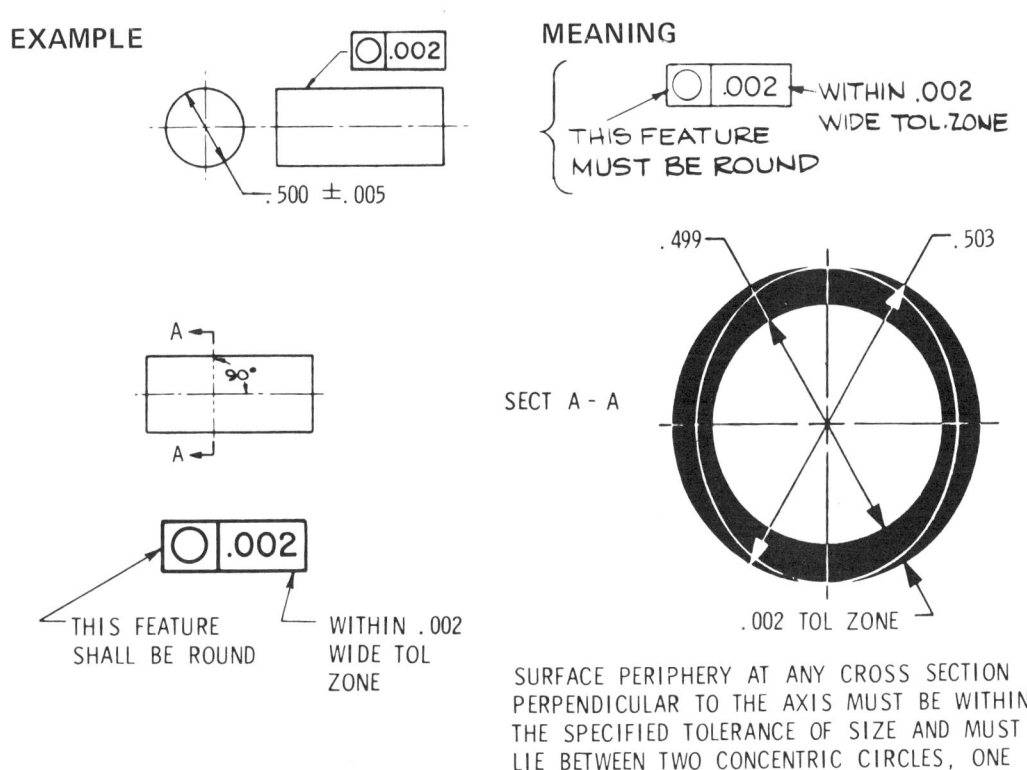

SURFACE PERIPHERY AT ANY CROSS SECTION PERPENDICULAR TO THE AXIS MUST BE WITHIN THE SPECIFIED TOLERANCE OF SIZE AND MUST LIE BETWEEN TWO CONCENTRIC CIRCLES, ONE HAVING A RADIUS .002 LARGER THAN THE OTHER

(ABOVE SIZES ARBITRARILY SELECTED FOR ILLUSTRATION)

ROUNDNESS OF A CONE

Example 1 on the next page illustrates a cone-shaped part for which a roundness tolerance of .003 is specified. As previously discussed, the periphery at any cross section perpendicular to the axis must be within the specified tolerance of size and must lie between the two concentric circles (one having a radius .003 larger than the other).

ROUNDNESS ○
ROUNDNESS OF A CONE

___

EXAMPLE 1

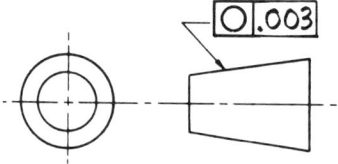

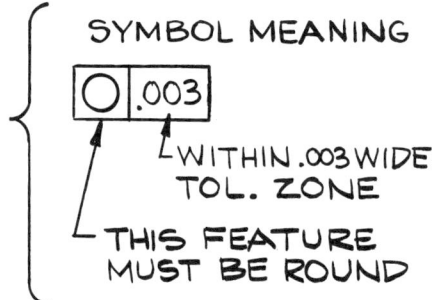

SYMBOL MEANING

WITHIN .003 WIDE TOL. ZONE

THIS FEATURE MUST BE ROUND

MEANING

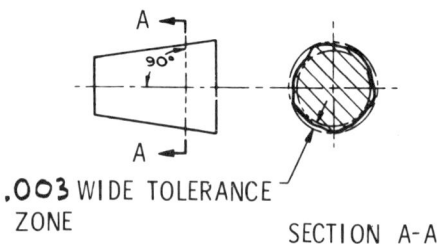

.003 WIDE TOLERANCE ZONE

SECTION A-A

THE PERIPHERY AT ANY CROSS-SECTION PERPENDICULAR TO THE AXIS MUST BE WITHIN THE SPECIFIED TOLERANCE OF SIZE AND MUST LIE BETWEEN TWO CONCENTRIC CIRCLES (ONE HAVING A RADIUS .003 LARGER THAN THE OTHER).

ROUNDNESS OF A SPHERE

Roundness of a spherical part is given the same basic interpretation (see Example 2 below) except that the tolerance control reference is to *any cross section passing through a common center* rather than to *any cross section perpendicular to the axis*, as in the conventional application of roundness tolerancing.

EXAMPLE 2

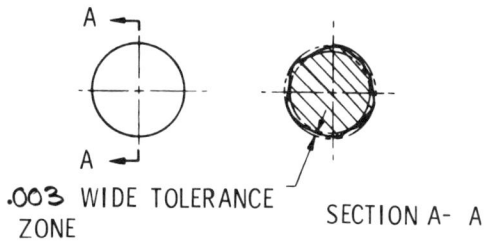

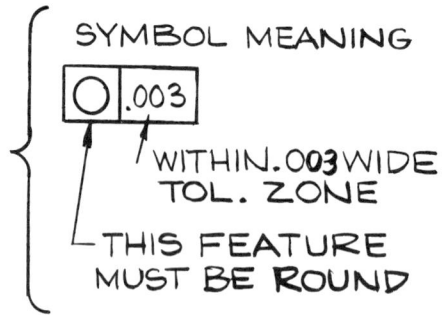

SYMBOL MEANING

WITHIN .003 WIDE TOL. ZONE

THIS FEATURE MUST BE ROUND

MEANING

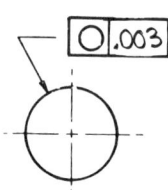

.003 WIDE TOLERANCE ZONE

SECTION A-A

THE PERIPHERY AT ANY CROSS-SECTION PASSING THROUGH A COMMON CENTER MUST BE WITHIN THE SPECIFIED TOLERANCE OF SIZE AND MUST BE BETWEEN TWO CON-CENTRIC CIRCLES (ONE HAVING A RADIUS .003 LARGER THAN THE OTHER). HENCE, THE SURFACE MUST LIE BETWEEN TWO CON-CENTRIC SPHERES SEPARATED .003 APART.

CYLINDRICITY

Definition. Cylindricity is the condition of a surface of revolution in which all points (elements) of the surface are equidistant from a common axis.

CYLINDRICITY TOLERANCE

A cylindricity tolerance specifies a tolerance zone bounded by two concentric cylinders within which the surface must lie.

CYLINDRICITY TOLERANCE APPLICATION

Limits of size exercise control of cylindricity within the size tolerance. This control is often adequate. However, where more refined form control is required, cylindricity tolerancing can be used. Note that in cylindricity, unlike roundness, the tolerance applies simultaneously to both circular and longitudinal elements of the entire surface.

The example illustrates a part with a cylindricity tolerance of .003. A cylindricity tolerance is interpreted as "on the radius" or, as in this case, within the .003 wide tolerance zone defined by two concentric cylinders .003 apart. A cylindricity tolerance can be considered roundness tolerancing extended to control the *entire* surface of a cylinder.

EXAMPLE

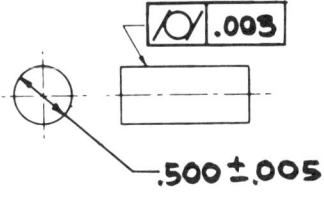

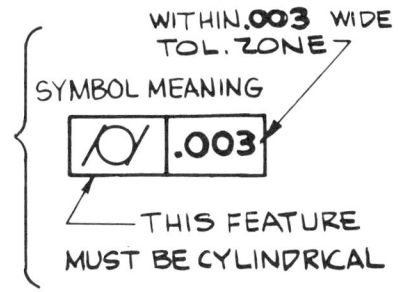

WITHIN **.003** WIDE TOL. ZONE

SYMBOL MEANING

THIS FEATURE MUST BE CYLINDRICAL

MEANING

THE FEATURE MUST BE WITHIN THE SPECIFIED TOLERANCE OF SIZE AND MUST LIE BETWEEN TWO CONCENTRIC CYLINDERS (ONE HAVING A RADIUS .003 LARGER THAN THE OTHER).

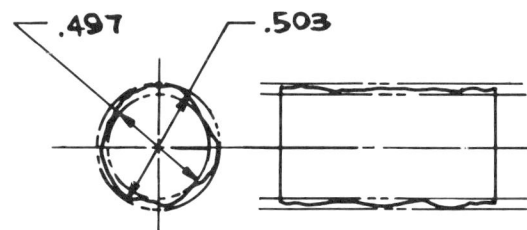

CYLINDRICITY /○/

Cylindricity tolerancing can be applied *only* to cylindrical forms. The leader from the feature control symbol may be directed to either view.

It should be noted that a cylindricity tolerance simultaneously controls roundness, straightness, and parallelism of the elements of the cylindrical surface.

A cylindricity tolerance applies to the *entire* cylindrical surface as opposed to the cross-sectional or diametral measurement considered in roundness. Also, in measuring cylindricity, the concentric cylinders defining the tolerance zone are always based on the size of the part produced. The part size in the example has been assumed to measure .503 at its largest diameter. This .503 is the diameter of the largest of the concentric cylinders defining the tolerance zone. The smaller of the concentric cylinders is .503 minus the amount of the cylindricity tolerance (.003 on R = .006 on dia), i.e., .497. The cylindricity tolerance zone is therefore .003 *wide* between the concentric cylinders. The entire part surface must then fall within this tolerance zone to be acceptable.

It should be noted that the cylindricity tolerance must always be contained *within* the part *size* tolerance range. The cylindricity tolerance cannot exceed the size tolerance limits.

As is seen, the cylindricity tolerance zone is established relative to the part size wherever it may fall in its *size* tolerance range. That is, the part size is first determined and its *cylindricity* is then defined as a refinement of the part *form* relative to that *size*.

If the largest measurement of the produced part had been near the *low* limit, for example at .497, the cylindricity tolerance could not have been more than .001 on R (= .002 on dia). The cylindricity tolerance cannot exceed *size* tolerance limits. Therefore, the form (cylindricity) variation could not be less than the low limit size of .495.

A cylindricity tolerance (similar to flatness) is a form control of a surface element as it compares to a perfect counterpart of itself. Therefore, no datum references are required nor proper. One can compare a cylindricity tolerance zone to a flatness tolerance zone by visualizing that flatness zone curled around a cylinder. A cylindricity tolerance does have a reference axis, but this is considered only a part of the perfect frame of reference (concentric cylinders and their common axis) for measurement, just as a flatness tolerance refers to a perfect plane of reference for its measurement.

Since cylindricity is a form tolerance controlling surface elements only, it cannot be modified to an MMC application.

EVALUATION OF ROUNDNESS AND CYLINDRICITY

Cylindricity is checked or measured by the same basic techniques that are used to check roundness except that cylindricity involves a cylindrical tolerance zone of uniform thickness over the entire surface and is based on a single size reference. This must be considered in any measuring procedure concerned with cylindricity. For example, in the following discussions of vee block, between centers, and polar graph methods, the measurements are made at cross sections only; whereas in cylindricity, one must consider the entire surface as controlled by the one tolerance zone.

Vee block or between centers methods are often used in open set-up inspection of roundness and cylindricity.

The vee block method provides an approximation or a rough check which may be adequate for some applications. Variables, such as the varying number and arrangement of lobes on the part surface and the angle of the vee block, may affect the resultant measurement sufficiently for the method to become rather inaccurate. In addition, the procedure itself contributes to inaccuracies, because one portion of the part surface (in the vee) is used as the basis for checking another portion of the same surface, thus compounding the chances for error. With certain lobe and vee block combinations, a visibly out-of-round part can be rotated with no evidence of error on the dial indicator (see next page).

When the vee block method is used, the total error read is roughly the error on the *diameter*; it must be halved, as an approximation, to compare the result with the specified roundness tolerance which is implied as the radial separation or total width between annular circles.

The between centers method more nearly establishes the reference axis for the geometric relationship. Any error in the centers, however, will affect the accuracy of the resultant measurement. Since the part is rotated about its nominal center or axis, the reading will be on the radius and representative of the total width zone between concentric circles. This method, however, is more correctly termed a runout relationship of a surface and centers and is technically not a roundness analysis.

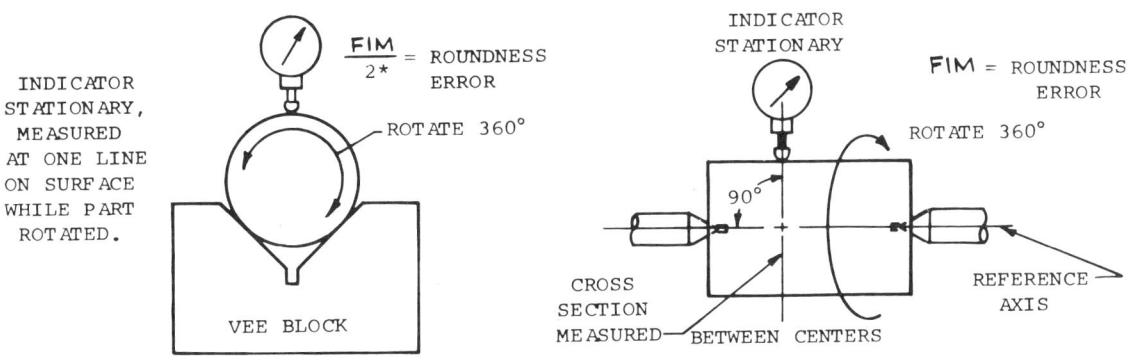

Figures 1 and 2 on the following page demonstrate that certain vee-block and lobed-part conditions do not register the correct part roundness or cylindricity error.

Obviously these parts have noticeable form errors. However, when the parts are rotated in the 60° vee-block, the true error will not register. In fact, in these hypothetical

EVALUATION OF ROUNDNESS AND CYLINDRICITY

examples of a five-lobed part and an oval shape, *no* error registers in any position. When the number of lobes of the part is known, certain vee-block angles can be used to obtain accurate diametral readings. However, the above indicated variables and the difficulty of accurately predicting conditions on any given part make this method impractical.

Figures 3 and 4 illustrate how "miking" a lobed part (particularly a part with an uneven number of lobes) will not pick up error. The parts shown, although obviously not true in form, register the same measurements at any diametrical location. Suppose the parts in Fig. 1 (five lobes) and Fig. 3 (three lobes) were "miked" at 1.999 since they are intended to go through the 2.000 ⌀ hole shown in Fig. 5. The parts will not pass through the hole although they were "miked" at the lower size measurement of 1.999.

Figures 6 and 7 illustrate the parts shown in Figs. 1 and 3 evaluated on the basis of radial values. Using appropriate measuring techniques, the error can be determined directly.

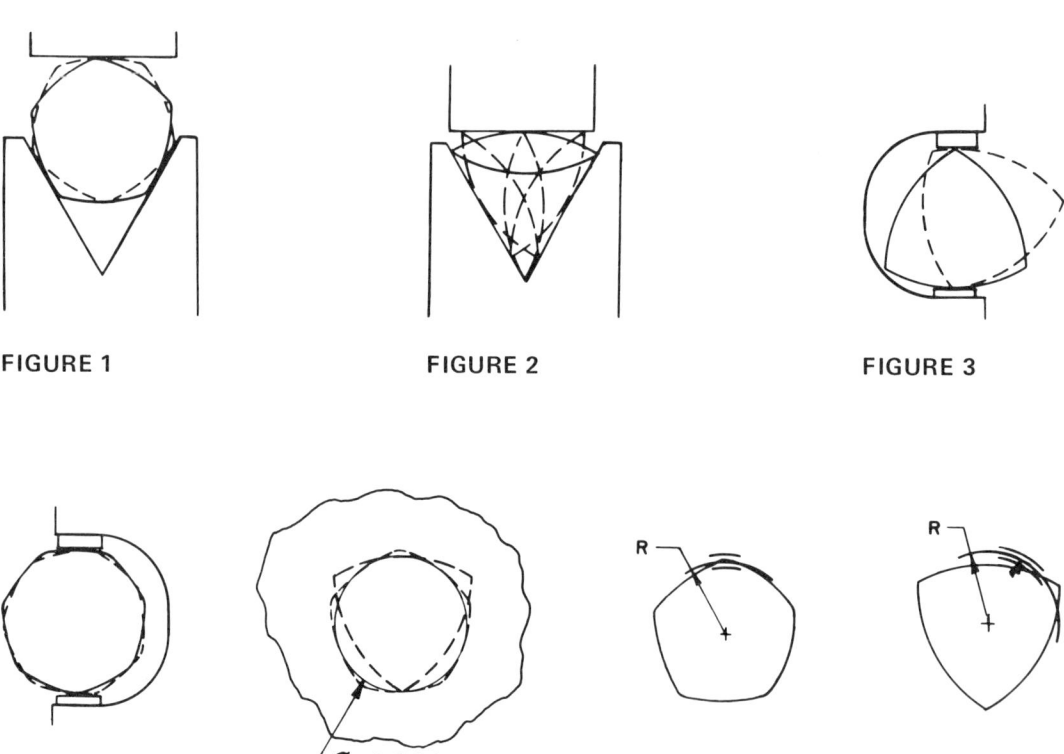

FIGURE 1   FIGURE 2   FIGURE 3

⌀ 2.000 HOLE

FIGURE 4   FIGURE 5   FIGURE 6   FIGURE 7

PRECISION METHOD OF EVALUATING ROUNDNESS AND CYLINDRICITY

To evaluate roundness and cylindricity precisely, one must relate the part surface periphery to the geometry of a perfectly round or cylindrical form as constructed from a reference axis. Several kinds of special gaging equipment utilizing optical, mechanical, electronic, and pneumatic principles are available. One method uses an electronic probe which travels around the periphery of the part while the part is chuck-mounted on an extremely accurate spindle and transcribes an enlarged profile of the part periphery on a polar graph. This profile is then compared with a transparent overlay gage which contains circles at various increments. Note that the final basis for comparison is the part profile only and the reference axis is merely a means of constructing the geometry for measurement. Although more costly and time consuming than other methods, this method utilizes geometric relationships which more directly evaluate roundness and cylindricity where necessary.

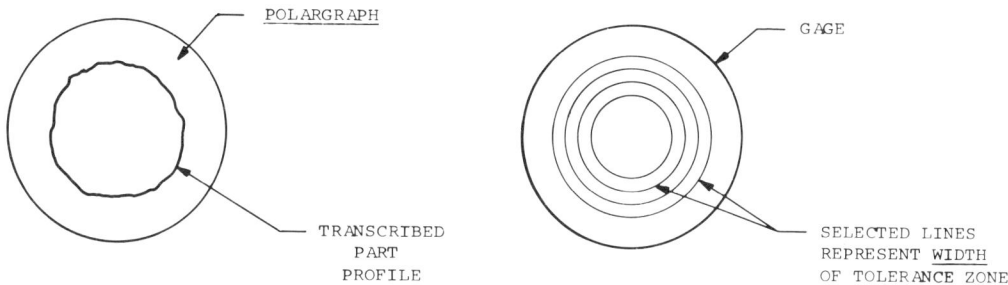

OTHER METHODS OF EVALUATING ROUNDNESS AND CYLINDRICITY

Due to the increasing sophistication of design requirements, manufacturing processes, and measuring processes, such control as roundness and cylindricity may be accomplished by yet further methods.

Such methods as "Least Mean Square," which determines the center of a round form by mathematical formula,

$$X = \frac{2\Sigma X}{N}, \; Y = \frac{2\Sigma Y}{N}, \; R = \frac{\Sigma R}{N} \; ,$$

or a computer program based on ordinate and radius measurements may be used. Other methods, such as "minimum circumscribed circle," "maximum inscribed circle," and "minimum radial separation," which utilize precision spindle techniques can be used in appropriate circumstances.

More specific callout of the drawing requirement indicating one of the above methods (or stylus tip radius, cycles per revolution, etc.) may occasionally be necessary.

DATUMS

BASIS FOR RELATING FEATURES TO ONE ANOTHER

MEANING OF DATUM

Flatness, straightness, roundness, cylindricity (and occasionally profile) tolerancing are characteristics or controls which are applied to single features. It might be said that in order to define these controls, the feature is compared to a perfect counterpart of itself with a stated tolerance to indicate the variation permissible from perfect for that feature. It might also be said that the feature serves as its own "datum"; a datum being a feature from which relationships originate. The following definition will further clarify the meaning and intention for a datum:

A DATUM is a point, line, plane, cylinder, axis, etc. which, for purposes of computation or reference, is assumed to be theoretically exact, and from which the location of part features may be established (or related). Datums are established by, or relative to, actual part features or surfaces.

DATUM VS. DATUM SURFACE AND DATUM FEATURES

A distinction must be made between a "datum" (theoretically exact — as on a drawing), which represents design requirement and function, and the actual "datum surface" or "datum feature" on the produced part.

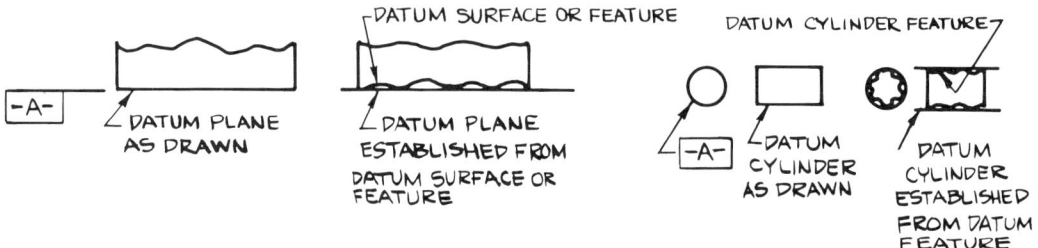

DATUMS REQUIRED TO RELATE FEATURES

When relating form, attitude (orientation) or location tolerances to features in relation to one another, a datum (or datums), must be used. Datum precedence also must be considered when necessary. (See top of following page.)

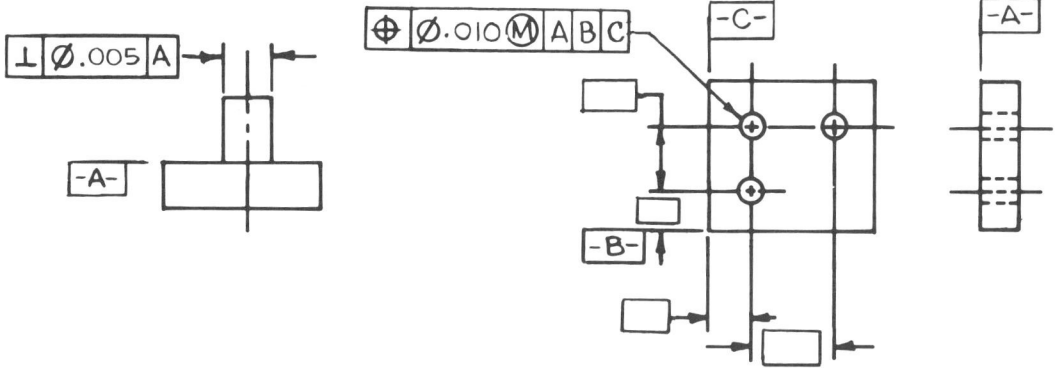

Further examples of datum application using form, attitude, and location characteristics are given in the following sections.

TOLERANCES OF FORM-RELATED FEATURES USING DATUMS

Tolerances of form used on related features, which require a datum involves the characteristics below:

(PERPENDICULARITY)   (ANGULARITY)   (PARALLELISM)

These characteristics are used to describe form tolerances of *single surface, element,* or *size* features and are always related to a *datum.*

See following pages for details of application.

PERPENDICULARITY ⊥
(SQUARENESS, NORMALITY)

Definition. Perpendicularity is the condition of a surface, median plane, or axis which is at exactly 90° to a datum plane or axis.

PERPENDICULARITY TOLERANCE

A perpendicularity tolerance specifies:

1) a tolerance zone defined by two parallel planes perpendicular to a datum plane within which

    a) the surface of a feature must lie (see Fig. 1);
    b) the median plane of a feature must lie (see Fig. 2);

2) a tolerance zone defined by two parallel planes perpendicular to a datum axis within which the axis of a feature must lie (see Fig. 3);

3) a cylindrical tolerance zone perpendicular to a datum plane within which the axis of a feature must lie (see Fig. 4);

4) a tolerance zone defined by two parallel, straight lines perpendicular to a datum plane or datum axis within which an element of the surface must lie (see Fig. 5 — radial perpendicularity).

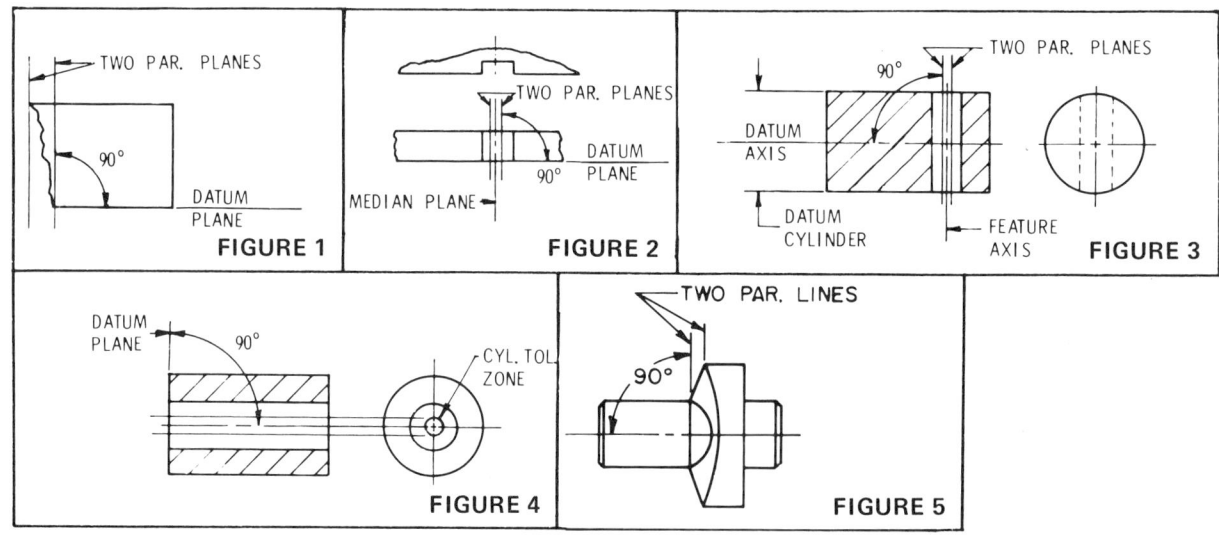

PERPENDICULARITY APPLICATION

The example below illustrates perpendicularity tolerance as applied to a surface.

EXAMPLE

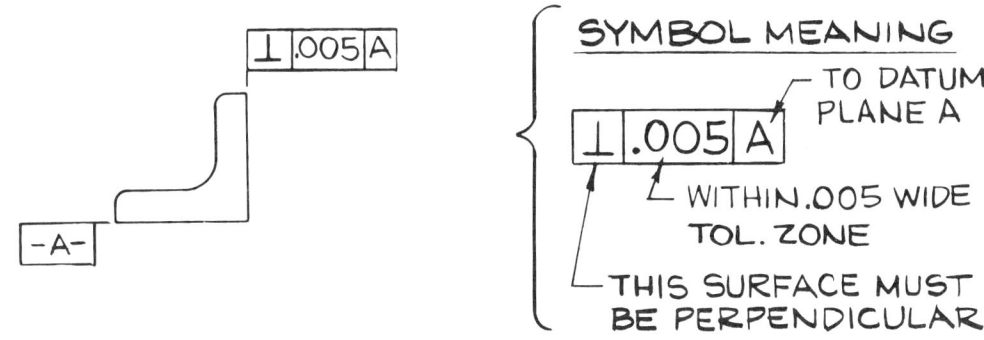

MEANING

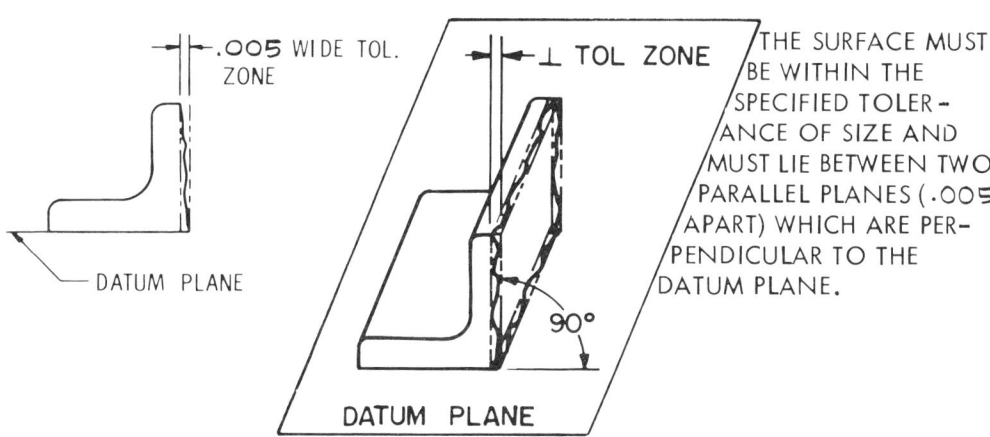

Note that the perpendicularity tolerance applied to a plane surface controls flatness if a flatness tolerance is not specified (that is, the flatness will be at least as good as the perpendicularity).

PERPENDICULARITY ⊥

The examples below and on the following pages show perpendicularity tolerancing under various conditions. Note also how MMC applications permit greater tolerance.

NONCYLINDRICAL FEATURE AT RFS, DATUM A PLANE

EXAMPLE

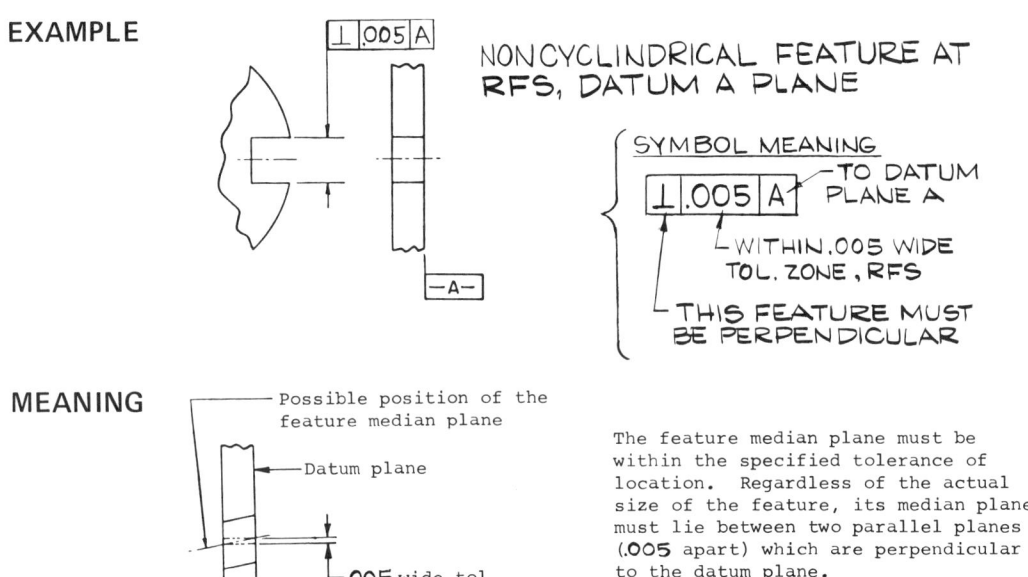

NONCYLINDRICAL FEATURE AT RFS, DATUM A PLANE

SYMBOL MEANING

⊥ .005 A — TO DATUM PLANE A

— WITHIN .005 WIDE TOL. ZONE, RFS

— THIS FEATURE MUST BE PERPENDICULAR

MEANING

Possible position of the feature median plane

Datum plane

.005 wide tol. zone

The feature median plane must be within the specified tolerance of location. Regardless of the actual size of the feature, its median plane must lie between two parallel planes (.005 apart) which are perpendicular to the datum plane.

NONCYLINDRICAL FEATURE AT MMC, DATUM A PLANE

EXAMPLE

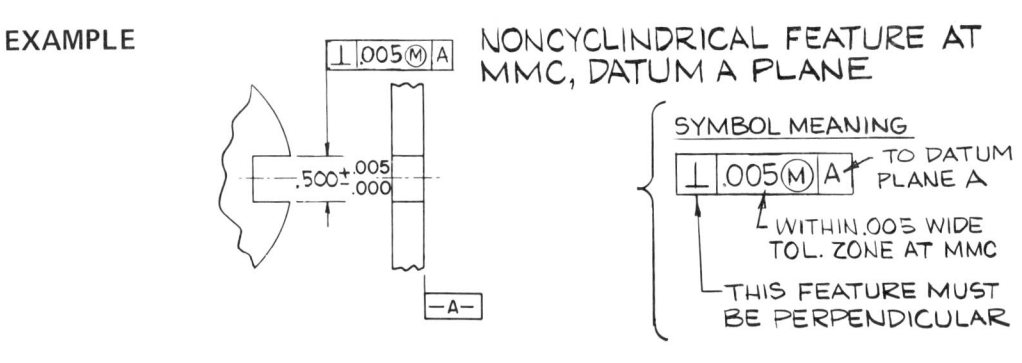

NONCYLINDRICAL FEATURE AT MMC, DATUM A PLANE

SYMBOL MEANING

⊥ .005 Ⓜ A — TO DATUM PLANE A

— WITHIN .005 WIDE TOL. ZONE AT MMC

— THIS FEATURE MUST BE PERPENDICULAR

MEANING

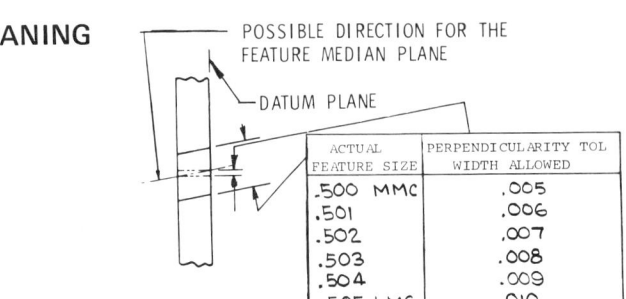

POSSIBLE DIRECTION FOR THE FEATURE MEDIAN PLANE

DATUM PLANE

ACTUAL FEATURE SIZE	PERPENDICULARITY TOL WIDTH ALLOWED
.500 MMC	.005
.501	.006
.502	.007
.503	.008
.504	.009
.505 LMC	.010

THE FEATURE MEDIAN PLANE MUST BE WITHIN THE SPECIFIED TOLERANCE OF LOCATION. WHEN THE FEATURE IS AT MAXIMUM MATERIAL CONDITION (.500) THE MAXIMUM PERPENDICULARITY TOLERANCE IS .005 WIDE. WHERE THE FEATURE IS LARGER THAN ITS SPECIFIED MINIMUM SIZE, AN INCREASE IN THE PERPENDICULARITY TOLERANCE IS ALLOWED.

CYLINDRICAL FEATURE AT RFS, DATUM A CYLINDER RFS

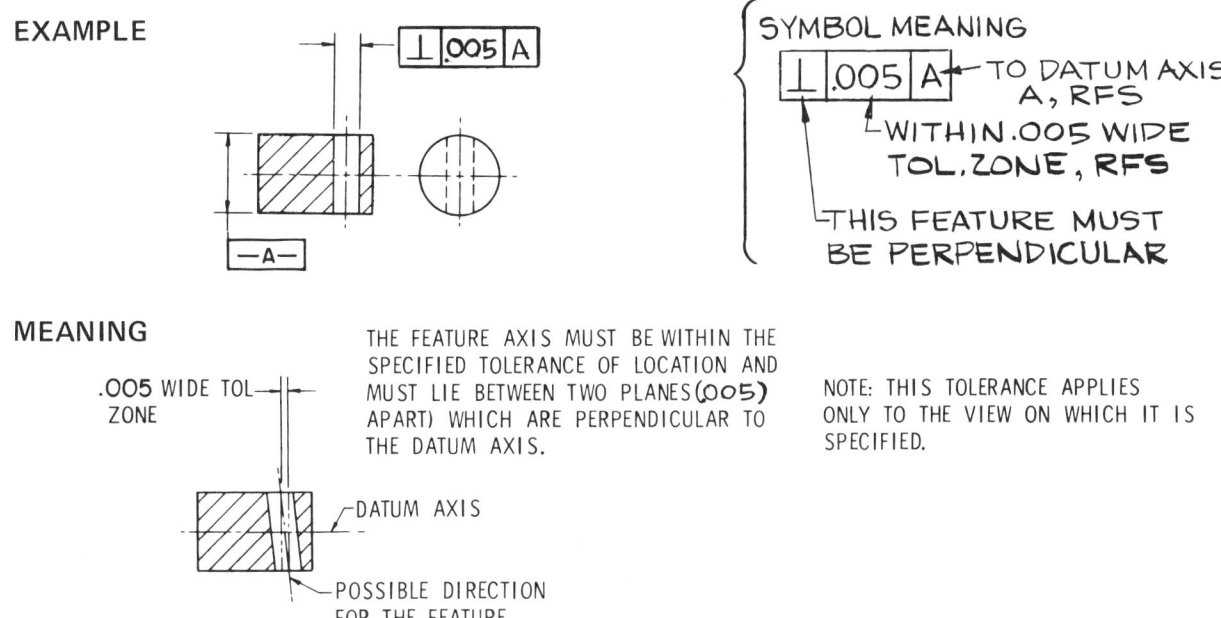

EXAMPLE

⊥ | .005 | A

SYMBOL MEANING

⊥ | .005 | A — TO DATUM AXIS A, RFS
— WITHIN .005 WIDE TOL. ZONE, RFS
— THIS FEATURE MUST BE PERPENDICULAR

MEANING

.005 WIDE TOL ZONE

DATUM AXIS

POSSIBLE DIRECTION FOR THE FEATURE AXIS

THE FEATURE AXIS MUST BE WITHIN THE SPECIFIED TOLERANCE OF LOCATION AND MUST LIE BETWEEN TWO PLANES (.005) APART) WHICH ARE PERPENDICULAR TO THE DATUM AXIS.

NOTE: THIS TOLERANCE APPLIES ONLY TO THE VIEW ON WHICH IT IS SPECIFIED.

CYLINDRICAL FEATURE AT MMC, DATUM A PLANE

Note that the ∅ symbol is required to indicate a diameter (cylindrical) tolerance zone.

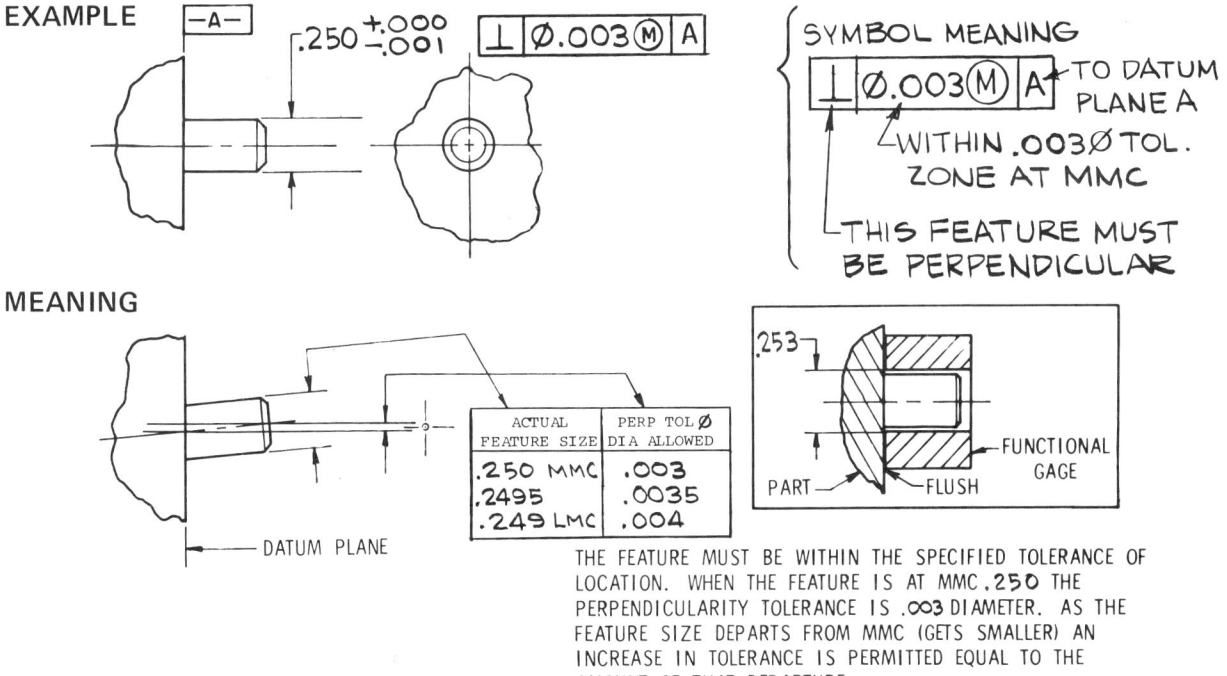

EXAMPLE

−A−
.250 +.000 −.001
⊥ | ∅.003 Ⓜ | A

SYMBOL MEANING

⊥ | ∅.003 Ⓜ | A — TO DATUM PLANE A
— WITHIN .003∅ TOL. ZONE AT MMC
— THIS FEATURE MUST BE PERPENDICULAR

MEANING

DATUM PLANE

ACTUAL FEATURE SIZE	PERP TOL ∅ DIA ALLOWED
.250 MMC	.003
.2495	.0035
.249 LMC	.004

.253

PART — FLUSH — FUNCTIONAL GAGE

THE FEATURE MUST BE WITHIN THE SPECIFIED TOLERANCE OF LOCATION. WHEN THE FEATURE IS AT MMC .250 THE PERPENDICULARITY TOLERANCE IS .003 DIAMETER. AS THE FEATURE SIZE DEPARTS FROM MMC (GETS SMALLER) AN INCREASE IN TOLERANCE IS PERMITTED EQUAL TO THE AMOUNT OF THAT DEPARTURE.

PERPENDICULARITY ⊥

CYLINDRICAL FEATURE AT MMC, DATUM A PLANE

Note that the ⌀ symbol is required to indicate a diameter (cylindrical) tolerance zone.

EXAMPLE

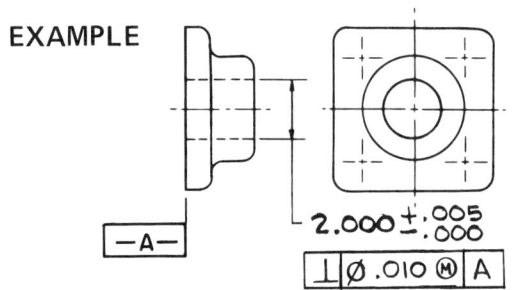

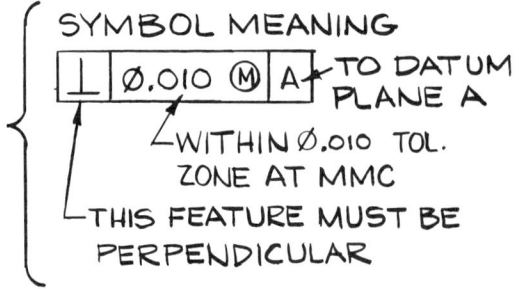

SYMBOL MEANING

⊥ | ⌀.010 Ⓜ | A ← TO DATUM PLANE A

WITHIN ⌀.010 TOL. ZONE AT MMC

THIS FEATURE MUST BE PERPENDICULAR

MEANING

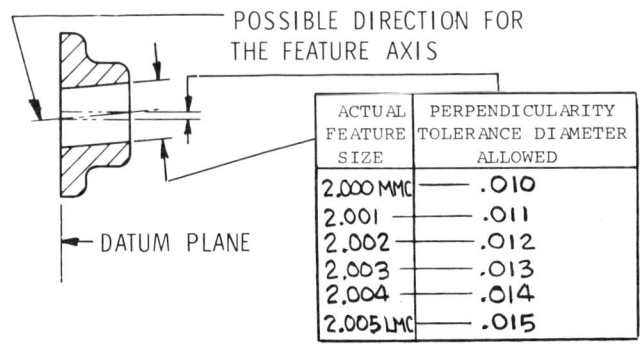

POSSIBLE DIRECTION FOR THE FEATURE AXIS

DATUM PLANE

ACTUAL FEATURE SIZE	PERPENDICULARITY TOLERANCE DIAMETER ALLOWED
2.000 MMC	.010
2.001	.011
2.002	.012
2.003	.013
2.004	.014
2.005 LMC	.015

WHEN THE FEATURE IS AT MMC (2.000), THE MAXIMUM PERPEN-DICULARITY TOLERANCE IS .010 DIAMETER. AS THE FEATURE SIZE DEPARTS FROM MMC (GETS LARGER), AN INCREASE IN TOL-ERANCE IS PERMITTED EQUAL TO THE AMOUNT OF THAT DEPARTURE.

RADIAL PERPENDICULARITY

EXAMPLE

⊥ | .001 | A
EACH RADIAL ELEMENT

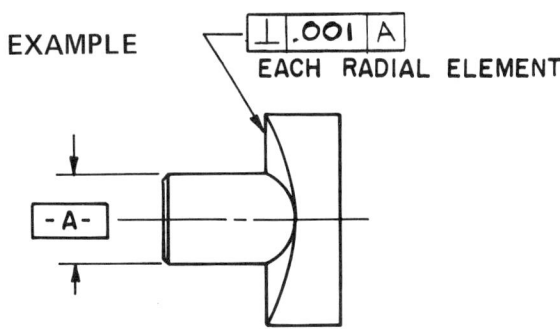

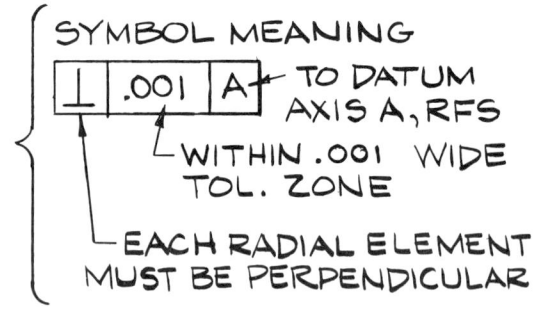

SYMBOL MEANING

⊥ | .001 | A ← TO DATUM AXIS A, RFS

WITHIN .001 WIDE TOL. ZONE

EACH RADIAL ELEMENT MUST BE PERPENDICULAR

MEANING

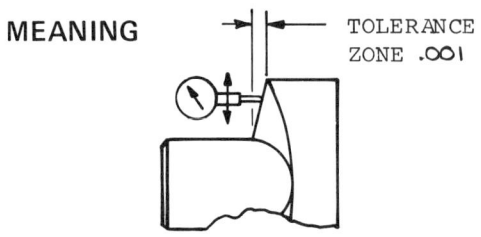

TOLERANCE ZONE .001

TRAVEL OF THE INDICATOR IS IN A RADIAL DIRECTION WITH PART HELD STATIONARY. EACH RADIAL ELEMENT OF THE SURFACE MUST BE WITHIN THE SPECIFIED TOLER ANCE OF SIZE AND MUST LIE BETWEEN TWO PARALLEL LINES (.001 APART) WHICH ARE PERPENDICULAR TO THE AXIS OF DIAMETER A.

Where "element" control is required and the perfect form at MMC requirement of Rule 1 is not applicable, the method below may be used (refer also to p. 23).

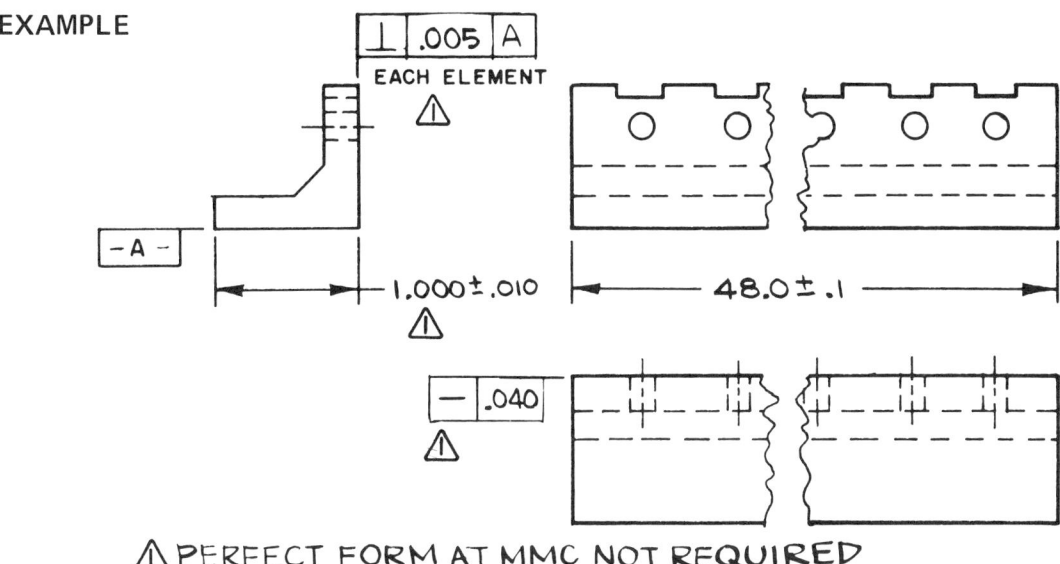

EXAMPLE

⚠ PERFECT FORM AT MMC NOT REQUIRED

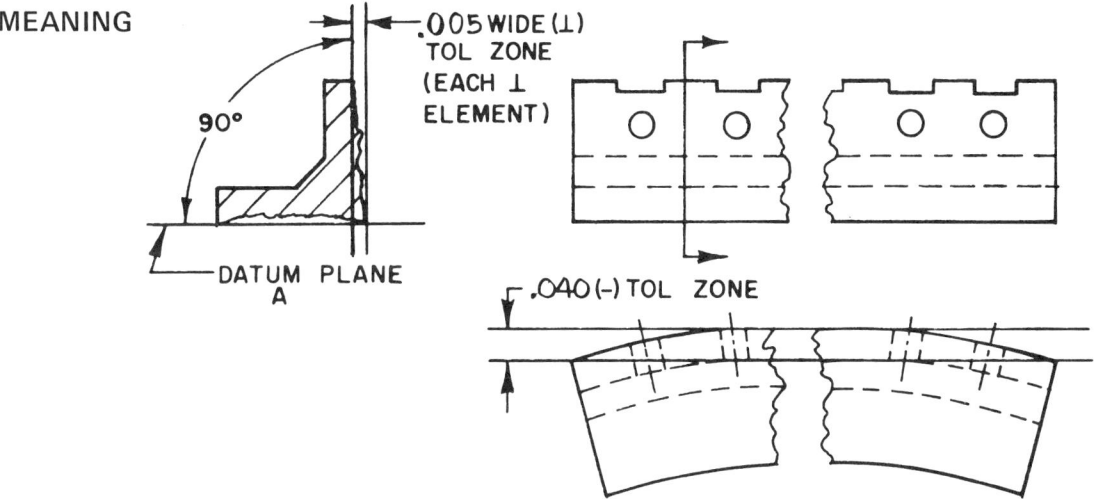

MEANING

Each perpendicular element of the surface at any location along the entire length of the surface must lie between two parallel lines (.005 apart) which are perpendicular to datum plane A. Each longitudinal element of the surface must lie between two parallel straight lines (.040 apart). The part perfect form of the MMC envelope may be exceeded but the stated tolerances are the maximum form error permissible on the surface.

PERPENDICULARITY ⊥

When perpendicularity tolerancing is critical, it may be necessary to limit the *tolerance* deviation to an amount equal to the feature *size* deviation from MMC. This assumes that the part form must be perfect at MMC size and that the virtual condition (size) can be no greater than that at MMC. The only permissible form tolerance must be acquired from the variation in part size (see Example 1 below) in the increase of the feature hole size.

Example 2 limits the tolerance acquired from the feature size increase to a MAX amount.

ZERO TOLERANCE AT MAXIMUM MATERIAL CONDITION

Note that the ⌀ symbol is required to indicate a diameter (cylindrical) tolerance zone.

EXAMPLE 1

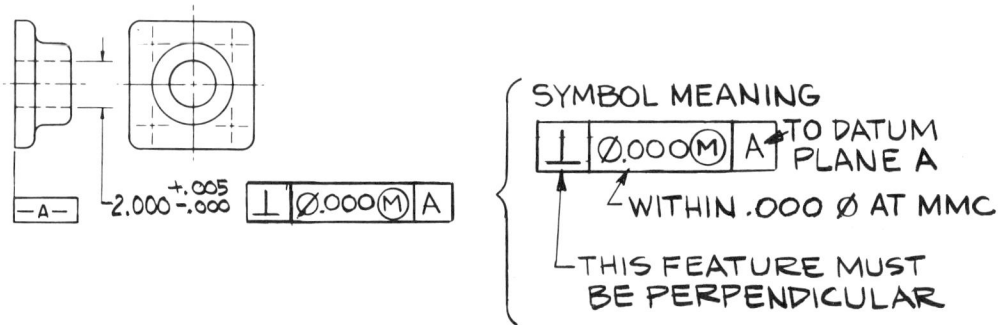

MEANING

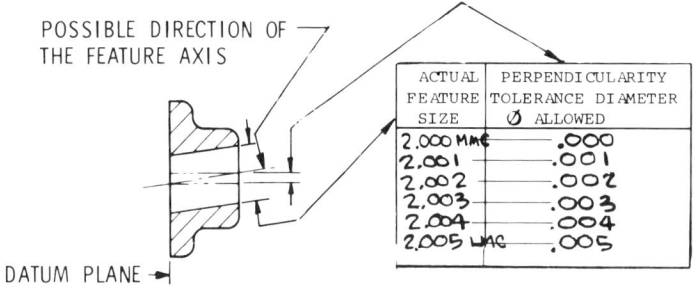

WHEN THE FEATURE IS AT MMC 2.000 ITS AXIS MUST BE PERPENDICULAR TO THE DATUM PLANE. AS THE FEATURE SIZE DEPARTS FROM MMC (GETS LARGER) AN INCREASE IN TOL. IS PERMITTED EQUAL TO THE AMOUNT OF THAT DEPARTURE.

ZERO TOLERANCE AT MMC, MAX DEVIATION

EXAMPLE 2

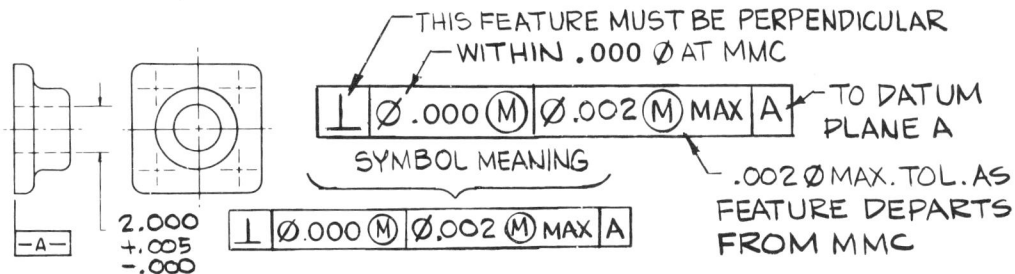

THIS FEATURE MUST BE PERPENDICULAR
WITHIN .000 Ø AT MMC

⊥ | Ø.000 Ⓜ | Ø.002 Ⓜ MAX | A

SYMBOL MEANING

TO DATUM PLANE A

.002 Ø MAX. TOL. AS FEATURE DEPARTS FROM MMC

2.000
+.005
-.000

-A-

⊥ | Ø.000 Ⓜ | Ø.002 Ⓜ MAX | A

MEANING

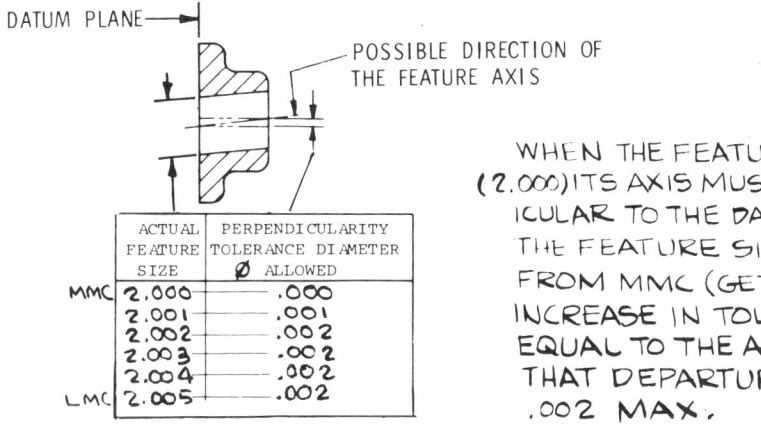

DATUM PLANE

POSSIBLE DIRECTION OF THE FEATURE AXIS

ACTUAL FEATURE SIZE	PERPENDICULARITY TOLERANCE DIAMETER Ø ALLOWED
MMC 2.000	.000
2.001	.001
2.002	.002
2.003	.002
2.004	.002
LMC 2.005	.002

WHEN THE FEATURE IS AT MMC (2.000) ITS AXIS MUST BE PERPENDICULAR TO THE DATUM PLANE. AS THE FEATURE SIZE DEPARTS FROM MMC (GETS LARGER) AN INCREASE IN TOL. IS PERMITTED EQUAL TO THE AMOUNT OF THAT DEPARTURE. UP TO .002 MAX.

ANGULARITY ∠

Definition. Angularity is the condition of a surface, axis, or median plane which is at the specified angle (other than 90°) from a datum plane or axis.

ANGULARITY TOLERANCE

Angularity tolerance is the distance between two parallel planes, inclined at the specified angle to a datum plane or axis, within which the toleranced surface, axis, or median plane must lie.

ANGULARITY APPLICATION

The example shows a part with a surface angular requirement. Note that the symbol is interpreted as "This surface must be at 45° angle in relation to datum plane A within .005 wide tolerance zone in relation to datum plane A."

The interpretation shows how the tolerance zone is established. Note that the angular tolerance zone is at 45° BASIC (exact) from the datum plane A. To be acceptable, the entire angular surface must fall within this tolerance zone. The angular surface must be contained within the limits of part size.

Note in the lower portion of the interpretation that the tolerance zone is also affected and established by the surface itself. That is, the surface extremities actually determine one plane of the tolerance zone as it bottoms out while the plane is inclined at an exact 45° angle with reference to the datum plane.

PLANE SURFACES

EXAMPLE

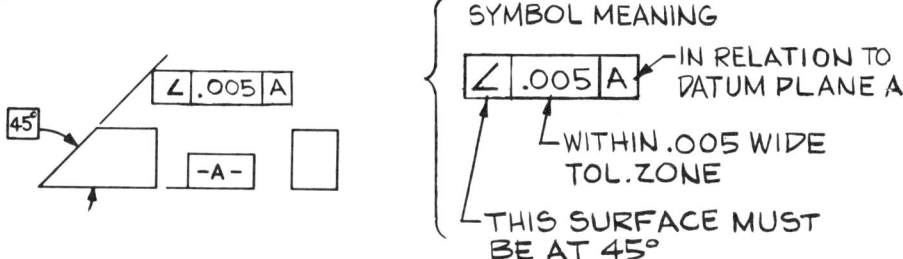

SYMBOL MEANING

∠ .005 A

IN RELATION TO DATUM PLANE A

WITHIN .005 WIDE TOL. ZONE

THIS SURFACE MUST BE AT 45°

MEANING

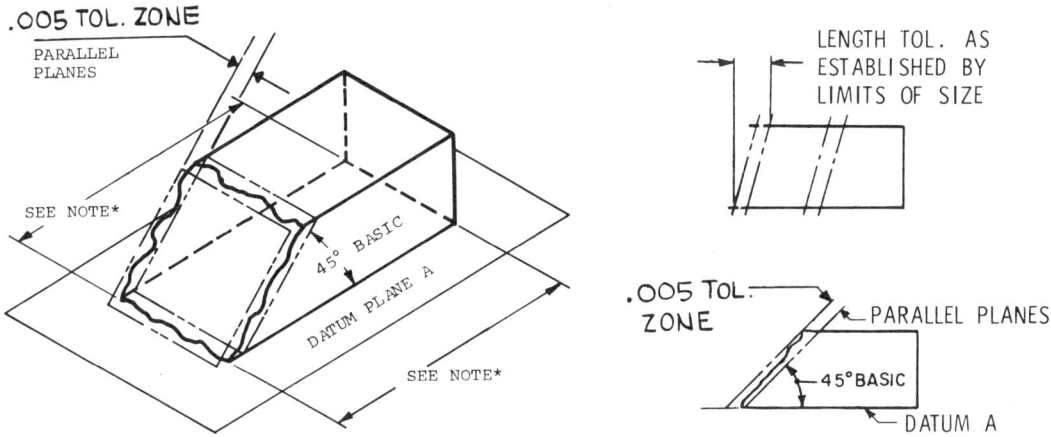

.005 TOL. ZONE

PARALLEL PLANES

SEE NOTE*

45° BASIC

DATUM PLANE A

SEE NOTE*

LENGTH TOL. AS ESTABLISHED BY LIMITS OF SIZE

.005 TOL. ZONE

PARALLEL PLANES

45° BASIC

DATUM A

NOTE*

Part must be within *size* tolerance limits. The angular tolerance zone, composed of two parallel planes .005 apart, is 45° BASIC to the datum plane A. This tolerance zone is established by contact of the outermost of the two planes with the extremities of the angular surface and with the other plane parallel and inward at .005 distance. The entire surface must fall within this tolerance zone to be acceptable. Actually, this also controls the *flatness* of the surface to .005. Note that the angular surface extremities must be within both size and angular tolerance. See the examples below.

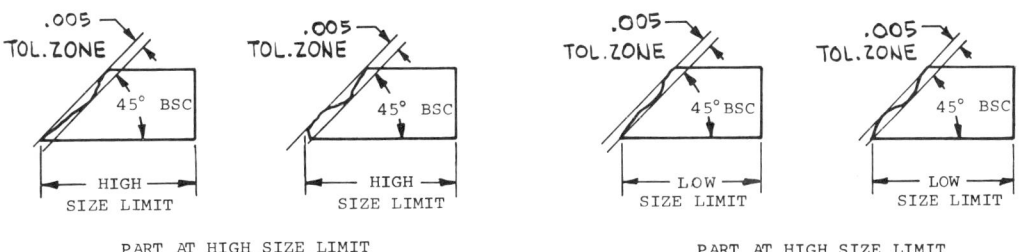

.005 TOL. ZONE — 45° BSC — HIGH SIZE LIMIT

.005 TOL. ZONE — 45° BSC — HIGH SIZE LIMIT

PART AT HIGH SIZE LIMIT

.005 TOL. ZONE — 45° BSC — LOW SIZE LIMIT

.005 TOL. ZONE — 45° BSC — LOW SIZE LIMIT

PART AT HIGH SIZE LIMIT

ANGULARITY ∠

Where angularity of a feature with *size* is to be specified, the median plane or axis of the feature becomes the basis for control. In this case, a *form* tolerance characteristic is used as a postional tolerance control.

Figure 1 illustrates an RFS application of a slot in an angular relationship to three datums. The produced slot median or center plane must be within the .005 total wide tolerance zone regardless of the actual size of the slot or datum. The tolerance zone extremities are defined as two parallel planes parallel to the median plane and perpendicular to datum A, with the vertex of the median plane angle established by the centers of datums B and C, RFS.

In this instance, the form characteristic, *angularity*, is used in a manner which can be considered a positional relationship. Therefore the position characteristic could also have been used. When position is used, the datums and the tolerance require Ⓢ modifiers to state the requirement as RFS, e.g.

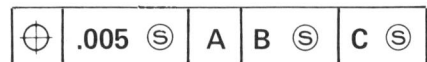

In our example, the RFS angularity specification appears most direct and simple although the multiple datum references may tend to complicate the use of the angularity characteristic. Position and its conventional relationship to a datum system may be considered preferred in this case.

Figure 2 illustrates an MMC application of a slot in an angular relationship to three datums. The produced slot median plane must line within the .005 total wide tolerance zone when the slot and datums B and C are at MMC sizes. The tolerance zone extremities are defined as two parallel planes parallel to the median plane and perpendicular to datum A, with the vertex of the median plane angle established by the centers of datums B and C at MMC size. As the inside slot departs from MMC size (gets larger), an amount equal to that departure is added to the total wide tolerance zone of the slot. At the slot high limit the total wide tolerance zone is .011. As the datums B and C depart from MMC size, further tolerance is added to the total tolerance zone of the slot. Due to the geometry of this part, the amount of added tolerance available relative to the datum departures from MMC size may be difficult to determine. Use of a functional gage will automatically provide the permissible amount of additional tolerance based on actually produced feature size. In this instance the form characteristic, angularity, is again used in a manner which can be considered a positional relationship. Therefore the position characteristic could also have been used.

Figure 3 illustrates an RFS application of a hole in an angular relationship with three datums. The produced hole axis must lie within the .005 total wide tolerance zone regardless of the actual size of the hole or the datums. The tolerance zone extremities are defined as two parallel planes parallel to the axis and perpendicular to datum A, with the vertex of the axis angle established by the centers of datums B and C, RFS.

It should be noted that this type of angularity tolerancing controls the feature axis only in angular orientation of the view shown. Postion control relating the feature to appropriate datums may often provide a more suitable control of a feature of this type.

MEDIAN PLANE

EXAMPLE (RFS)

FIGURE 1

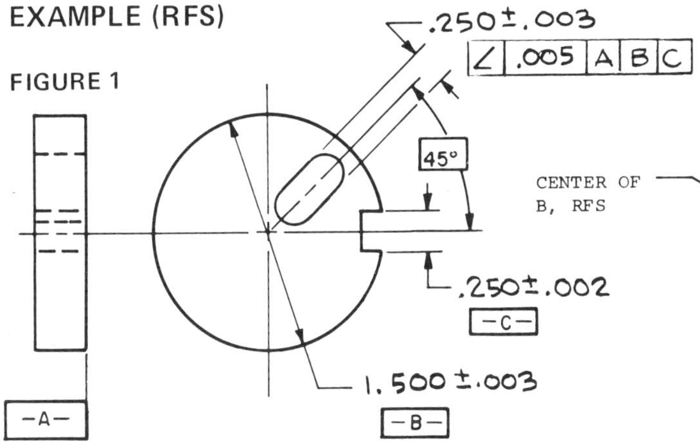

.250 ±.003

∠ | .005 | A | B | C

45°

.250 ±.002

-C-

1.500 ±.003

-B-

-A-

MEANING

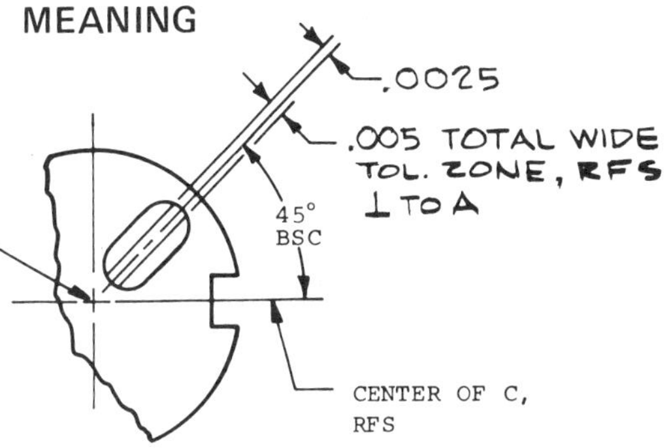

.0025

.005 TOTAL WIDE TOL. ZONE, RFS ⊥ TO A

45° BSC

CENTER OF B, RFS

CENTER OF C, RFS

EXAMPLE (MMC)

FIGURE 2

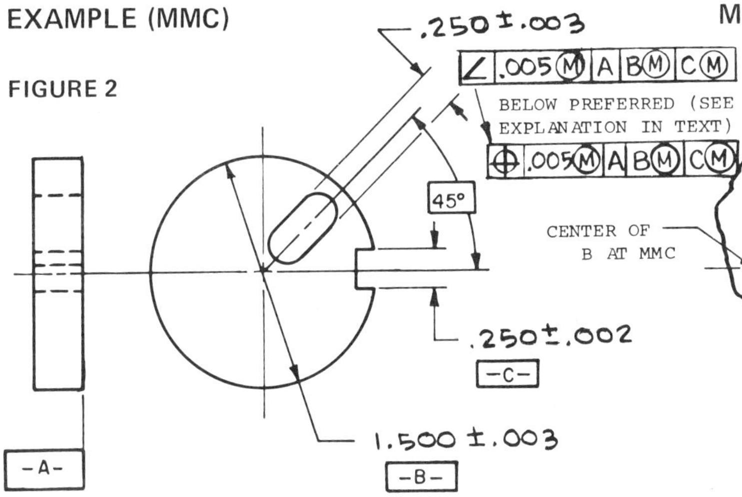

.250 ±.003

∠ | .005 Ⓜ | A | B Ⓜ | C Ⓜ

BELOW PREFERRED (SEE EXPLANATION IN TEXT)

⊕ | .005 Ⓜ | A | B Ⓜ | C Ⓜ

45°

CENTER OF B AT MMC

.250 ±.002

-C-

1.500 ±.003

-B-

-A-

MEANING

.0025 (AT MMC)
.0055 AT HIGH LIMIT

.005 TOTAL WIDE TOL. ZONE AT MMC, .011 TOTAL WIDE TOL ZONE AT HIGH LIMIT SIZE OF SLOT * ⊥ TO A

45° BSC

CENTER OF C AT MMC

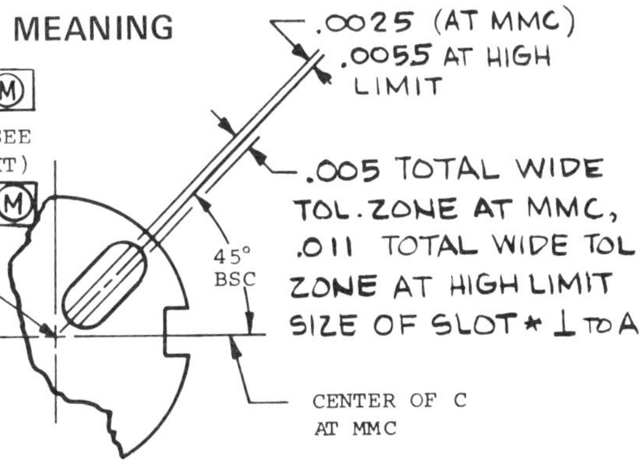

* SLOT TOL ZONE HAS FURTHER IN-CREASE RELATIVE TO DEPARTURE FROM MMC SIZE OF DATUMS B & C.

AXIS

EXAMPLE (RFS)

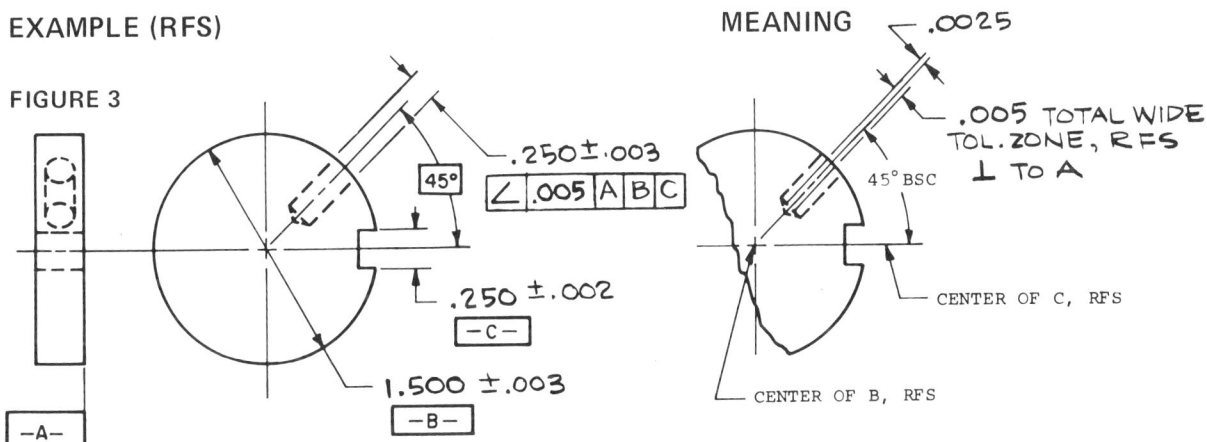

FIGURE 3

.250 ±.003

∠ | .005 | A | B | C

45°

.250 ±.002

-C-

1.500 ±.003

-B-

-A-

MEANING

.0025

.005 TOTAL WIDE TOL. ZONE, RFS ⊥ TO A

45° BSC

CENTER OF C, RFS

CENTER OF B, RFS

PARALLELISM //

Definition. Parallelism is the condition of a surface or axis which is equidistant at all points from a datum plane or axis.

PARALLELISM TOLERANCE

A parallelism tolerance specifies:

1) a tolerance zone defined by two planes or lines parallel to a datum plane (or axis) within which the considered feature (axis or surface) must lie (see Figs. 1 and 2);

2) a cylindrical tolerance zone parallel to a datum axis within which the axis of the feature under consideration must lie (see Fig. 3).

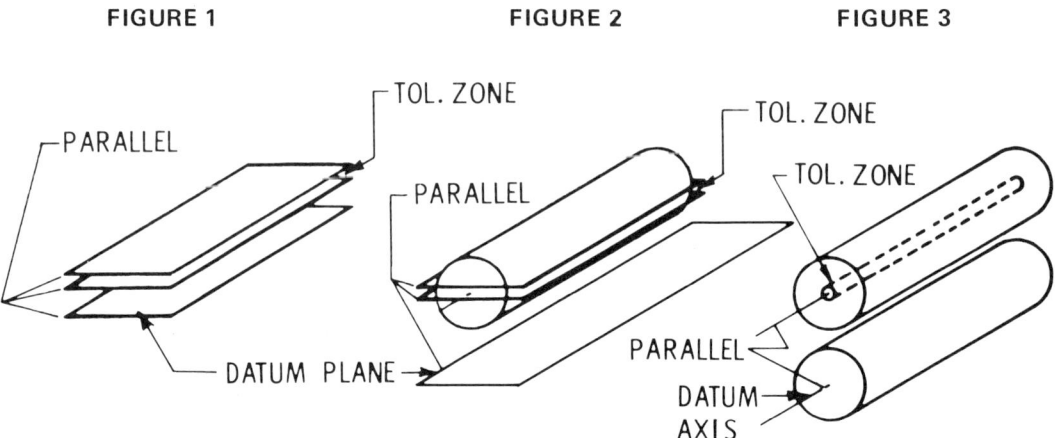

FIGURE 1   FIGURE 2   FIGURE 3

PARALLELISM APPLICATION

Note in the following example that the bottom surface has been selected as the datum and the top surface is to be parallel to datum plane A within .002.

The MEANING beneath the example clarifies the symbol: it reads, ''This feature must be parallel within .002 to datum plane A.

The lower example illustrates the tolerance zone and the manner in which the surface must fall within the tolerance zone to be acceptable. Note that the tolerance zone is established parallel to the datum plane A. Note also that the parallelism tolerance, when applied to a plane surface, controls flatness if a flatness tolerance is not specified (that is, the implied flatness will be *at least* as good as the parallelism).

SURFACE TO DATUM PLANE

EXAMPLE

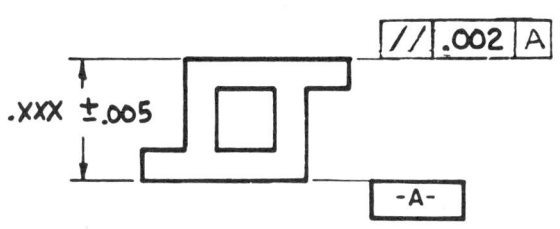

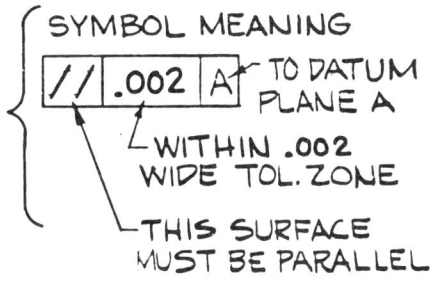

SYMBOL MEANING

// .002 A ← TO DATUM PLANE A
└ WITHIN .002 WIDE TOL. ZONE
└ THIS SURFACE MUST BE PARALLEL

MEANING

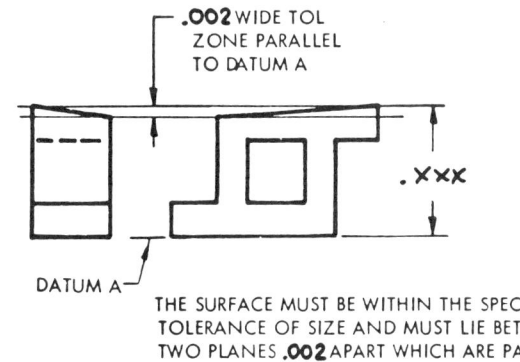

.002 WIDE TOL ZONE PARALLEL TO DATUM A

.XXX

DATUM A

THE SURFACE MUST BE WITHIN THE SPECIFIED TOLERANCE OF SIZE AND MUST LIE BETWEEN TWO PLANES .002 APART WHICH ARE PARALLEL TO THE DATUM PLANE.

EXAMPLE   MEANING

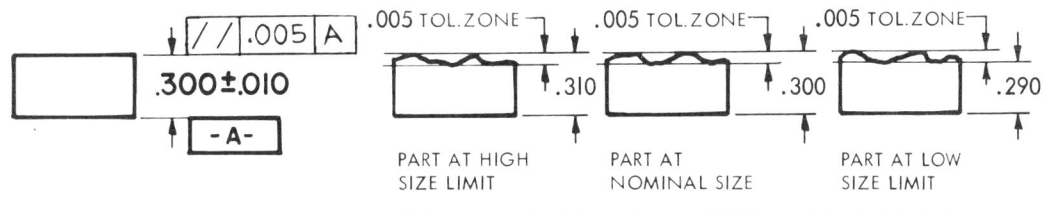

.005 TOL. ZONE ┐   .005 TOL. ZONE ┐   .005 TOL. ZONE ┐

.310   .300   .290

PART AT HIGH SIZE LIMIT   PART AT NOMINAL SIZE   PART AT LOW SIZE LIMIT

(POSSIBLE VARIATION OF PARALLELISM TOLERANCE ZONE WITHIN SIZE TOLERANCE RANGE)

PARALLELISM //

PARALLELISM — FEATURES OF SIZE

This page and the one following cover the application of parallelism tolerancing to features of size. Note how the MMC applications give the possibility of greater tolerance.

FEATURE AT RFS, DATUM A PLANE

EXAMPLE

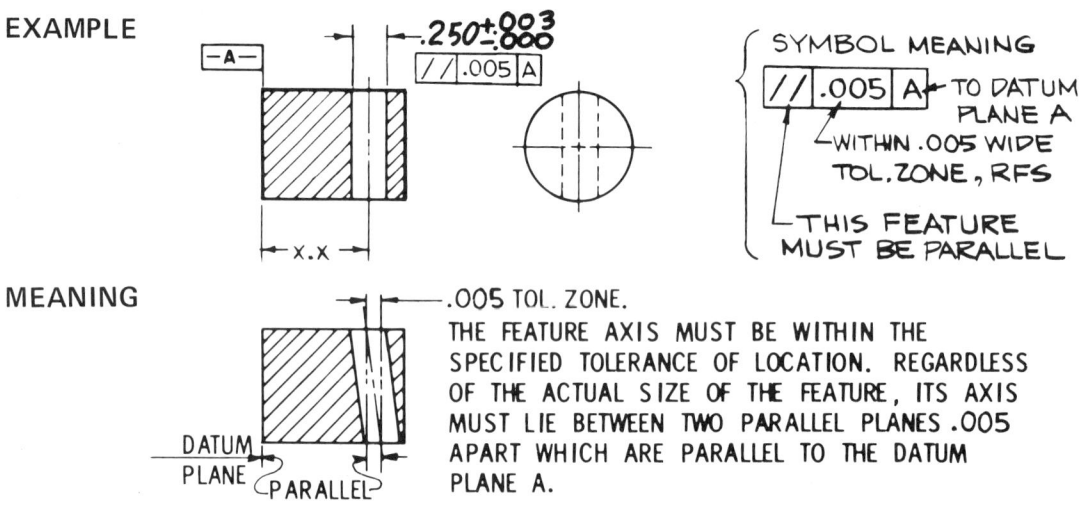

SYMBOL MEANING

$\boxed{//\ \boxed{.005}\ \boxed{A}}$ — TO DATUM PLANE A

— WITHIN .005 WIDE TOL. ZONE, RFS

— THIS FEATURE MUST BE PARALLEL

MEANING

.005 TOL. ZONE.

THE FEATURE AXIS MUST BE WITHIN THE SPECIFIED TOLERANCE OF LOCATION. REGARDLESS OF THE ACTUAL SIZE OF THE FEATURE, ITS AXIS MUST LIE BETWEEN TWO PARALLEL PLANES .005 APART WHICH ARE PARALLEL TO THE DATUM PLANE A.

FEATURE AT MMC, DATUM A PLANE

EXAMPLE

SYMBOL MEANING

$\boxed{//\ \boxed{.005\ \text{Ⓜ}}\ \boxed{A}}$ — TO DATUM PLANE A

— WITHIN .005 WIDE TOL. ZONE AT MMC

— THIS FEATURE MUST BE PARALLEL

MEANING

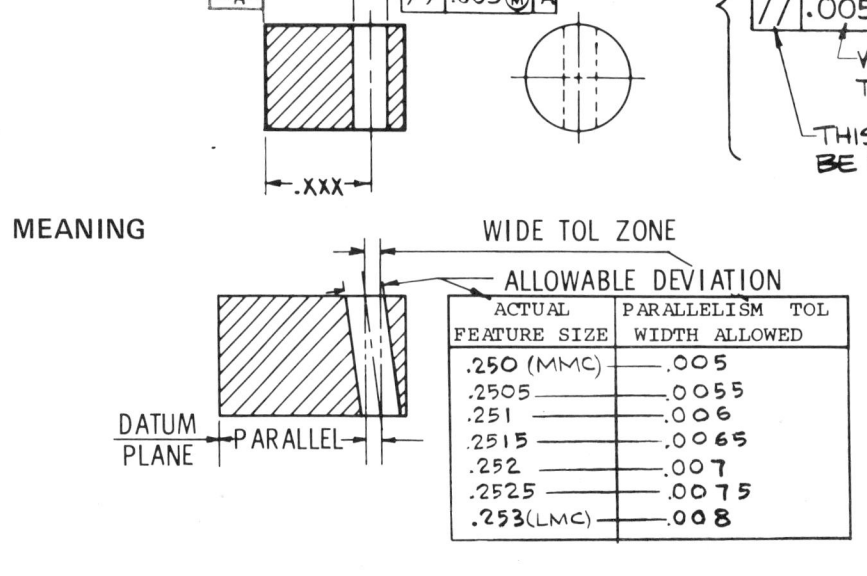

WIDE TOL ZONE

ALLOWABLE DEVIATION

ACTUAL FEATURE SIZE	PARALLELISM TOL WIDTH ALLOWED
.250 (MMC)	.005
.2505	.0055
.251	.006
.2515	.0065
.252	.007
.2525	.0075
.253 (LMC)	.008

FEATURE AXIS MUST BE WITHIN SPECIFIED TOL. OF LOCATION & HOLE (AT MMC) AXIS MUST LIE BETWEEN TWO PARALLEL PLANES .005 APART WHICH ARE PARALLEL TO THE DATUM PLANE A.

WHEN FEATURE SIZE DEPARTS FROM MMC (GETS LARGER), AN INCREASE IN THE PARALLELISM TOL. IS ALLOWED EQUAL TO THE AMOUNT OF THAT DEPARTURE.

FEATURE AT RFS, DATUM FEATURE AT RFS

(Note that the ⌀ symbol is required to indicate a diameter (cylindrical) tolerance zone.)

EXAMPLE

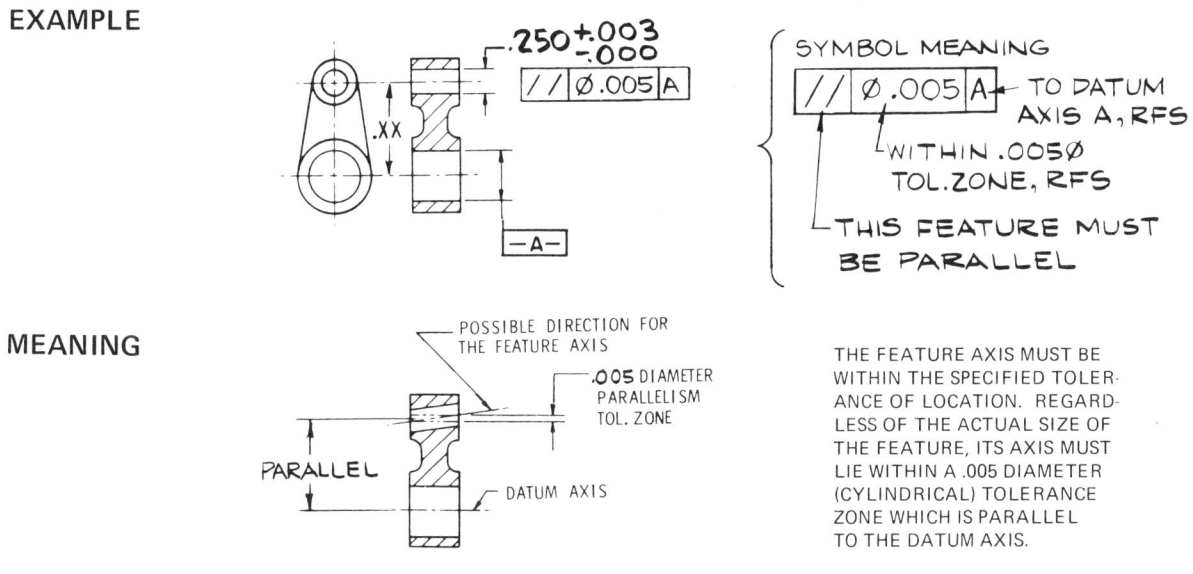

SYMBOL MEANING

// | ⌀ .005 | A ← TO DATUM AXIS A, RFS

WITHIN .005⌀ TOL. ZONE, RFS

THIS FEATURE MUST BE PARALLEL

MEANING

POSSIBLE DIRECTION FOR THE FEATURE AXIS

.005 DIAMETER PARALLELISM TOL. ZONE

PARALLEL

DATUM AXIS

THE FEATURE AXIS MUST BE WITHIN THE SPECIFIED TOLERANCE OF LOCATION. REGARDLESS OF THE ACTUAL SIZE OF THE FEATURE, ITS AXIS MUST LIE WITHIN A .005 DIAMETER (CYLINDRICAL) TOLERANCE ZONE WHICH IS PARALLEL TO THE DATUM AXIS.

FEATURE AT MMC, DATUM FEATURE AT RFS

(Note that the ⌀ symbol is required to indicate a diameter (cylindrical) tolerance zone.)

EXAMPLE

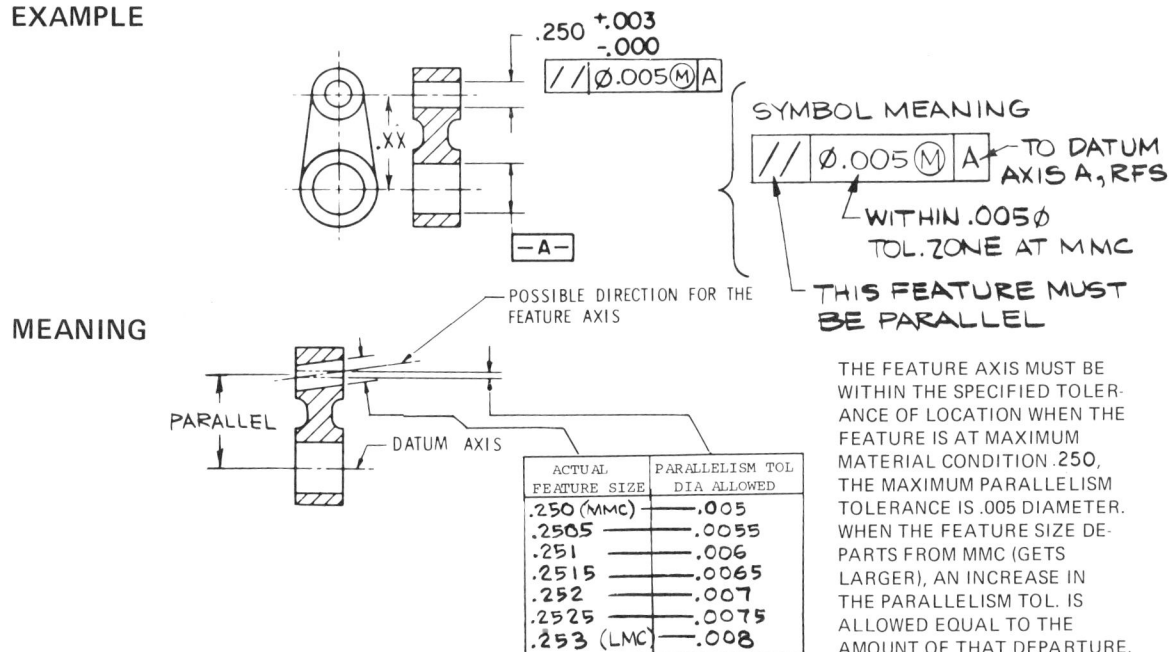

SYMBOL MEANING

// | ⌀ .005 Ⓜ | A ← TO DATUM AXIS A, RFS

WITHIN .005⌀ TOL. ZONE AT MMC

THIS FEATURE MUST BE PARALLEL

MEANING

POSSIBLE DIRECTION FOR THE FEATURE AXIS

PARALLEL

DATUM AXIS

ACTUAL FEATURE SIZE	PARALLELISM TOL DIA ALLOWED
.250 (MMC)	.005
.2505	.0055
.251	.006
.2515	.0065
.252	.007
.2525	.0075
.253 (LMC)	.008

THE FEATURE AXIS MUST BE WITHIN THE SPECIFIED TOLERANCE OF LOCATION WHEN THE FEATURE IS AT MAXIMUM MATERIAL CONDITION .250, THE MAXIMUM PARALLELISM TOLERANCE IS .005 DIAMETER. WHEN THE FEATURE SIZE DEPARTS FROM MMC (GETS LARGER), AN INCREASE IN THE PARALLELISM TOL. IS ALLOWED EQUAL TO THE AMOUNT OF THAT DEPARTURE.

TOLERANCES OF FORM — PROFILE TOLERANCING

APPLIED TO INDIVIDUAL FEATURES (NO DATUM)
OR RELATED FEATURES (USING DATUMS)

Profile tolerancing is of two varieties and involves the characteristics below. According to the design requirement, these characteristics may be applied to an individual feature, such as a *single surface* or *element*, or to related features, such as a *single surface* or *element* relative to a datum or datums.

(PROFILE OF A LINE) (PROFILE OF A SURFACE)

See following pages for details of application.

PROFILE
METHOD OF SPECIFYING

Definition. Profile tolerancing is a method used to specify a uniform amount of variation of a surface or line elements of a surface.

PROFILE TOLERANCE

A profile tolerance (either bilateral or unilateral) specifies a tolerance zone, always* intended and measured normal to the basic profile (applicable to the view in which drawn) at all points of the profile, within which the specified part surface profile or line profile must lie.

PROFILE TOLERANCE APPLICATION

Profile tolerancing is an effective method for controlling lines, arcs, irregular surfaces, or other unusual part profiles. Profile tolerances are usually applied to surface features but may also be applied to a line (element on a feature surface). In either case, these requirements must be specified in association with the desired profile in a plane of projection (view) on the drawing as follows:

a) An APPROPRIATE VIEW or section is drawn which shows the desired basic profile in true shape.
b) The profile is defined by basic dimensions. This dimensioning may be in the form of located radii and angles, or it may consist of coordinate dimensioning to points on the the profile.
c) Depending on design requirements, the tolerance may be divided bilaterally to both sides of the true profile or applied unilaterally to either side of the true profile.

---

*Unless otherwise specified.

Where an equally disposed bilateral tolerance is intended, it is only necessary to show the feature control symbol with a leader directed to the surface. For an unequally disposed or unilateral tolerance, phantom lines are drawn parallel to the true profile to clearly indicate the tolerance zone inside or outside the true profile. One end of a dimension line is extended to the feature control symbol.

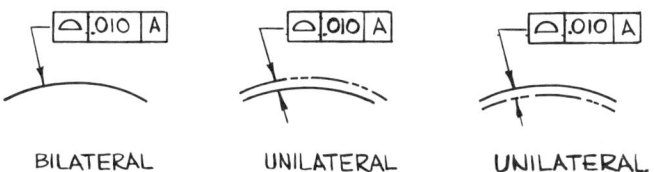

BILATERAL   UNILATERAL   UNILATERAL

d) Other appropriate dimensions, as well as the applicable feature control symbol, are added. The SYMBOL should be applied in a view in which the surface or lines to be controlled is shown as a profile and which pictorially represents the desired feature orientation and relationship.

e) If some limits on a drawing are expressed by a profile tolerance and others by regular tolerance dimensions, the extent or limitation of the profile tolerance must be clearly indicated by reference letters applied at the extremities of the profile controlled portion and a notation (for example, BETWEEN A & B) added with the profile tolerance symbol or notes. Where a profile tolerance applies all around the profile of the part, the notation "ALL AROUND" is placed beneath the feature control symbol.

Common surface profile tolerancing includes a combination of both FORM and SIZE controls, with the profile basic dimensions amounting to overall dimensions as well. In this situation, the standard limits of size interpretation, which require that "form" control must be contained within "size" control (i.e. Rule 1), do *not* apply.

Surface profile tolerancing may also be associated with conventional size dimensions and tolerances. The profile itself, however, must be controlled by basic dimensions and the profile tolerance zone. Under this condition the standard limits of size interpretation (Rule 1) do apply and the profile tolerance zone must be contained *within* the *size* tolerance zone. If, for example, the bilateral profile method is used, some portions of the profile tolerance zone may be sacrificed if the controlled feature is at its extreme size limit at that point of the profile.

Since profile tolerancing is a control of *surface form,* no modifier (MMC) can be used on the feature controlled. Where a datum reference is used, it is usually intended to apply RFS. However, there may be unique applications permitting the use of an MMC datum to expedite gaging provided there is no detrimental effect on the design requirement (e.g., a functional shaft size as related to a cam profile).

GAGING

The example meaning shows the basic gaging principle of measurement *normal to the basic profile.* Gage traverse or part rotation techniques could be used. Where overall size permits, optical comparator (or similar) techniques are usually the most economical and effective method of inspecting this type of part. These techniques make it

PROFILE

possible to compare a blown-up (10X, 20X, etc.) projected shadow or profile of the part with an optical gage chart containing accurately scribed profile tolerance zone limits.

Note, however, that shadow image of the part profile which is seen on the comparator screen will be the extreme profile and may not represent the entire surface. Focus adjustment and surface illumination methods (available on some comparators) may be able to account for the entire surface. If necessary, further inspection may be required to account for surface irregularities which are below the extreme profile or not made visible by the method used.

A larger part exceeding the normal limitations of the optical chart size might be checked in segments or with the use of tracer and reticle.

TWO TYPES OF PROFILE TOLERANCE

In practice, a profile tolerance may be applied to either an entire surface or to individual line element profiles taken at various cross sections through the part. The two types or methods of controlling profile are:

a) Profile of a surface ⌒

The tolerance zone established by profile of a surface tolerance is a three-dimensional zone or total control across the entire length and width or circumference of the feature; it may be applied to parts having a constant cross section or to parts having a surface of revolution. Usually profile of a surface requires datum references.

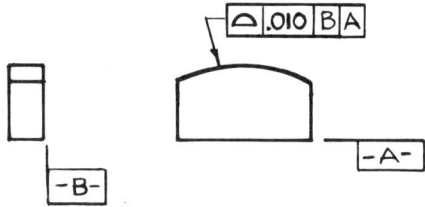

b) Profile of a line ⌒

The tolerance zone established by profile of a line tolerance is a two-dimensional zone extending along the length of the considered feature; it may apply to the profiles of parts having a varying cross section such as a propeller, aircraft wing, nose cone, or to random cross sections of parts where it is not desired to control the entire surface as a single entity. Profile of a line may, or may not, require datum references.

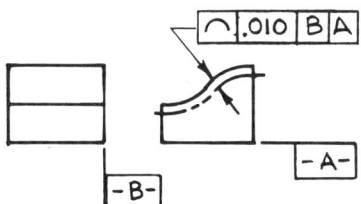

PROFILE OF A SURFACE ⌒

EXAMPLE

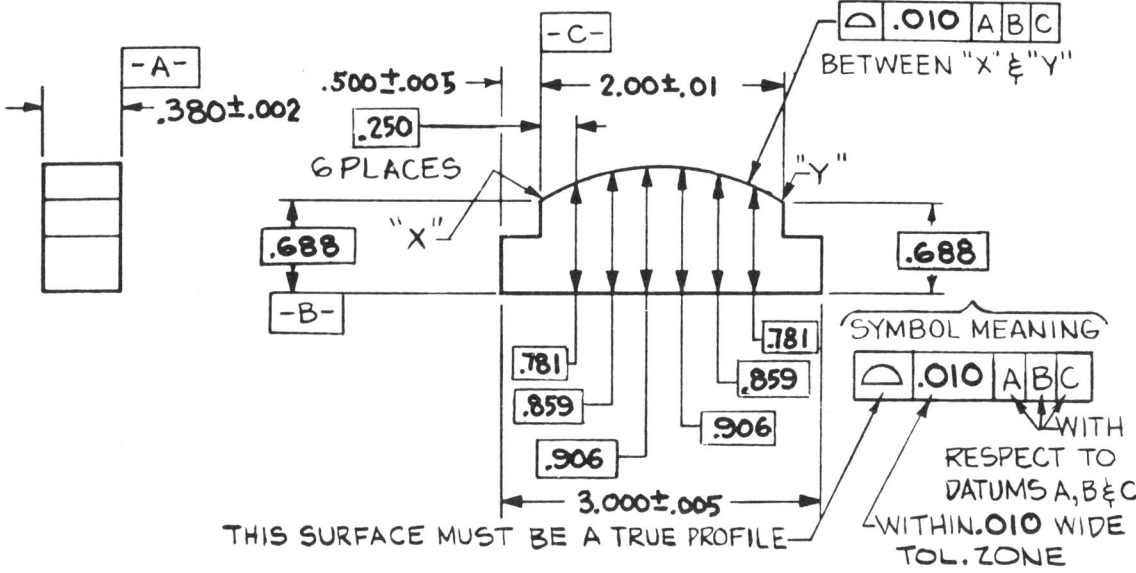

THIS SURFACE MUST BE A TRUE PROFILE

SYMBOL MEANING

WITH RESPECT TO DATUMS A, B & C

WITHIN .010 WIDE TOL. ZONE

MEANING

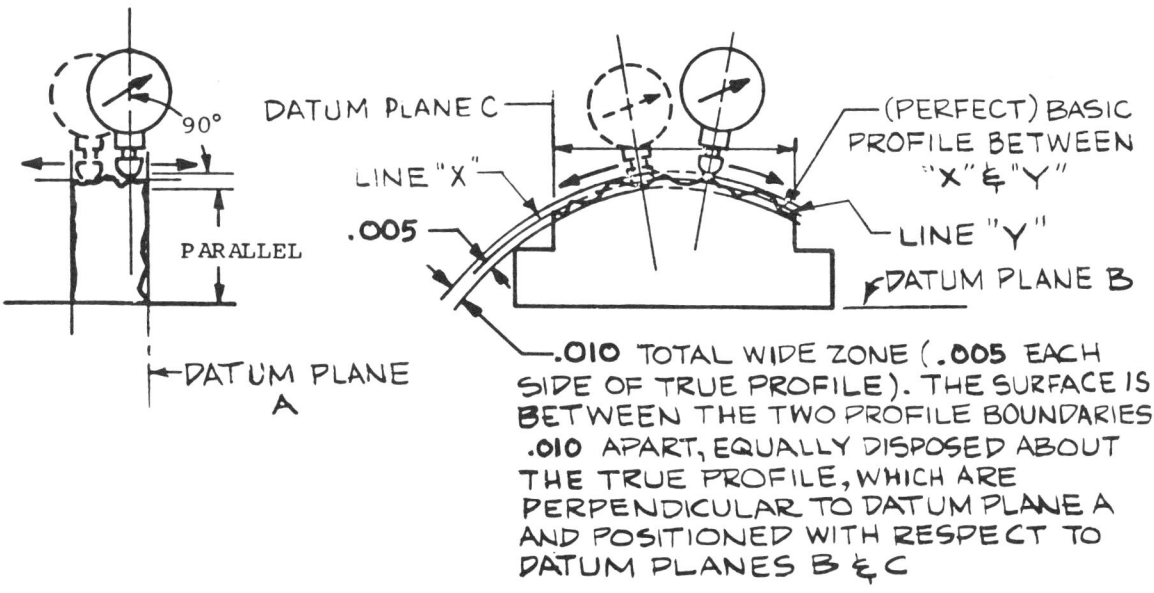

.010 TOTAL WIDE ZONE (.005 EACH SIDE OF TRUE PROFILE). THE SURFACE IS BETWEEN THE TWO PROFILE BOUNDARIES .010 APART, EQUALLY DISPOSED ABOUT THE TRUE PROFILE, WHICH ARE PERPENDICULAR TO DATUM PLANE A AND POSITIONED WITH RESPECT TO DATUM PLANES B & C

PROFILE
PROFILE OF A SURFACE △

Where the profile tolerance is to be used in controlling the entire surface profile, the words "ALL AROUND" should be added to the profile tolerance symbol. *

ZONE TOLERANCE AROUND ENTIRE PROFILE

EXAMPLE

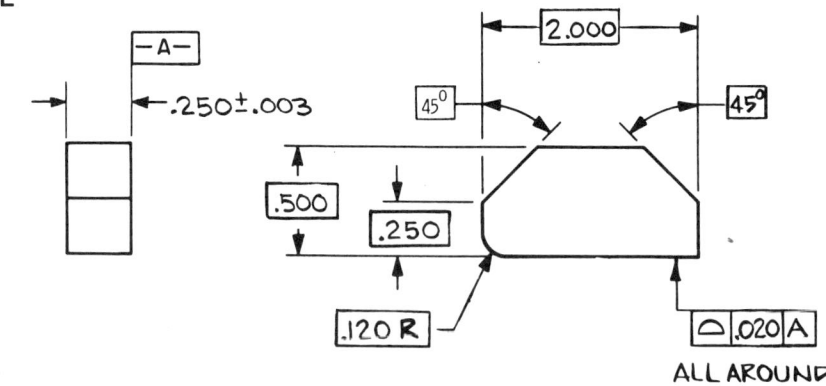

MEANING

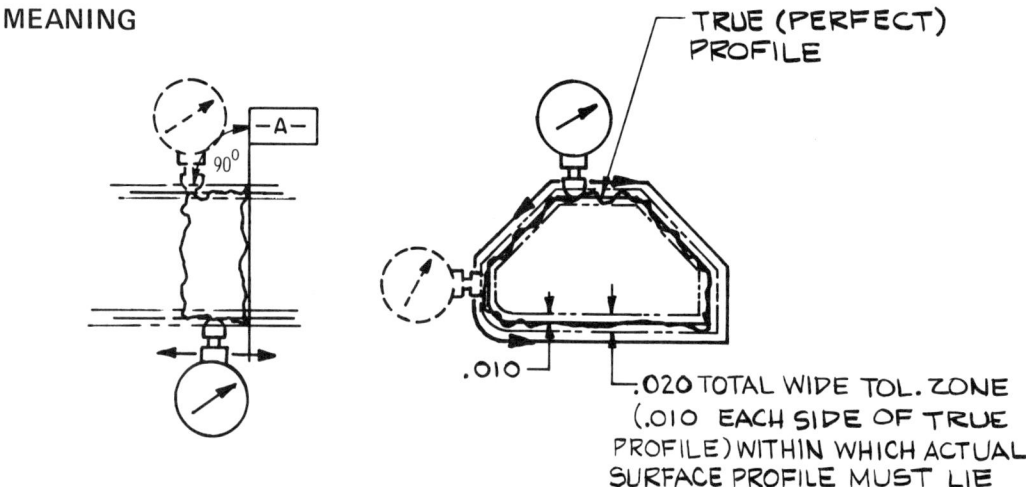

GAGING

The example illustrates the basic gaging principle of measurement *normal to the true profile* as oriented from datum A. Where overall size permits, optical "comparator" (or similar) techniques are usually the most economical and effective method of inspecting this type of part (see also GAGING of preceding example).

Where the profile tolerance is to be used in controlling all, or large portions, of the surface profile, specific notations (such as angular degrees on a cam) should be added with the profile tolerance symbol or note.

This type of tolerance control is particularly useful on cams or similar parts.

---

* Future standards may use new symbol for ALL AROUND: ⊕─| △ | .020 | A | .

DIFFERENT ZONE TOLERANCE AROUND ENTIRE PROFILE

EXAMPLE

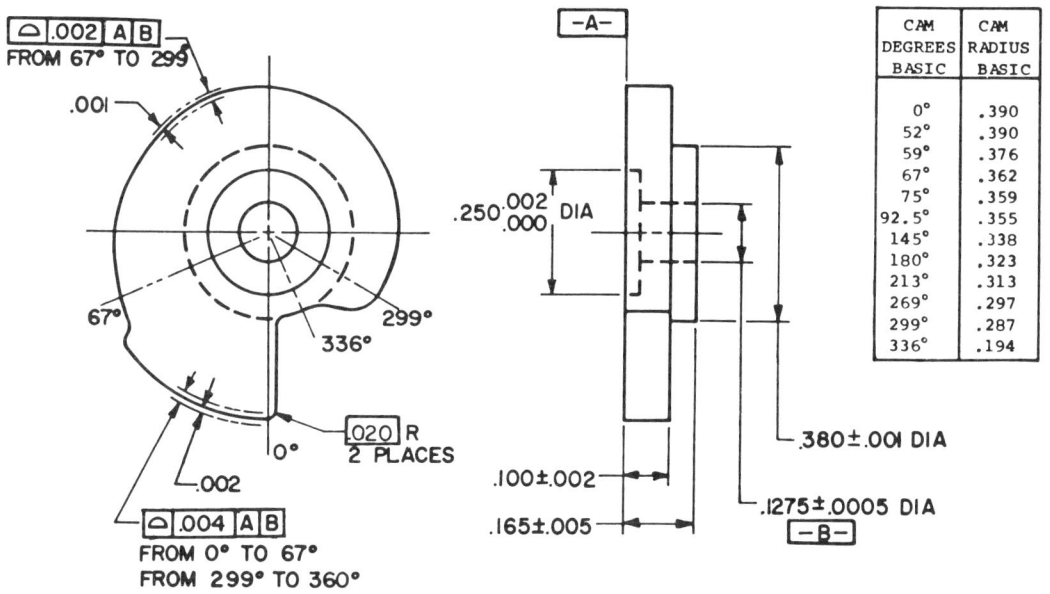

CAM DEGREES BASIC	CAM RADIUS BASIC
0°	.390
52°	.390
59°	.376
67°	.362
75°	.359
92.5°	.355
145°	.338
180°	.323
213°	.313
269°	.297
299°	.287
336°	.194

MEANING

GAGING

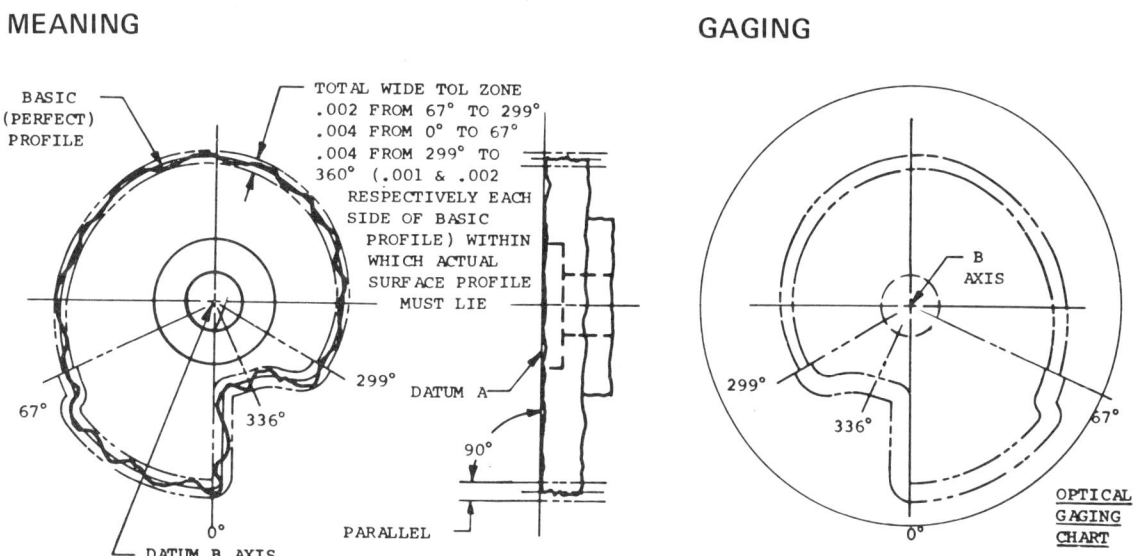

Optical comparator (or similar) techniques are ideal for this part. A transparent chart made to 20X, 50X, etc. size with accurately scribed lines defining the tolerance zones permits a direct surface profile check of the magnified part with the chart. If the surface profile of the part lies within the scribed tolerance zones when oriented to the datum B axis and while bottoming on datum A, it is acceptable.

The profile of a surface controls the entire surface, or all the elements of the surface, within a uniform tolerance zone as established from the true profile. However, where line elements of a surface are to be specifically controlled or controlled as a refinement of size or surface profile control, the profile of a line characteristic may be used. Line profile control is applied in a manner similar to the application of surface profile.

The example at the right illustrates line profile control as a refinement of size. As with surface profile, line profile must be shown in the drawing view in which the control applies. The tolerance zone is established in the same way as surface profile. However, its tolerance zone is disposed about each element of the surface. Therefore the tolerance zone applies for the full length of each element of the surface (in the view in which it is shown), but only for the width or height at a cutting plane bisecting the element. This cutting plane must be considered to be perpendicular or parallel to the part orientation within its size tolerances. Where datum references are specified, the orientation is with respect to the datums to give a more specific relationship. Each actual line element of the controlled profile may vary within its prescribed tolerance zone, but the element-to-element control may vary within the entire size tolerance. It may be seen however that some line profile tolerance could be sacrificed as the actual profile reaches its maximum size limit (Rule 1 applies).

Line profile control is illustrated in the example. Accuracy and consistency of the individual elements at the part surface are critical, but variation from element to element is less critical. More lenient control within the size tolerances is adequate in the direction perpendicular to the line profile. The requirement to hold an accurate line profile on an irregularly shaped part within a more lenient size control is similar to holding straightness within a more lenient size control on a square, rectangular, or cylindrical part.

Surface profile and line profile may be applied to the same feature, where applicable—for example, in situations in whch the line elements are to be controlled more closely than the surface as a whole. Combined surface and line profile control may be used and specified as shown in the figure below.

EXAMPLE

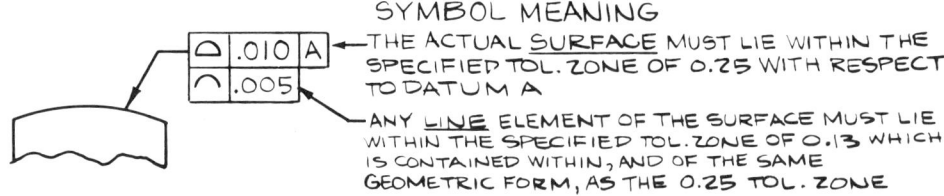

SYMBOL MEANING

THE ACTUAL SURFACE MUST LIE WITHIN THE SPECIFIED TOL. ZONE OF 0.25 WITH RESPECT TO DATUM A

ANY LINE ELEMENT OF THE SURFACE MUST LIE WITHIN THE SPECIFIED TOL. ZONE OF 0.13 WHICH IS CONTAINED WITHIN, AND OF THE SAME GEOMETRIC FORM, AS THE 0.25 TOL. ZONE

EXAMPLE

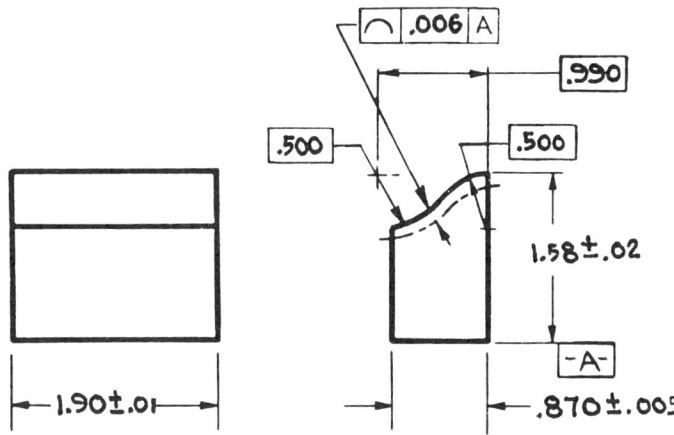

MEANING

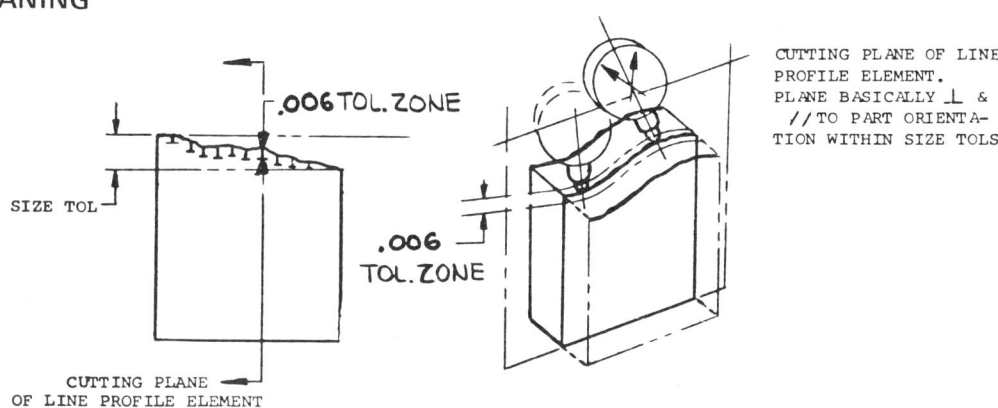

Any profile line element of the surface (elements in the view in which the symbol is shown and element represented by the basic profile) must be within the specified .006 tolerance zone which is equally disposed about the basic profile. The *profile* tolerance zone must be contained within the *size* tolerance zone.

GAGING

The illustration above shows the basic gaging principle of measurement *normal to the basic profile* and in an imaginary cutting plane which bisects the line element while oriented basically perpendicular and parallel within the size tolerances.

Dial indicator, master template, and optical comparator (or similar) techniques can be used (see also GAGING in preceding examples).

COPLANAR SURFACES

Profile tolerancing may be used to specify coplanarity of two or more surfaces where it is desired to treat these surfaces as a single interrupted or noncontinuous surface. In such an application, a control is provided similar to that achieved by a flatness tolerance applied to a single plane surface. Coplanarity is the condition of two or more surfaces having all elements in one plane.

As shown in Example 1, the profile of a surface tolerance establishes a tolerance zone defined by two parallel planes within which the considered surfaces must lie. No datum reference is stated (as in the case of flatness) since the orientation of the tolerance zone is established from the contact of the part against a reference standard; the datum is established by the considered surfaces themselves.

Where more than two surfaces are involved, it may be desirable to identify which specific surface (or surfaces) is to be used as a datum and to establish the tolerance zone. Where necessary, datum target methods could also be used. Where a datum (or datums) is used, it is understood that the tolerance zone established applies to all coplanar surfaces, including the datum surfaces, unless otherwise specified as shown in Example 2.

OFFSET SURFACES

As a variation of coplanar surfaces, the principles described above may be extended to offset surfaces as shown in the example. In this instance, the variation from coplanar is a desired amount (.105) and is stated by a basic dimension.

COPLANAR SURFACES

EXAMPLE 1

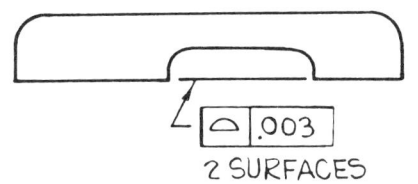

2 SURFACES

MEANING

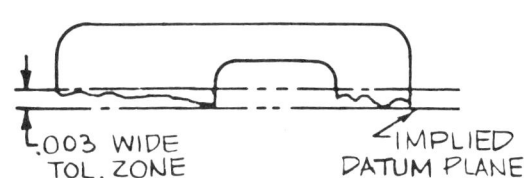

.003 WIDE TOL. ZONE

IMPLIED DATUM PLANE

THE SURFACES MUST BE WITHIN THE SPECIFIED TOL. OF SIZE AND MUST BOTH LIE BETWEEN TWO PARALLEL PLANES .003 APART.

EXAMPLE 2

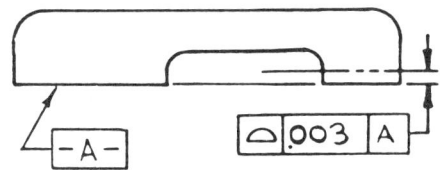

MEANING

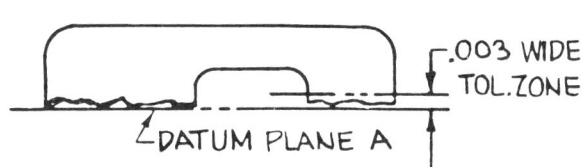

.003 WIDE TOL. ZONE

DATUM PLANE A

THE SURFACE MUST BE WITHIN THE SPECIFIED TOL. OF SIZE AND MUST LIE BETWEEN TWO PARALLEL PLANES .003 APART AS ESTABLISHED RELATIVE TO DATUM PLANE A

OFFSET SURFACES

EXAMPLE

MEANING

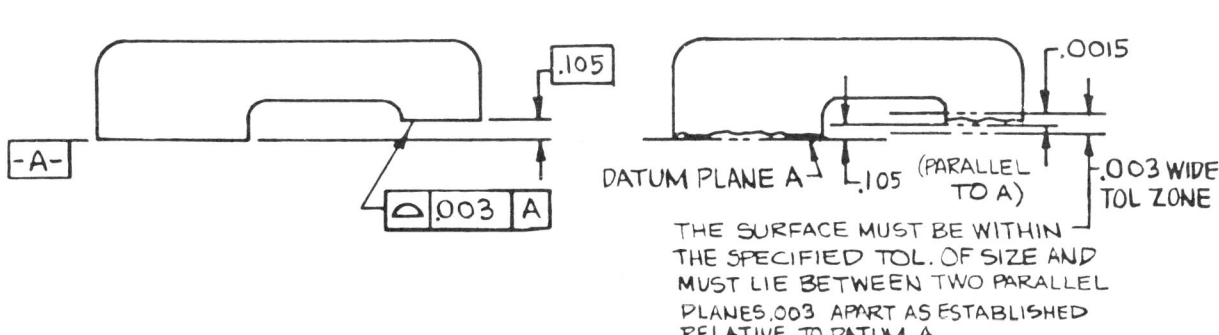

.0015

DATUM PLANE A   .105 (PARALLEL TO A)   .003 WIDE TOL ZONE

THE SURFACE MUST BE WITHIN THE SPECIFIED TOL. OF SIZE AND MUST LIE BETWEEN TWO PARALLEL PLANES .003 APART AS ESTABLISHED RELATIVE TO DATUM A

75

RUNOUT TOLERANCES — RELATED FEATURES USING DATUMS

A runout tolerance is a relationship between surfaces or features, therefore, a datum (or datums) is required.

Runout tolerance may be applied in two different ways using the characteristics shown below.

(CIRCULAR RUNOUT)   TOTAL
                    (TOTAL RUNOUT)*

See following pages for details of application.

RUNOUT ↗ (CIRCULAR AND TOTAL)

Definition. Runout is the composite deviation from the desired form of a part surface of revolution during full rotation (360°) of the part on a datum axis.

RUNOUT TOLERANCE

Runout tolerance states how far an actual surface or feature is permitted to deviate from the desired form implied by the drawing during full rotation (360°) of the part on a datum axis.

RUNOUT APPLICATION

Runout tolerancing is a method used to control the composite surface effect of one or more features of a part relative to a datum axis. Runout tolerance is applicable to rotating parts in which this composite surface control is based on the part function and design requirement. A runout tolerance always applies on an RFS basis; namely, size variation has no affect upon the runout tolerance compliance.

Each considered feature must be within its individual runout tolerance when rotated 360° about the datum axis. The tolerance specified for a controlled surface is the total tolerance or full indicator movement (FIM) in terms of common inspection criteria. Former terms, full indicator reading (FIR) and total indicator reading (TIR) have the same meaning as FIM.

As is seen at right, the basis of runout tolerance control is the datum axis of the part. Surfaces controlled may be those constructed *around* a datum axis, or those constructed at *right angles* to a datum axis. As is also seen, the datum axis is established from a datum feature.

---

* Anticipated new symbol in future standards for total runout, ↗↗.

BASIS OF CONTROL OF RUNOUT TOLERANCE

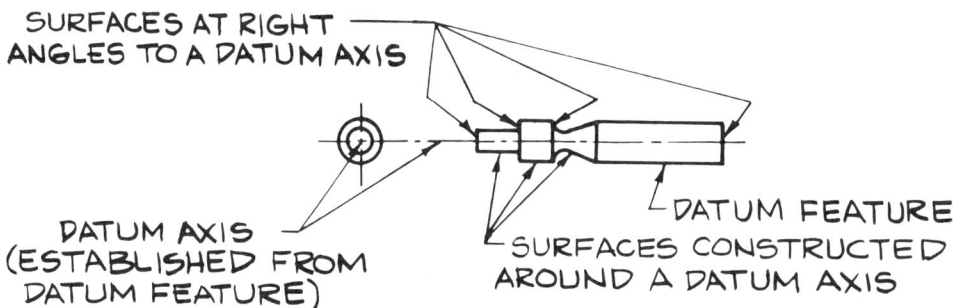

SURFACES AT RIGHT ANGLES TO A DATUM AXIS

DATUM AXIS (ESTABLISHED FROM DATUM FEATURE)

DATUM FEATURE

SURFACES CONSTRUCTED AROUND A DATUM AXIS

TYPES OF FEATURES APPLICABLE TO A RUNOUT TOLERANCE

CIRCULAR RUNOUT

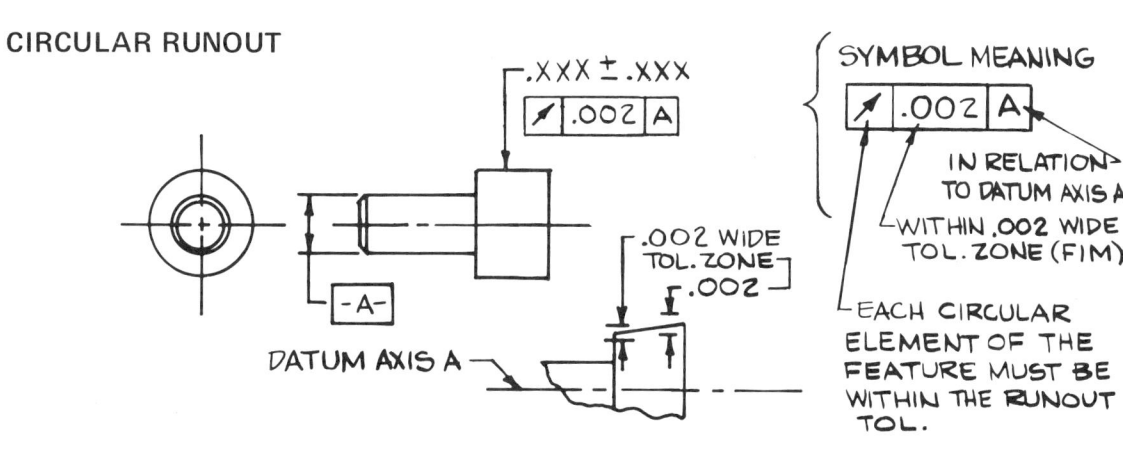

.XXX ± .XXX

↗ .002 A

-A-

DATUM AXIS A

.002 WIDE TOL. ZONE

.002

SYMBOL MEANING

↗ .002 A

IN RELATION TO DATUM AXIS A

WITHIN .002 WIDE TOL. ZONE (FIM)

EACH CIRCULAR ELEMENT OF THE FEATURE MUST BE WITHIN THE RUNOUT TOL.

TOTAL RUNOUT

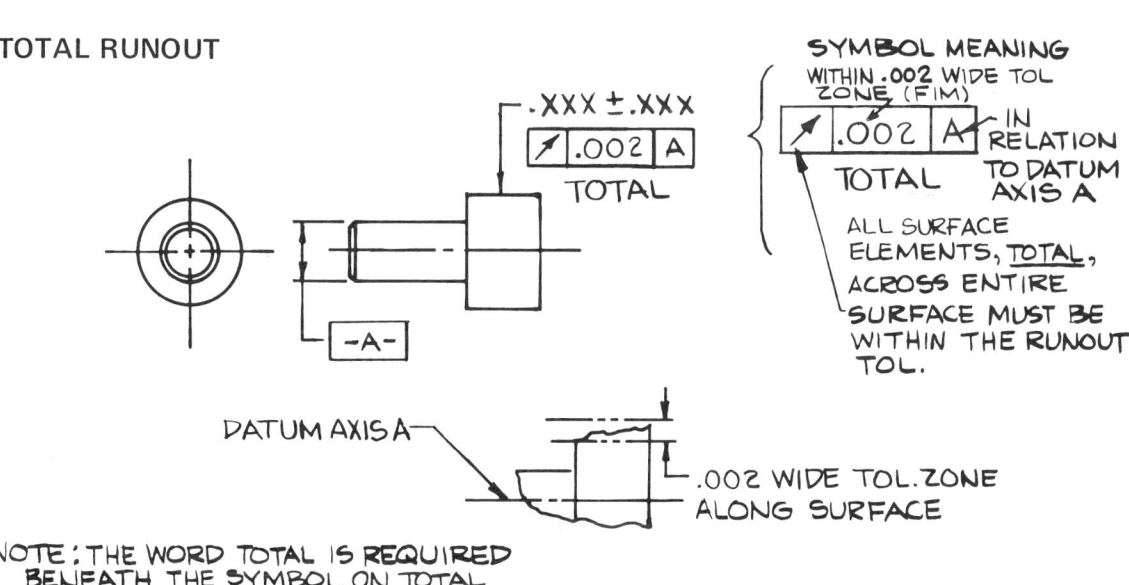

.XXX ± .XXX

↗ .002 A

TOTAL

-A-

DATUM AXIS A

.002 WIDE TOL. ZONE ALONG SURFACE

SYMBOL MEANING

WITHIN .002 WIDE TOL ZONE (FIM)

↗ .002 A

TOTAL

IN RELATION TO DATUM AXIS A

ALL SURFACE ELEMENTS, TOTAL, ACROSS ENTIRE SURFACE MUST BE WITHIN THE RUNOUT TOL.

NOTE : THE WORD TOTAL IS REQUIRED BENEATH THE SYMBOL ON TOTAL RUNOUT

RUNOUT ⌁ (CIRCULAR AND TOTAL)

The datum axis may be established by a single diameter (cylinder) of sufficient length, two diameters with sufficient axial separation, or a diameter and a face surface which is at right angles to it. Features selected as datums should, as much as possible, be functional to the part requirement (e.g., bearing mounting diameters, etc.).

TYPES OF RUNOUT CONTROL

The two types of runout control are *circular* runout and *total* runout. Selection of the proper type is based upon the design requirement and manufacturing considerations. The fundamental difference between the two types are illustrated on p. 77 with detailed examples following in this section. Note that with either method, the collective or composite control of various form variations of the part provides a more direct representation of part functions, integrates manufacturing operations, and minimizes inspection setup requirements.

COAXIAL FEATURES — SELECTION OF PROPER CONTROL

There are three methods of controlling interrelated coaxial features:

1. RUNOUT TOLERANCE (circular or total) (RFS)
2. POSITION TOLERANCE   (MMC)
3. CONCENTRICITY TOLERANCE   (RFS)

Any of the above methods provide effective control. However it is important to select the *most appropriate* one to both meet the design requirements and provide the most economical manufacturing conditions. (See also details of above sections).

Below are recommendations to assist in selecting the proper control:

If the need is to control only CIRCULAR cross sectional elements in a composite relationship to the datum axis, RFS, e.g., multi-diameters on a shaft, use

CIRCULAR RUNOUT   **EXAMPLE:**

(This method controls any composite error effect of roundness and circular cross-sectional profile variations)

If the need is to control the TOTAL cylindrical or profile surface in composite relative to the datum axis' RFS, e.g., multi-diameters on a shaft, bearing mounting diameters, etc., use

EXAMPLE: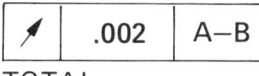

TOTAL RUNOUT   TOTAL   TOTAL

(This method controls any composite error effect of roundness, cylindricity, straightness, angularity, and parallelism.)

Note: Runout is always implied as an RFS application. It cannot be applied on an MMC basis, since an MMC situation involves functional interchangeability or assemblability (probably of mating parts), in which case POSITION tolerance would be used. See below.

If the need is to control the total cylindrical or profile surface and its axis in composite location relative to the datum axis on an MMC basis, e.g., on mating parts to assure interchangeability or assemblability, use

POSITION

EXAMPLE:

If the need is to control the *axis* of one or more features in composite relative to a *datum axis*, RFS, e.g., to control balance of a rotating part, use

EXAMPLE: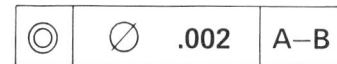

CONCENTRICITY

Note: Concentricity is always implied as an RFS application. Variations in size (departure from MMC size, out-of-roundness , out-of-cylindricity, etc.) do not in themselves conclude *axis* error.

CIRCULAR RUNOUT ⚡ *

PART MOUNTED ON FUNCTIONAL DIAMETER (DATUM)

Circular runout provides a composite control of circular elements of a surface. Circular runout is normally a less complex requirement than total runout. The tolerance is applied independently at any circular cross section or measuring position on the part as it rotates through 360°. Circular runout should be considered when the part function and manufacturing requirements are satisfied by this type of control. Where more complete control of *all* elements in composite is necessary, total runout should be considered.

Circular runout controls composite variations of roundness and cross-sectional form variations of the surface at each circular element where applied to surfaces constructed around a datum axis. It controls circular elements of the surface (wobble) where applied to surfaces constructed at right angles (perpendicular) to a datum axis.

The examples at right utilize circular runout control. The upper example applies circular runout to the angular surface controlling the individual circular elements of the part as it rotates.

When circular runout is to be applied at specific locations, it must be so stated on the drawing. The lower example at right illustrates this application.

Please see following pages and examples in this section for further applications of and considerations on circular runout and for circular runout and total runout used on the same part.

---

*It should be noted that in previous U.S.A. standard, ANSI-Y14.5-1966, circular runout was indicated by the addition of the word "CIRCULAR" beneath the feature control symbol while "total" runout was implied by the use of the arrow symbol ⚡ only. The standard meanings have now been reversed to be compatible to ISO (International Standards) and worldwide practices. Anticipated new symbol for total runout in future standards (ANSI, ISO), ⚡⚡.

EXAMPLE

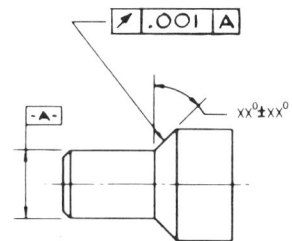

SYMBOL MEANING

┌─────┬──────┬───┐
│ ↗ │ .001 │ A │
└─────┴──────┴───┘

← IN RELATION TO DATUM AXIS A

└ WITHIN .001 WIDE TOL. ZONE (FIM)

└ EACH CIRCULAR ELEMENT MUST BE WITHIN THE RUNOUT TOLERANCE.

MEANING

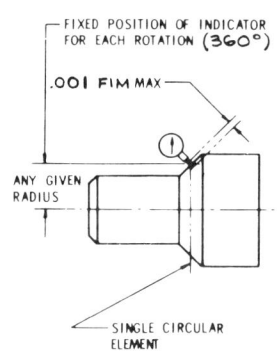

THE FEATURE MUST BE WITHIN THE SPECIFIED TOLERANCE OF SIZE AT ANY MEASURING POSITION, EACH CIRCULAR ELEMENT OF THE SURFACE MUST BE WITHIN .001 FULL INDICATOR MOVEMENT WHEN THE PART IS ROTATED ONE FULL ROTATION ABOUT THE SPECIFIED DATUM AXIS WITH THE INDICATOR FIXED IN A POSITION NORMAL TO THE SURFACE. (THIS DOES NOT CONTROL FORM OF THE TOTAL SPECIFIED SURFACE AREA, BUT ONLY CONTROLS THE RUNOUT OF EACH CIRCULAR ELEMENT.)

EXAMPLE

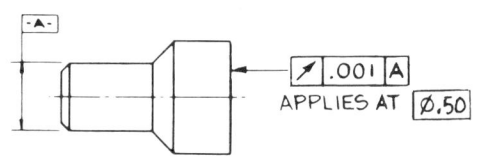

SYMBOL MEANING

┌─────┬──────┬───┐
│ ↗ │ .001 │ A │
└─────┴──────┴───┘

← IN RELATION TO DATUM AXIS A

└ WITHIN .001 WIDE TOL. ZONE (FIM)

└ THE CIRCULAR ELEMENT AT ∅.50 MUST BE WITHIN THE RUNOUT TOLERANCE.

MEANING

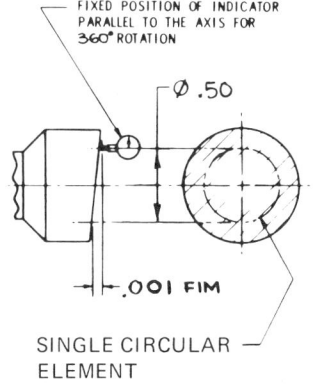

THE CIRCULAR ELEMENT OF THE SURFACE AT ∅ .50 MUST BE WITHIN .001 FULL INDICATOR MOVEMENT WHEN THE PART IS ROTATED ONE FULL ROTATION ABOUT THE SPECIFIED DATUM AXIS WITH THE INDICATOR FIXED IN A POSITION PARALLEL TO THE AXIS. (THIS DOES NOT CONTROL PERPENDICULARITY, BUT CONTROLS ONLY THE LATERAL RUNOUT (WOBBLE) OF EACH CIRCULAR ELEMENT AT THE SPECIFIED SURFACE LOCATION.)

CIRCULAR RUNOUT ⟋

PART MOUNTED ON TWO FUNCTIONAL DIAMETERS (DATUMS)

Where multiple diameters of a rotating part are to be controlled relative to a datum axis, the datum axis can be established by the features which will provide the functional mounting of the part at assembly.

Two diameters (cylinders) are used on the part at right to establish the datum axis of relationship. These diameters could represent the bearing mounting features and thus they establish the datum axis of rotation. Note that two datum features are selected (C and D). By stating the two datums simultaneously with a dash line between them, the relationship of all the other features so designated relate to their common datum axis C-D. The precedence of datums C and D are equal in this case.

Any two surfaces given runout tolerances about the datum axis are to be individually within their stated runout tolerance; collectively they are related to each other within the sum of their indicator readings.

A runout tolerance specified for the datum feature has no effect on the considered features related to it.

PART MOUNTED ON TWO FUNCTIONAL DIAMETERS (DATUMS)

EXAMPLE

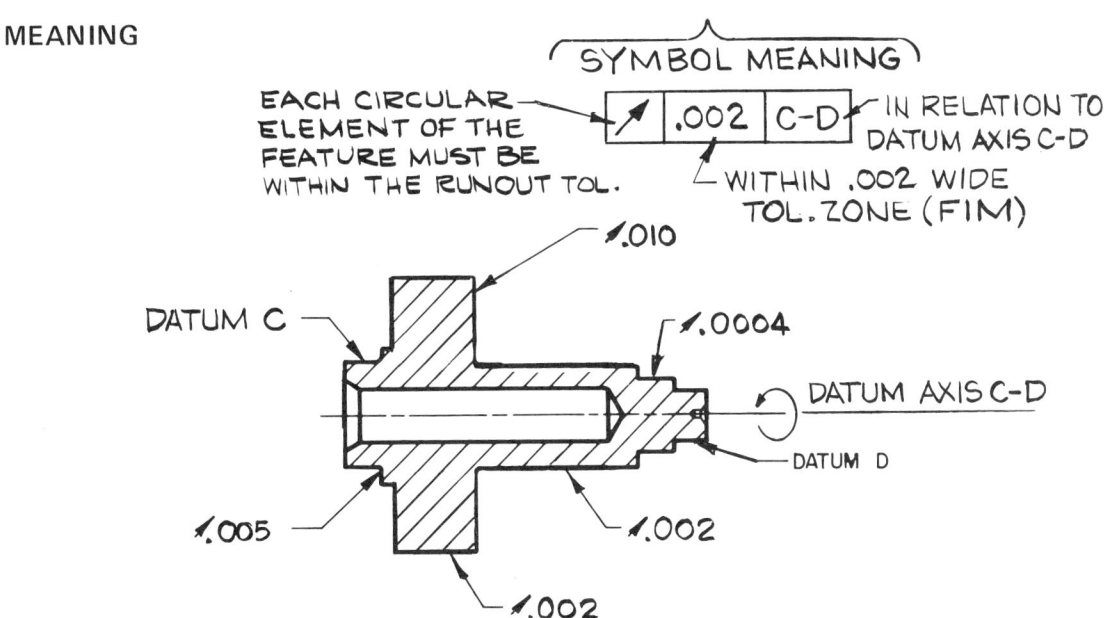

(Size dimensions & tolerances
not shown for simplicity)

MEANING

WHEN MOUNTED ON DATUMS C AND D, DESIGNATED SURFACES MUST
BE WITHIN CIRCULAR RUNOUT ( ) TOLERANCE SPECIFIED, RFS

CIRCULAR RUNOUT ✗ AND TOTAL RUNOUT ✗*

PART MOUNTED ON TWO FUNCTIONAL DIAMETERS (DATUMS) INCLUDING RUNOUT TOLERANCE ON DATUMS.

The part shown at right extends the principles of the previous example. It is a shaft of multi-diameters about a common datum axis C–H with each feature, *including the datum features,* stating an individual runout tolerance. It utilizes both circular runout and total runout control.

Total runout controls composite surface variations of roundness, cylindricity, parallelism, straightness, angularity, taper, and profile of a surface where applied to features constructed around a datum axis. Total runout applied to surfaces constructed perpendicular to a datum axis, controls composite variations of perpendicularity and flatness. The tolerance is applied *simultaneously* at all circular and profile measuring position as the part is rotated through 360°.

Due to the design requirements involved in this example, certain diameters (the datums) must be given total runout control, whereas the remaining diameters and face surfaces may be controlled with circular runout.

Since in the end assembly of this part, (mount to bearings), the relationship to both datums simultaneously is the result desired, all the features including each datum (C and H) must meet their individual runout tolerances. Note that the runout tolerance of each datum (C and H) individually is relative to their common axis C–H.

Multiple leaders may be used in controlling two or more features with a common runout tolerance as is shown. Runout tolerance may be specified individually or in groups, as is convenient, without affecting the runout tolerance.

---

* It should be noted that in previous U.S.A. standard, ANSI Y14.5–1966, total runout was implied by the use of the arrow symbol ✗ only while "circular" runout was indicated by the addition of the word "CIRCULAR" beneath the symbol. The standard meanings have now been reversed to be compatible with ISO (International Standards) and worldwide practices. Proposed in ISO standards is the symbol (⟋⟋) for total runout which may be found in future U.S. and ISO standards.

PART MOUNTED ON TWO FUNCTIONAL DIAMETERS (DATUMS)

EXAMPLE

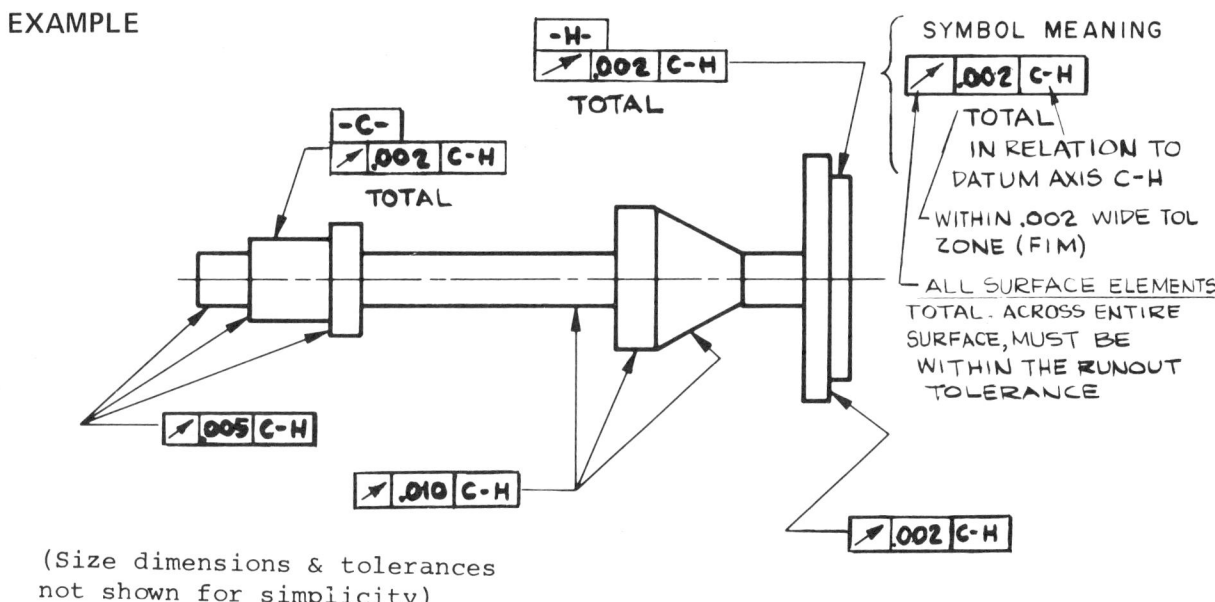

(Size dimensions & tolerances
not shown for simplicity)

MEANING

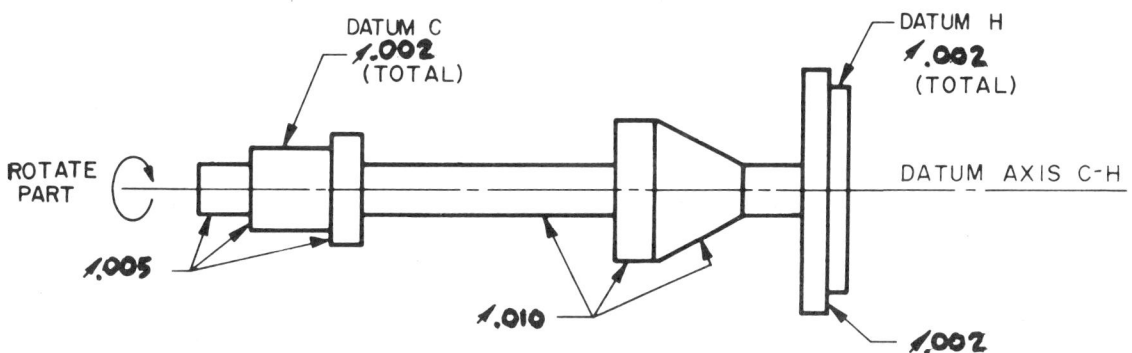

WHEN MOUNTED ON DATUMS C AND H, DESIGNATED SURFACES MUST BE WITHIN
CIRCULAR AND TOTAL RUNOUT (✓) TOLERANCE SPECIFIED, RFS

PART MOUNTED ON TWO FUNCTIONAL DIAMETERS (DATUMS) INCLUDING RUNOUT AND CYLINDRICITY TOLERANCE ON DATUM

The part shown at the right further extends the principles of the previous examples. It is a shaft of multi-diameters about a common datum axis C–D, with each feature including the datum features stating an individual runout tolerance. In addition to having both circular and total runout control, each of the datum features has a cylindricity tolerance.

It should be noted that the datum features of this part have controls of three orders of magnitude: size, total runout, and cylindricity (form). In this example, an accurate fit to bearings as controlled by size and form was required in addition to the runout relationship of the datum features to the datum axis (C–H) of rotation.

Features with a specific relationship to another feature rather than to the common datum axis, may be so indicated by addition of appropriate datum references (such as datum E on page at right). Note that the inside diameter on the right end of the part relates the runout requirement to datum E.

PART MOUNTED ON TWO FUNCTIONAL DIAMETERS (DATUMS)

EXAMPLE

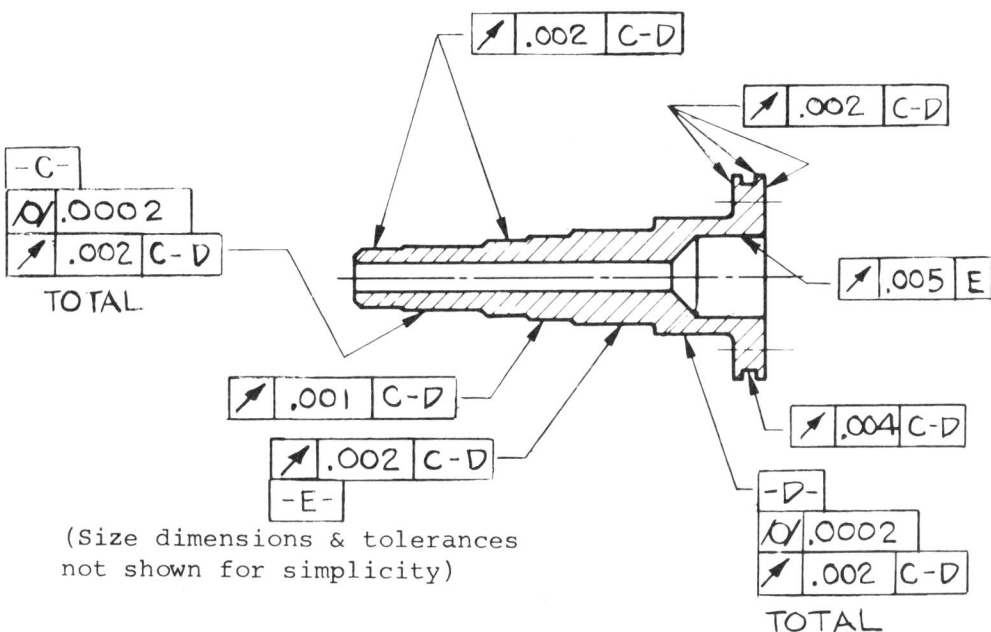

(Size dimensions & tolerances
not shown for simplicity)

MEANING

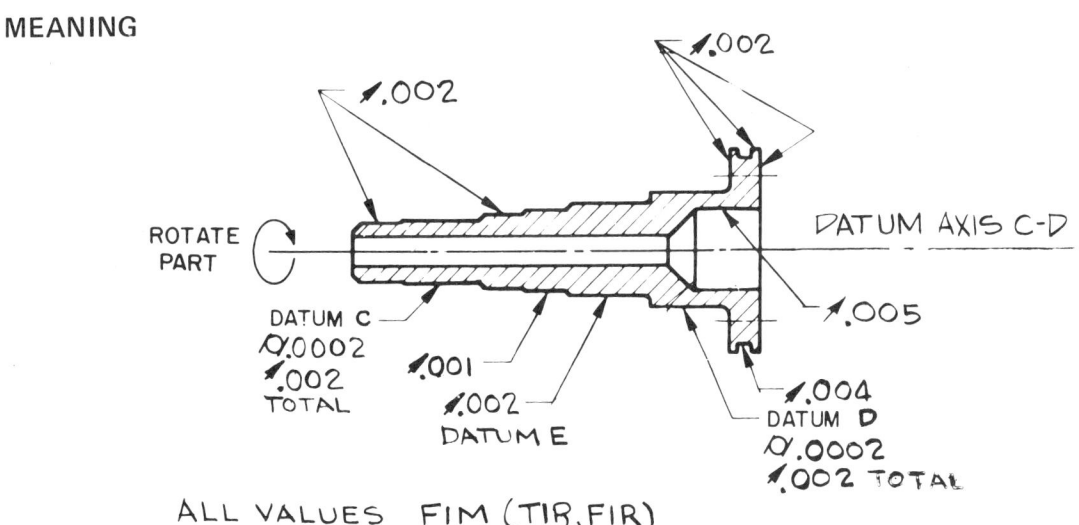

ALL VALUES FIM (TIR, FIR)

WHEN MOUNTED ON DATUMS C AND D, DESIGNATED SURFACES MUST BE
WITHIN CIRCULAR RUNOUT (↗) TOLERANCE SPECIFIED. DATUMS C AND
D MUST ALSO BE WITHIN TOTAL RUNOUT (↗) TOLERANCE SPECIFIED AND
CYLINDRICAL WITHIN .0002 RFS.

WHEN MOUNTED ON DATUM E, DESIGNATED SURFACE MUST BE WITHIN
CIRCULAR RUNOUT (↗) TOLERANCE SPECIFIED.

PART MOUNTED ON FUNCTIONAL FACE SURFACE (DATUM)
AND DIAMETER (DATUM)

Runout tolerancing may be applied to features of rotation where the feature datum references are an axis and a face surface perpendicular to the axis. In such an application, datum precedence is usually considered necessary. The influence of the appropriate feature as a primary datum is determined relative to the design requirement.

In the example at the right, the face surface perpendicular to the axis of rotation is considered the functionally important surface (bottoms on the inner race of bearing) and is thus selected as the primary datum A. Secondary datum diameter (cylinder) B is in contact with the extremities of datum feature B. With the part in this orientation and located functionally on its datums, all related features must meet their individual runout tolerances when the part is rotated about datum axis B.

Note that the primary datum (A) and secondary datum (B) features are indicated with that precedence by separate enclosures in the feature control symbol box (reading left to right).

PART MOUNTED ON FUNCTIONAL FACE SURFACE (DATUM)
AND DIAMETER (DATUM)

EXAMPLE

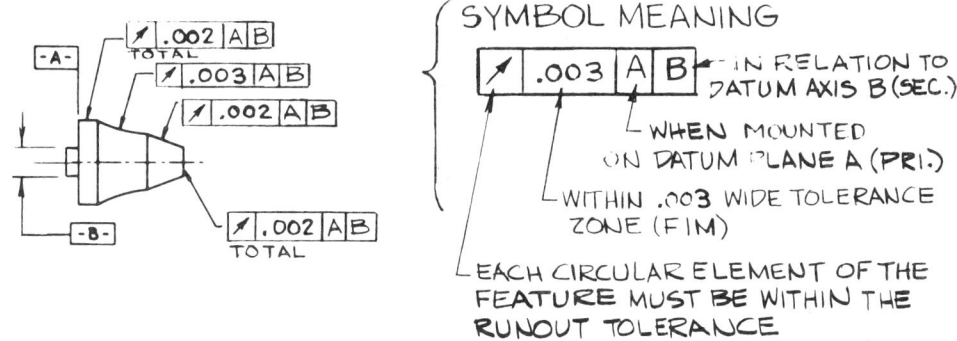

MEANING

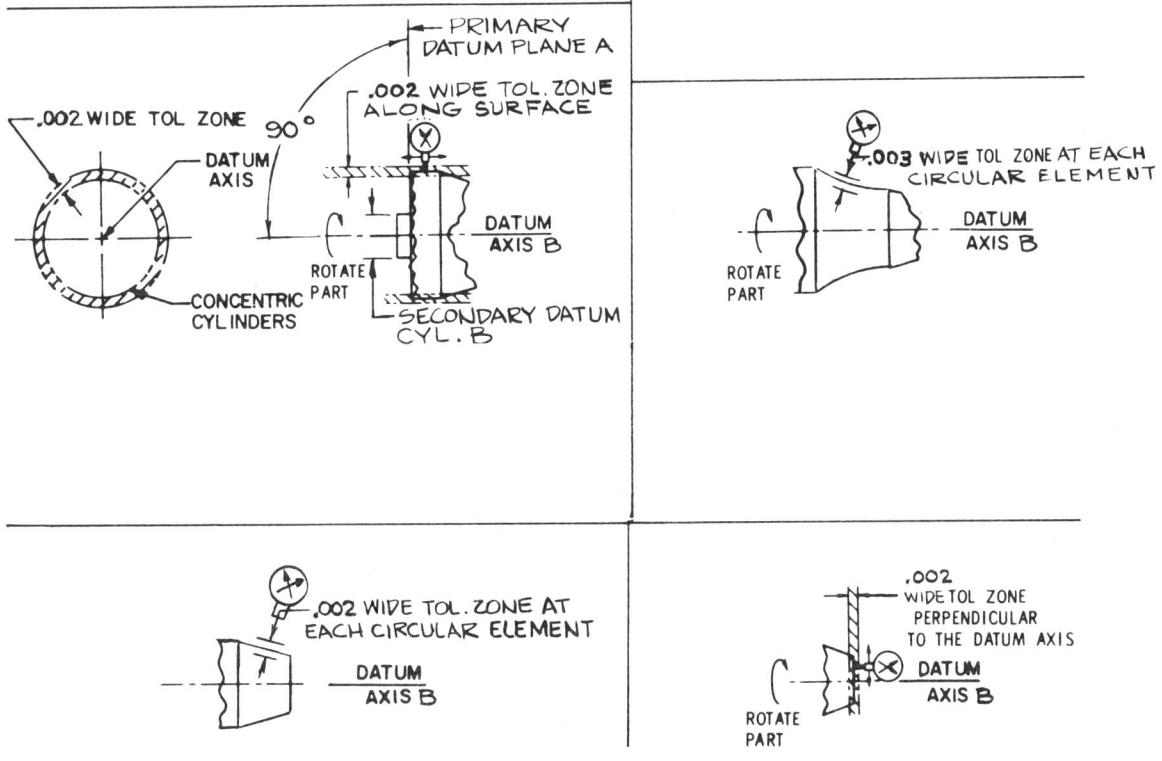

CIRCULAR RUNOUT ⤢ AND FLATNESS ▱

PART MOUNTED ON FUNCTIONAL FACE SURFACE (DATUM) AND DIAMETER (DATUM)

The part shown at right extends the principles of runout tolerancing. It illustrates a situation in which the large flat mounting surface requires flatness control to assure desired accuracy of the primary datum surface as it mounts and orients the remainder of the part to proper functional relationships. In this instance it can be assumed that this method was most representative of part end function.

The remainder of the part relates to the datum system C and D, with C being the primary datum for attitude of the part, and D the secondary datum for establishing axis of rotation. Note again that the other features *including datum D* relate to the datum system C and D.

PART MOUNTED ON LARGE FLAT SURFACE (DATUM) AND DIAMETER (DATUM)

EXAMPLE

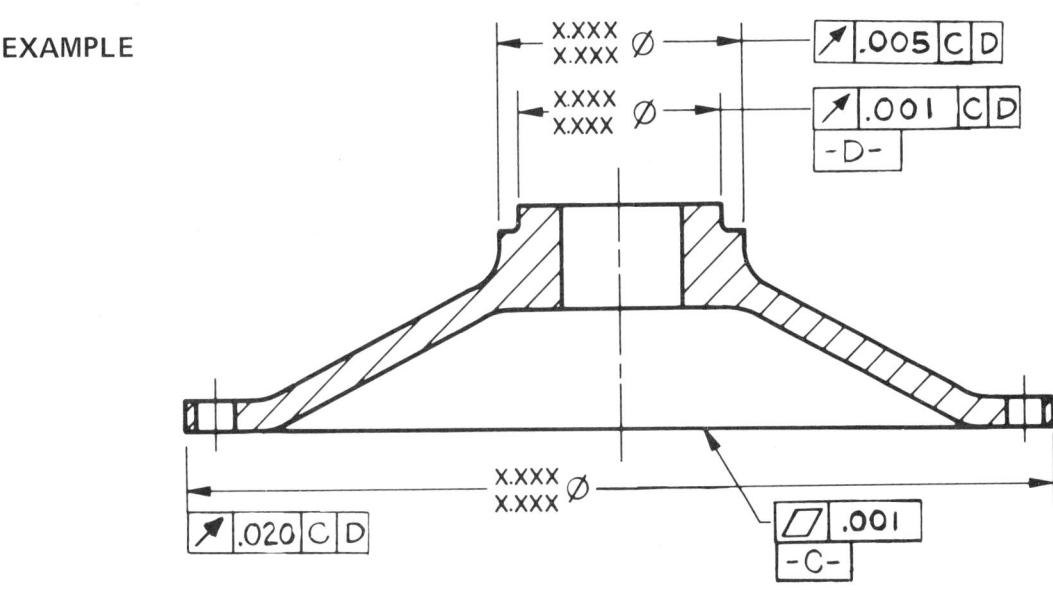

MEANING

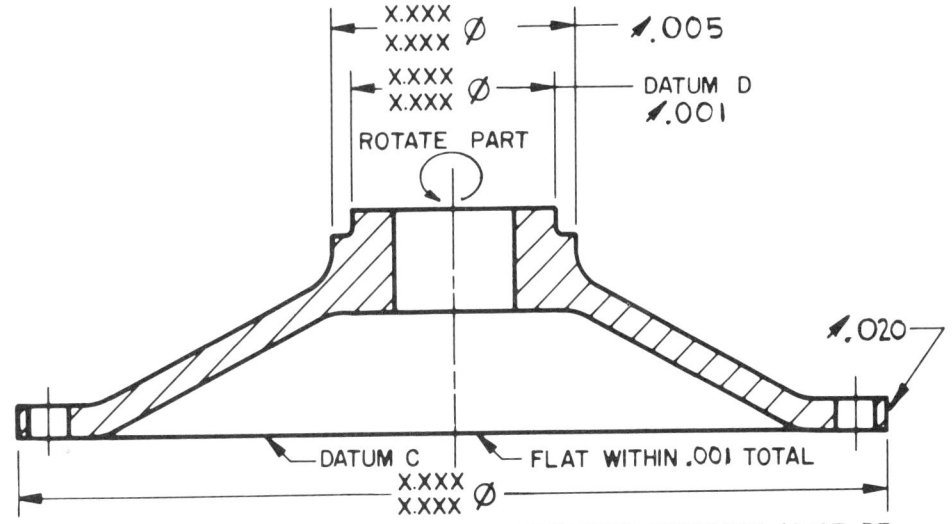

WHEN MOUNTED ON DATUMS C AND D, DESIGNATED SURFACES MUST BE WITHIN CIRCULAR RUNOUT (↗) TOLERANCE SPECIFIED, RFS

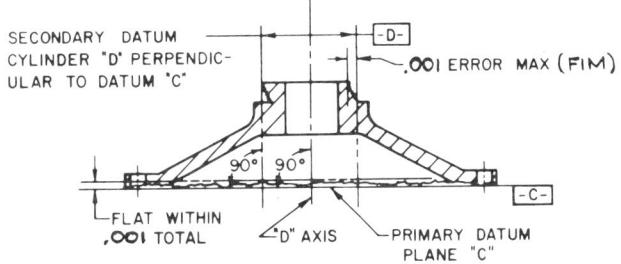

TOTAL RUNOUT ⚡ TOTAL
RUNOUT APPLICATION AND ANALYSIS

Typical total runout applications, inspection analysis, and meaning are discussed below and in the following text. An understanding of the fundamentals involved in these examples will provide the reader with the basis for total runout characteristic and use. The methods and principles shown may be readily adapted to circular runout as well, except for the substitution, where appropriate in the text and illustrations, of the circular elements (only) principles. In checking total runout, the datum feature, or features, are normally centered or positioned in an appropriate inspection device (e.g., between centers, collet, chuck, mandrel, vee block, or other centering device) to establish the datum axis from which the total runout relationships are to occur.

PART MOUNTED ON MACHINING CENTERS

EXAMPLE

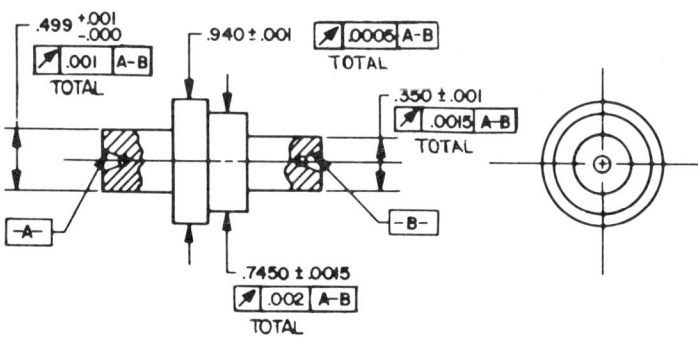

MEANING

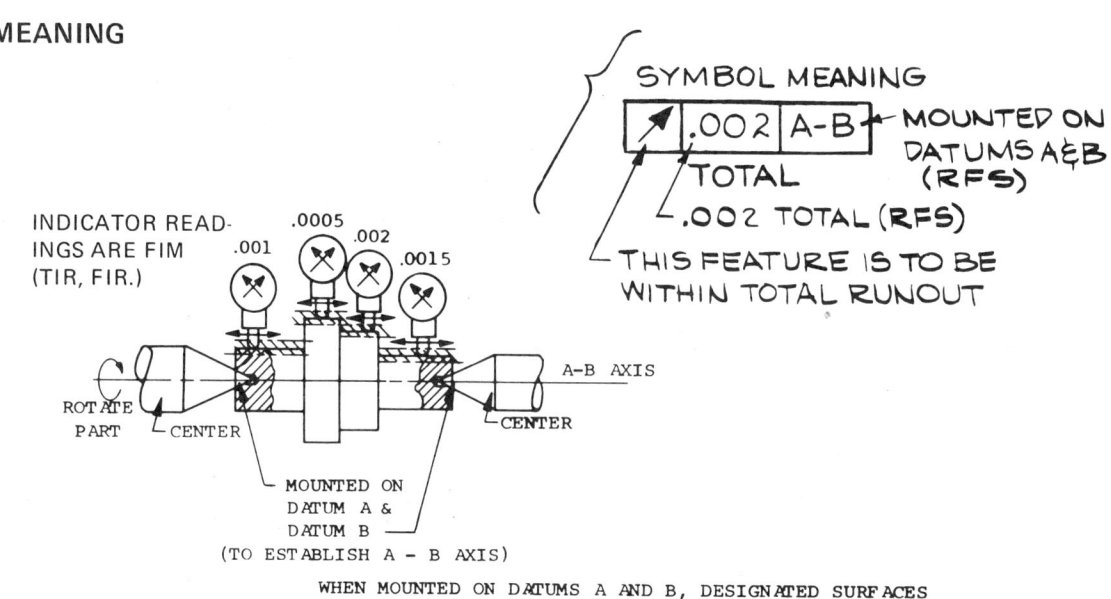

WHEN MOUNTED ON DATUMS A AND B, DESIGNATED SURFACES
MUST BE WITHIN TOTAL RUNOUT (⚡) TOLERANCE SPECIFIED.

PART MOUNTED ON TWO FUNCTIONAL DIAMETERS

EXAMPLE

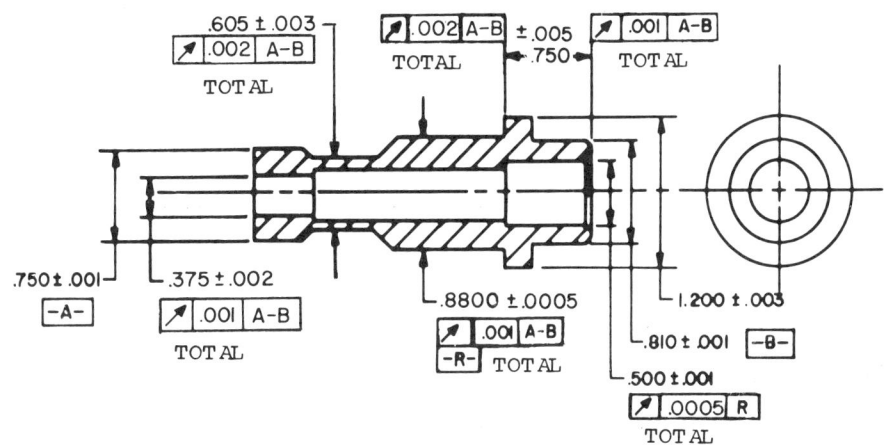

MEANING 1

INDICATOR READINGS ARE TOTAL (FIR, TIR)

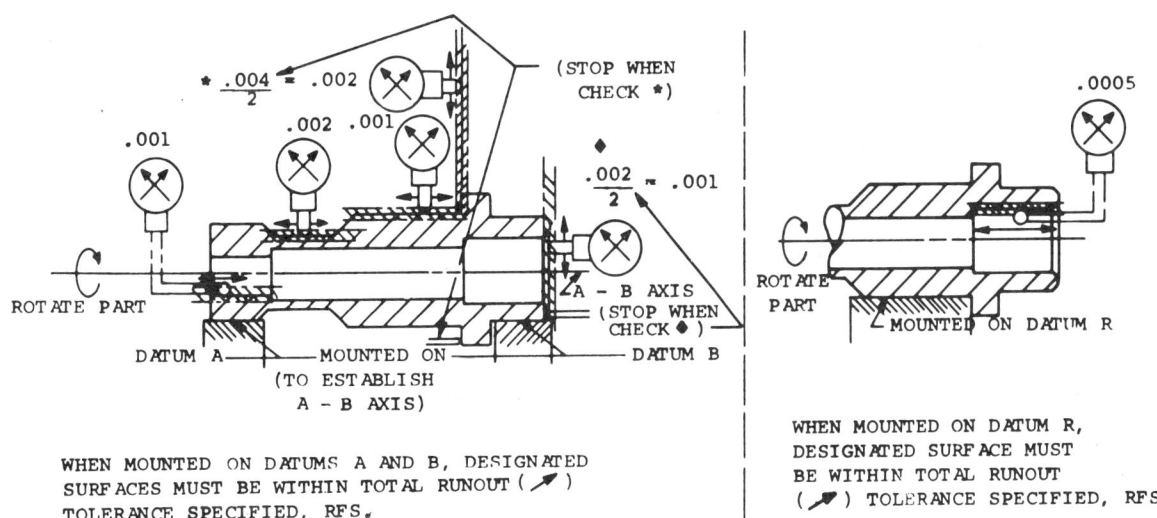

WHEN MOUNTED ON DATUMS A AND B, DESIGNATED
SURFACES MUST BE WITHIN TOTAL RUNOUT (↗)
TOLERANCE SPECIFIED, RFS.

RUNOUT REQUIREMENT RELATING
TO A - B AXIS

WHEN MOUNTED ON DATUM R,
DESIGNATED SURFACE MUST
BE WITHIN TOTAL RUNOUT
(↗) TOLERANCE SPECIFIED, RFS.

RUNOUT REQUIREMENT RELATIVE
TO R AXIS

TOTAL RUNOUT ↗ TOTAL
TOTAL SURFACE RUNOUT (WOBBLE)*

Note that when flat or face surface runout (wobble) is part of the specification as in the case of the example on the preceding page, the runout analysis must consider the need for restricting end movement of the part while checking it. If collets or other holding devices are applied to establish the datum axis, end movement of the part is controlled. Also, specified secondary or tertiary datums (specified faces, shoulders, etc.) may indicate the end location or "stop" surface. However, in the absence of the above conditions and when the flat or face surface runout is to be checked, one should. if possible, "stop" on the surface being checked (see MEANING 1 on the preceding page and continued below), in order to avoid adding the error of *another* surface to the reading. When this is done however, dependent on the location or placement of the stop, the dial indicator runout reading may need to be *halved* (divided by 2) before comparison with the stated part drawing runout tolerance to determine whether the requirement has been met. Referring to the illustration below, note that to use the fixed factor 2 the stop should be as close as possible to the O.D. of the surface being checked. As the placement of the stop approaches the axis of the mounting datum diameters (see also MEANING 3), the factor approaches a 1:1 ratio.

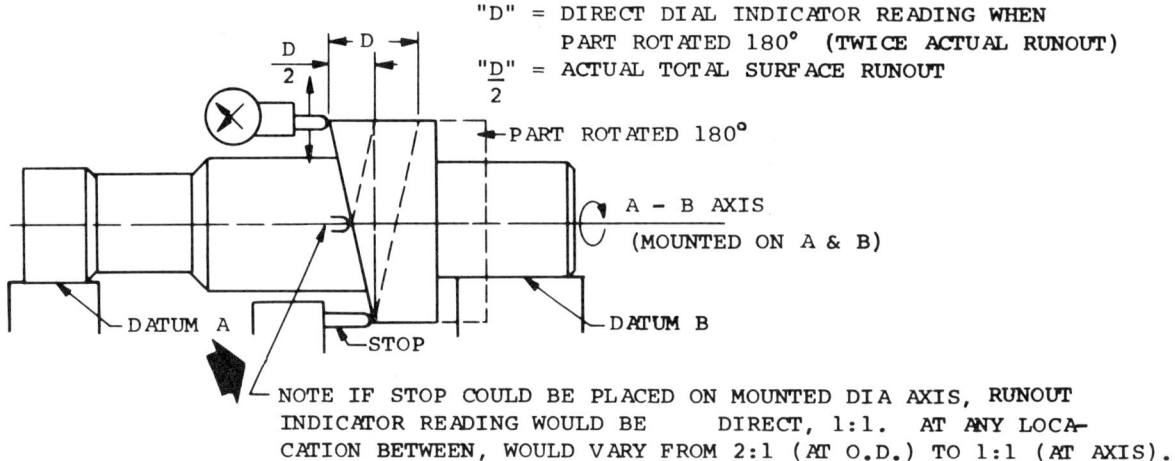

"D" = DIRECT DIAL INDICATOR READING WHEN
       PART ROTATED 180° (TWICE ACTUAL RUNOUT)

$\frac{"D"}{2}$ = ACTUAL TOTAL SURFACE RUNOUT

PART ROTATED 180°

A – B AXIS

(MOUNTED ON A & B)

DATUM A   STOP   DATUM B

NOTE IF STOP COULD BE PLACED ON MOUNTED DIA AXIS, RUNOUT INDICATOR READING WOULD BE DIRECT, 1:1. AT ANY LOCATION BETWEEN, WOULD VARY FROM 2:1 (AT O.D.) TO 1:1 (AT AXIS).

---

*Flat or face surface total runout (wobble) as illustrated can be considered to be identical to a "perpendicularity" requirement. However, runout is preferred on such an application which is associated with other runout requirements, so that a single setup can be used.

MEANING 2

As an alternative method a full end stop may be used. The end stop must, however, be extremely accurate at 90° (exact within gage tolerances) to the axis of the mounting diameters. See below.

In this case the dial indicator reading taken is direct and can be compared immediately with the stated part drawing runout (wobble) tolerance.

Note also that no error of the part surface or face which is used for "stopping" is introduced into the indicator reading, since the high point or extremity of this surface remains in consistent contact with the full stop gage as the part is rotated.

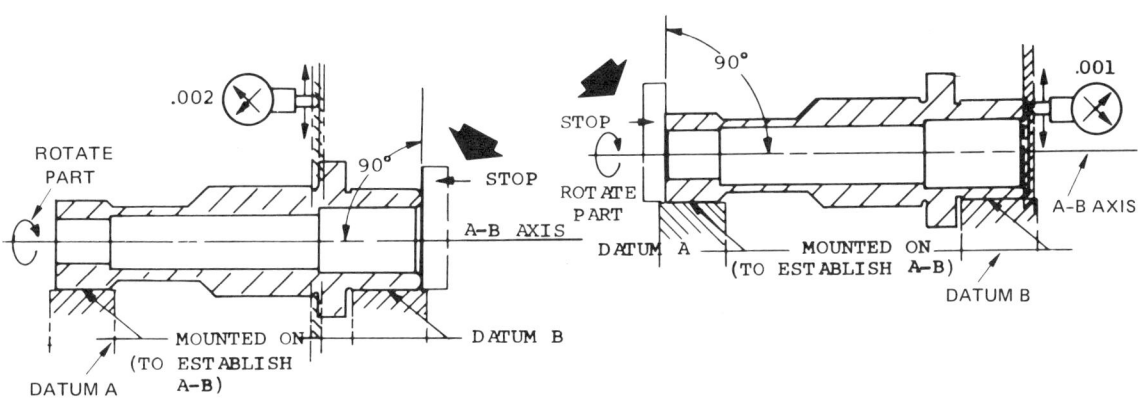

MEANING 3

If the part has a *closed end* and a stop can be placed at the *axis* of the mounting datum diameters (see illustration below), the runout requirement may be checked without regard for extreme accuracy of the stop. The dial indicator reading taken is direct and can be compared immediately with the stated part drawing runout (wobble) tolerance.

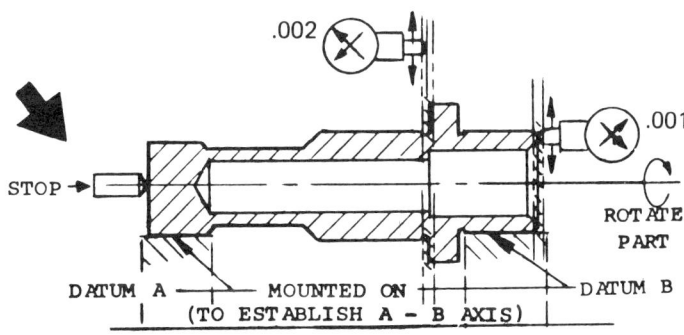

TOLERANCES OF LOCATION

TOLERANCES OF LOCATION STATE THE PERMISSIBLE VARIATION IN THE SPECIFIED LOCATION OF A FEATURE IN RELATION TO SOME OTHER FEATURE OR DATUM.

Tolerances of location refer to the geometric characteristics: position, concentricity, and symmetry.

In the course of the discussion on location tolerancing, more detail on maximum material condition, datums, basic dimensioning, and the interrelationship of location and form tolerancing will be introduced.

Location tolerances involve features of size and relationships of center lines, center planes, axes, etc. At least two features, one of which is a size feature, are required before location tolerancing is valid. Where function or interchangeability of mating part features is involved, the MMC principle may be introduced to great advantage. Perhaps the most widely used and best example of the application of this principle is position tolerancing.

The use of the position concept in conjunction with the maximum material condition concept provide some of the major advantages of the geometric tolerancing system.

POSITION ⊕

Definition. Position is a term used to describe the perfect (exact) location of a point, line, or plane (normally the center) of a feature in relationship with a datum reference or other feature.

POSITION TOLERANCE

A position tolerance is the total permissible variation in the location of a feature about its exact position. For cylindrical features (holes and bosses) the position tolerance is the *diameter (cylinder)* of the tolerance zone within which the axis of the feature must lie, the center of the tolerance zone being at the exact position. For other features (slots, tabs, etc.) the position tolerance is the total width of the tolerance zone within which the center plane of the feature must lie, the center plane of the zone being at the exact position.

POSITION THEORY

We shall now examine the position theory as typically applied to a part for purposes of function or interchangeability. As a means of describing this theory we shall first compare the position system with the bilateral or coordinate system.

Imagine a part with four holes in a pattern which must line up with a mating part to accept screws, pins, rivets, etc. to accomplish assembly, or four holes in a pattern to accept the pins, dowels, or studs of a mating part to accomplish assembly.

The top figure at the right shows the part with a hole pattern dimensioned and toleranced using a coordinate system. The bottom figure shows the same part dimensioned using the position system. Comparing the two approaches, we find the following differences:

1) The derived tolerance zones for the hole centers are square in the coordinate system and round in the position system.

2) The hole center location tolerance in the top figure is part of the coordinates (the 2.000 and 1.750 dimensions). In the bottom figure, however, the location tolerance is associated with the hole size dimension and is shown in the feature control symbol at the right. The 2.000 and 1.750 coordinates are retained in the position application, but are stated as BASIC or exact values.

For this comparison, the .005 square coordinate tolerance zone has been converted to an equivalent .007 position tolerance zone. The two tolerance zones are superimposed on each other in the enlarged detail.

The black dots represent possible inspected centers of this hole on eight separate piece parts. We see that if the coordinate zone is applied, only three of the eight parts are acceptable. However with the position zone applied, six of the eight parts appear immediately acceptable.

COORDINATE SYSTEM

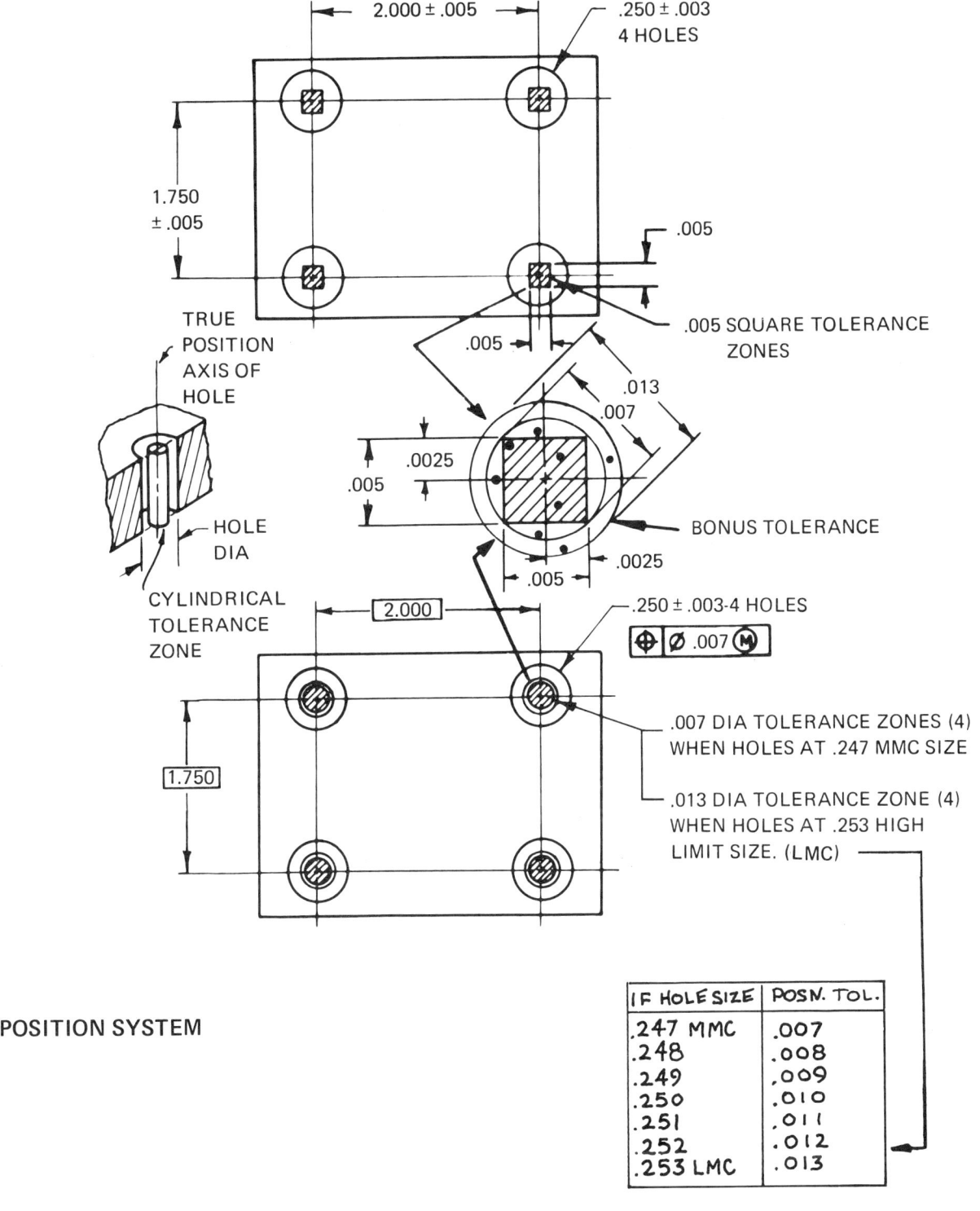

2.000 ± .005

.250 ± .003
4 HOLES

1.750
± .005

.005

.005 SQUARE TOLERANCE ZONES

.005

TRUE
POSITION
AXIS OF
HOLE

HOLE
DIA

CYLINDRICAL
TOLERANCE
ZONE

.013

.007

.0025

.005

BONUS TOLERANCE

.0025

.005

2.000

1.750

.250 ± .003-4 HOLES

⊕ | ⌀ .007 Ⓜ

.007 DIA TOLERANCE ZONES (4)
WHEN HOLES AT .247 MMC SIZE

.013 DIA TOLERANCE ZONE (4)
WHEN HOLES AT .253 HIGH
LIMIT SIZE. (LMC)

POSITION SYSTEM

IF HOLE SIZE	POSN. TOL.
.247 MMC	.007
.248	.008
.249	.009
.250	.010
.251	.011
.252	.012
.253 LMC	.013

POSITION

The position diameter shaped zone can be justified by recognizing that the .007 diagonal is unlimited in orientation. Also, a cylindrical hole should normally have a cylindrical tolerance zone.

A closer analysis of the representative black dots and their position with respect to the desired exact location clearly illustrates the fallacies of the coordinate system when applied to a part such as that illustrated.

The dot in the upper left diagonal corner of the square zone, and the dot on the left outside the square zone are in reality at nearly the same distance from the desired exact center. However, in terms of the square coordinate zone, the hole on the left is unacceptable by a wide margin, whereas the upper left hole is acceptable.

Note, then, that a hole produced off center under the coordinate system has *greater* tolerance if the shift is on the diagonal and not in the horizontal or vertical direction.

Realizing that the normal function of a hole relates to its mating feature in *any* direction (i.e., a hole vs. a round pin), we see that the square zone restriction seems unreasonable and incorrect. Thus the position tolerance zone, which recognizes and accounts for unlimited orientation of round or cylindrical features as they relate to one another, is more realistic and practical.

In normal applications of position principles, the tolerance is derived, of course, from the design requirement, *not* from converted coordinates. The maximum material sizes of the features (hole and mating component) are used to determine this tolerance.

Thus the .007 position tolerance of the example on the preceding page (see also Fig. 1) would normally be based on the MMC size of the hole (.247). As the hole size deviates from the MMC size, the position of the hole is permitted to shift off its "true position" beyond the original tolerance zone to the extent of that departure. The "bonus tolerance" of .013 illustrates the possible position tolerance should the hole be produced, for example, to its high limit size of .253. The tabulation on the preceding page shows the enlargement of the position tolerance zone as the hole size departs from MMC.

Although we have considered only one hole to this point in the explanation, the same reasoning applies to all the holes in the pattern. Note that position tolerancing is also a noncumulative type of control in which each hole relates to its own true or exact position and no error is accumulated from the other holes in the pattern.

Position tolerancing is usually applied on mating parts in cases where fit, function, and interchangeability are the considerations. It provides greater production tolerances, ensures design requirements, and provides the advantages of functional inspection practices as desired.

Functional gaging techniques, familiar to a large segment of industry through many years of application, are fundamentally based on the MMC position concept. It should be clearly understood, however, that functional gages are not mandatory in fulfilling MMC position requirements.

Functional gages are used and discussed in this text for the dual purpose of explaining the principles involved in position tolerancing and of introducing the functional gage technique as a valuable tool. A functional gage can be considered as a simulated master mating part at its worst condition.

Position, although a locational tolerance, also includes form tolerance elements in composite. For example, as shown in the illustration on the preceding page, perpendicularity is invoked as part of the control to the extent of the diameter zone, actually as a "cylindrical" zone, for the depth of the hole. Further, the holes in the pattern are parallel to one another within the position tolerance. Various other elements of form are included as a part of the composite functional control provided by position tolerancing. Addition of datum references and the relation of the hole pattern to surfaces and other features will be discussed in later text (see DATUM section).

POSITION SYSTEM

The example at the right further clarifies the position theory; two of the holes on the part shown in the previous example are enlarged to illustrate the actual effect of feature *size* variation on the positional location of the features.

POSITION THEORY

FIGURE 1

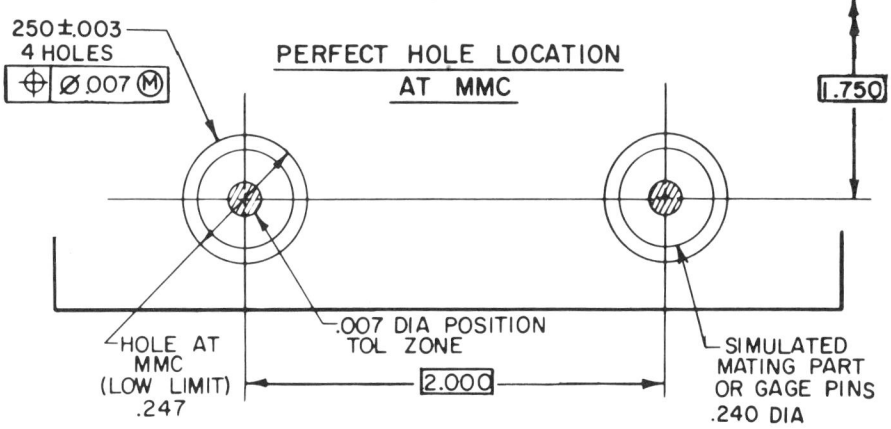

FIGURE 2

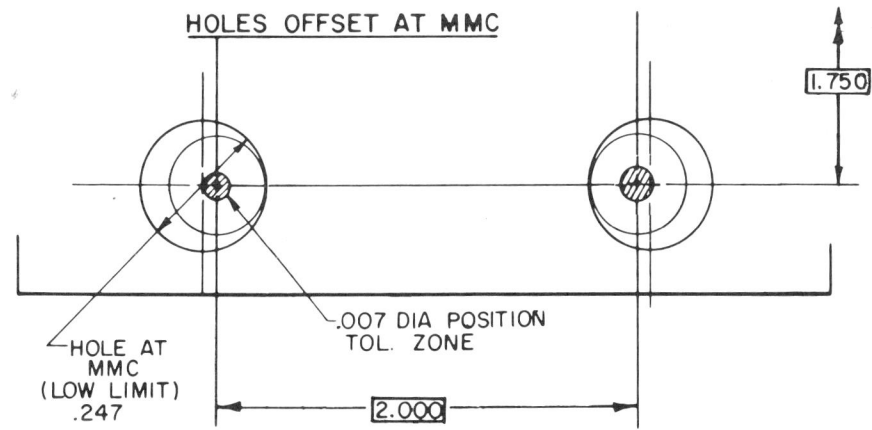

POSITION ⊕

Figure 1 shows the two .250 ± .003 holes at a MMC size (or the low limit of their size tolerance) of .247 and with their centers perfectly located in the .007 diameter position tolerance zone. The drawing illustrates the mating part situation represented by a functional or fixed pin. The gage pins are shown undersize an amount equal to the positional tolerance of .007; i.e., at .240 diameter. This represents the maximum permissible offset of the holes within their stated positional tolerance when the hole is at MMC size of .247.

Figure 2 shows the two .250 MMC holes offset in opposite directions to the maximum permissible limits of the .007 position tolerance zone. Note that we illustrate the worst condition: the edges of the holes are tangent to the diameters of the simulated mating part or gage pins. The holes are within tolerance and, as can be seen, would satisfactorily pass the simulated mating part condition as represented by the gage pins.

In Fig. 3 the .250 ± .003 holes have been produced to the opposite, or *high* limit (minimum or least material condition) size of .253. It can now be seen that when we retain the same offset and tangency of the holes and mating part of the gage pins as shown in Fig. 2, the produced centers of the holes are allowed to shift *beyond* the original .007 tolerance zone to a resulting .013 diameter tolerance zone still providing an acceptable situation.

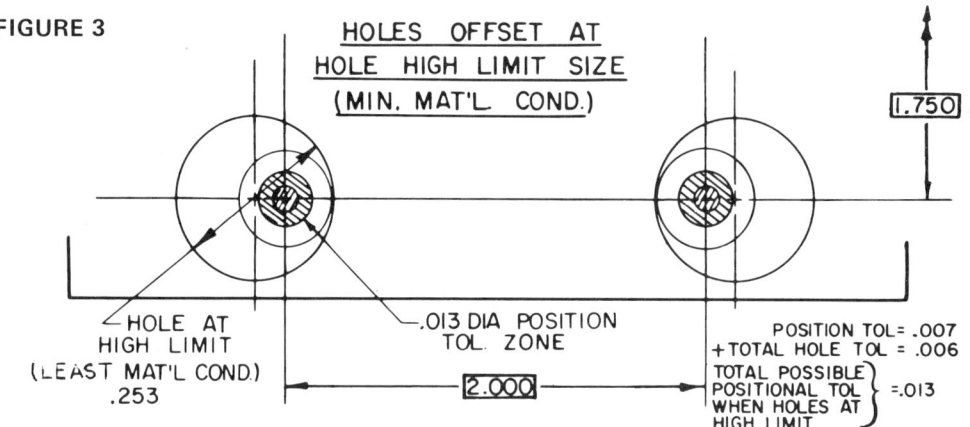

The foregoing illustrates the interrelationship of size and location tolerances which is utilized in position dimensioning and tolerancing.

Although in this example we have used only two of the holes, the same reasoning applies to *all* the holes in the pattern; similarly, each individual hole could be offset within its tolerance zone in any direction around 360° and provide an acceptable situation.

It should be noted that a functional or fixed pin gage such as the one used here to explain the position theory can be used *only* to check the *positional* location of the holes. *Positional* tolerance can be added as the holes increase in size or depart from MMC size within their size tolerance range. Hole *size* tolerance, however, must be held within the tolerances specified on the drawing and must be checked individually and separately from the positional check.

MATING PARTS — FLOATING FASTENER

Position tolerancing techniques are most effective and appropriate in mating part situations. The illustrations on page 105, in addition to demonstrating the calculations required, also emphasize the importance of decisions at the design stage to recognize and initiate the position principles.

The mating parts shown in the illustration are to be interchangeable. Thus the calculation of their position tolerances should be based on the two parts and their interface with the fastener in terms of MMC sizes.

The two parts are to be assembled with four screws. The holes in the two parts are to line up sufficiently to pass the four screws at assembly. Since the four screws ("fasteners") are separate components, they are considered to have some "float" with respect to one another. The colloquial term, "floating fastener" application, has been popularly used to describe this situation.

The calculations are shown in the upper right corner of the illustration. Also, note that, in this case, the same basic dimensions and position tolerances are used on both parts. They are, of course, separate parts and are on separate drawings.

The position tolerance calculations are based on the MMC sizes of the holes and the screws. The maximum material basis then sets the stage for maximum producibility, interchangeability, functional gaging (if desired), etc., at production. As seen from the illustration, part acceptance tolerances will increase as the hole sizes in the parts are actually produced and vary in size as a departure from MMC. From the .016 diameter tolerance calculated, the tolerance may increase to as much as .022 dependent upon the actually produced hole size. It should be noted that clearance between the mating features (in this case hole and screw) is the criterion for establishing the position tolerances.

Simultaneously with these production advantages, the design is protected since it has been based upon the realities of the hole and screw sizes as they interrelate at assembly and in their function. Thus, as parts are produced, assembly is ensured, and the design function is carried out specifically as planned.

A possible functional gage is also shown in the illustration on page 105. The .190 gage pin diameters are determined by the MMC size of the hole, .206, minus the stated position tolerance of .016. In our example, the same functional gage can be used on both parts. Functional gages are, of course, not required with position application, but they do, however, provide an effective method of evaluation where desired.

Referring to the position tolerance calculations, if more than two parts are assembled in a floating fastener application, we must determine the position tolerance to ensure that any two parts and the fastener will mate properly. Calculate each part to mate with the fastener using the illustrated formula and MMC sizes.

The calculations on the illustrated parts on page 105 show a balanced tolerance application in which the total permissible position tolerance of the holes on the two parts is the same, i.e., .016. The total position tolerance can however be distributed as desired: for example, if one part specifies only .010 of the .016 tolerance available for

POSITION ⊕

each part, .006 may be added to the calculated position tolerance of the mating part. (See example below.)

Distribution of the tolerances may, if desired, be adjusted as shown below:

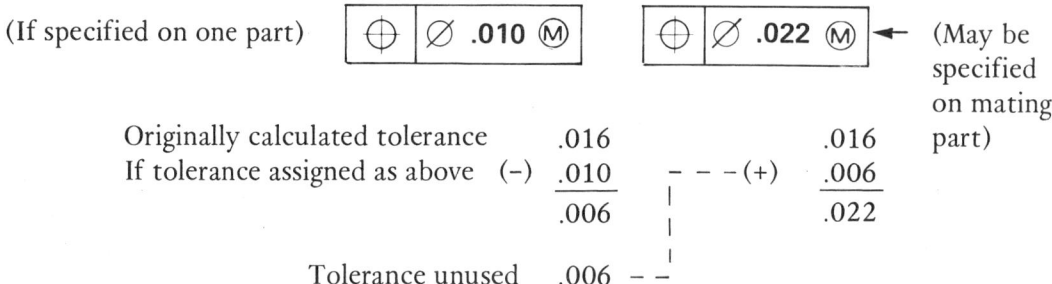

(If specified on one part) ⊕ | ⌀ .010 Ⓜ   ⊕ | ⌀ .022 Ⓜ ← (May be specified on mating part)

Originally calculated tolerance	.016	.016
If tolerance assigned as above (−)	.010	(+) .006
	.006	.022
Tolerance unused	.006	

The clearance holes on these parts are all specified as same size. Where they are specified as *different* sizes, the total position tolerance is equal to the *average* diametral clearance between mating holes and fasteners. An example is shown below:

MMC hole		.200
MMC fastener	(−)	.190
		.010

$$\frac{.010 + .016}{2} = .013$$

Other part		
MMC hole		.206
MMC fastener	(−)	.190
		.016

⊕ | ⌀ .013 Ⓜ
(Specified on both parts)

Or, each part can be calculated separately for the allowable positional displacement based on the difference between the MMC of the hole and fastener. If one part in our example had .203 ± .003 − 4 HOLES specified, the method below would be used:

MMC hole		.200
MMC fastener	(−)	.190
		.010

⊕ | ⌀ .010 Ⓜ

Other part		
MMC hole		.206
MMC fastener	(−)	.190
		.016

⊕ | ⌀ .016 Ⓜ

The position tolerance calculation method illustrated assumes the possiblility of a zero interference-zero clearance condition of the mating part features at extreme tolerance limits. Additional compensation of the calculated tolerance values should be considered as necessary relative to the particular application.

Formulas used as a basis for the position floating fastener calculations are:

To calculate position tolerance with fastener and hole size known:

$$T = H - F$$

where T = tolerance, H = MMC hole, and F = MMC fastener.

Where the hole size or fastener size is to be derived from an established position tolerance, the formula is altered to:

$$H = T + F$$
$$F = H - T$$

MATING PARTS — FLOATING FASTENER

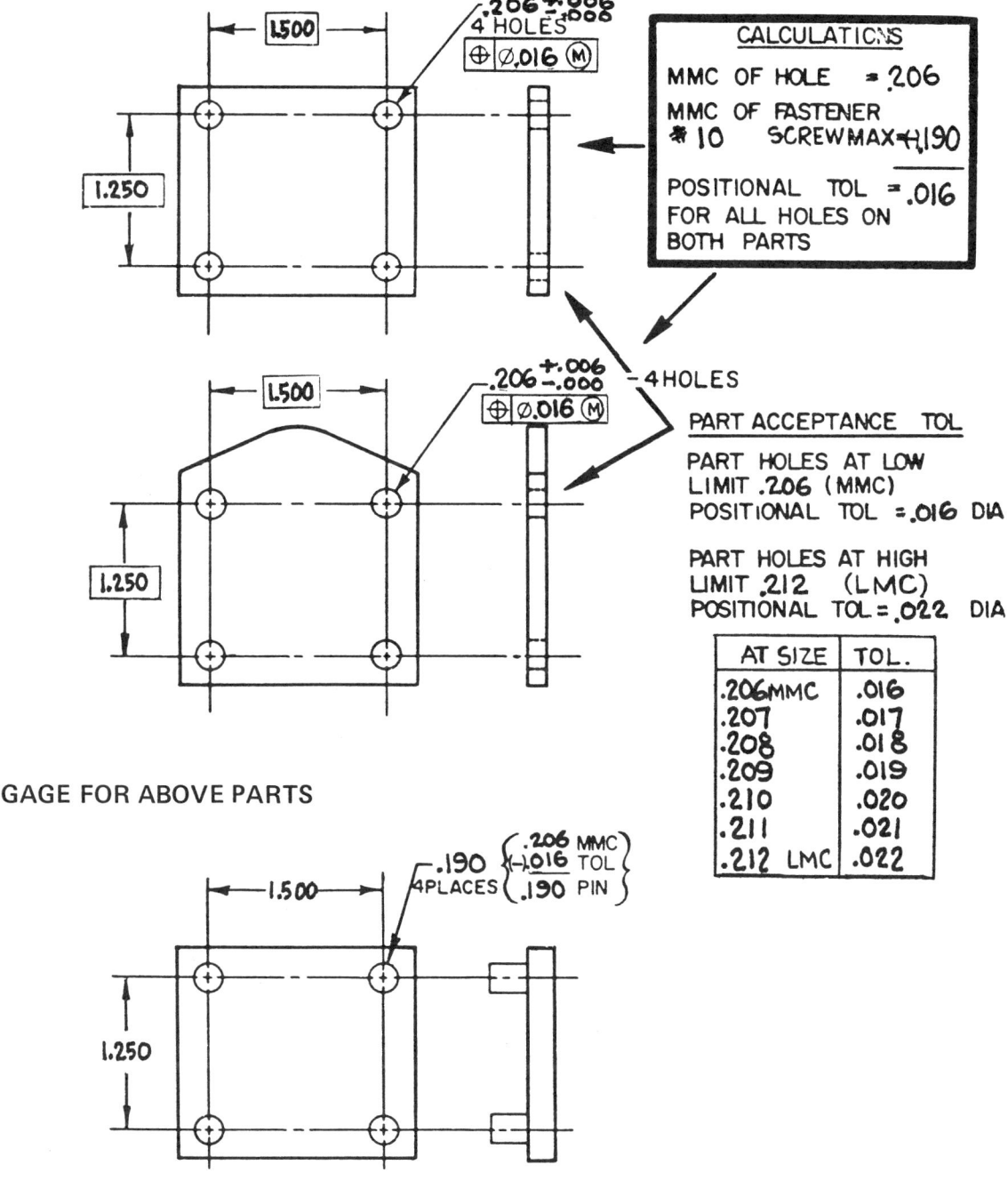

GAGE FOR ABOVE PARTS

When one of two mating parts has "fixed" features, such as the threaded studs in this example, the "fixed fastener" method is used in calculating position tolerances.

The term "fixed fastener" is a colloquialism popularly used to describe this application. Both the term and the technique are applied to numerous other manufacturing situations such as locating dowels and holes, tapped holes, etc.

The advantages of the MMC principle as described in the foregoing "floating fastener" application also apply here. However, with a "fixed fastener" application, the difference between the MMC sizes of mating features must be divided between the two features, since the total position tolerance must be shared by the two mating features. In this example, the two mating features (actually four of each in each pattern) are the studs and the clearance holes. The studs must fit through the holes at assembly.

Again, we see that the clearance of the mating features as they relate to each other at assembly determines the position tolerances. When one feature is to be assembled *within* another on the basis of the MMC sizes and "worst" condition of assembly, the clearance, or total tolerance, must be divided for assignment to *each* of the mating part features. In this case, the derived .016 was divided equally, with .008 diameter position tolerance assigned to each mating part feature (stud and hole). The total tolerance of .016 can be distributed to the two parts as desired, so long as the total is .016 (e.g., .010 + .006, .012 + .004, etc.). This decision is made at the design stage, however, and must be fixed on the drawing before release to production.

Application of the MMC principle to situations of this type guarantees functional interchangeability, design integrity, maximum production tolerance, functional gaging (if desired), and uniform understanding of the requirements.

As the part features of both parts are produced, any departure in size from MMC will increase the calculated position by an amount equal to that departure. Thus, for example, the position tolerance of the upper part could possibly increase up to .014, and that of the lower part up to .013 dependent upon the amount of departure from their MMC sizes. However, parts must actually be produced and sizes established before the *amount* of increase in tolerance can be determined.

Functional gages (shown below each part in the illustration) can be used for checking, and, although their use is not a must, they provide a very effective method of evaluation if desired. Note that the functional gages resemble the mating parts; as a matter of fact, functional gages simulate mating parts at their worst condition.

The functional gage pins of the upper part are determined by the MMC hole size minus the stated position tolerance. Gage tolerances are not shown, although they may be imagined to be on the order of $.1981 \, {}^{+.0002}_{-.0000}$ for pin size, and $\pm .0002$ on between pin locations. Local gage practices would prevail.

The functional gage on the lower part of the illustration contains holes instead of pins. The gage hole sizes are determined by the MMC (O.D.) size of the 10-32 pins plus the stated position tolerance. The tolerances are similar to those of the above pin gage. Tolerances on the order of $.1979 \, {}^{+.0000}_{-.0002}$ for hole size, and $\pm .0002$ between holes could be applied, depending on local gage practices.

It should be noted that the term MAJOR $\varnothing$ is used beneath the position callout on the lower part. In the absence of this special notation of exception, ANSI Y14.5

MATING PARTS — FIXED FASTENER

PART

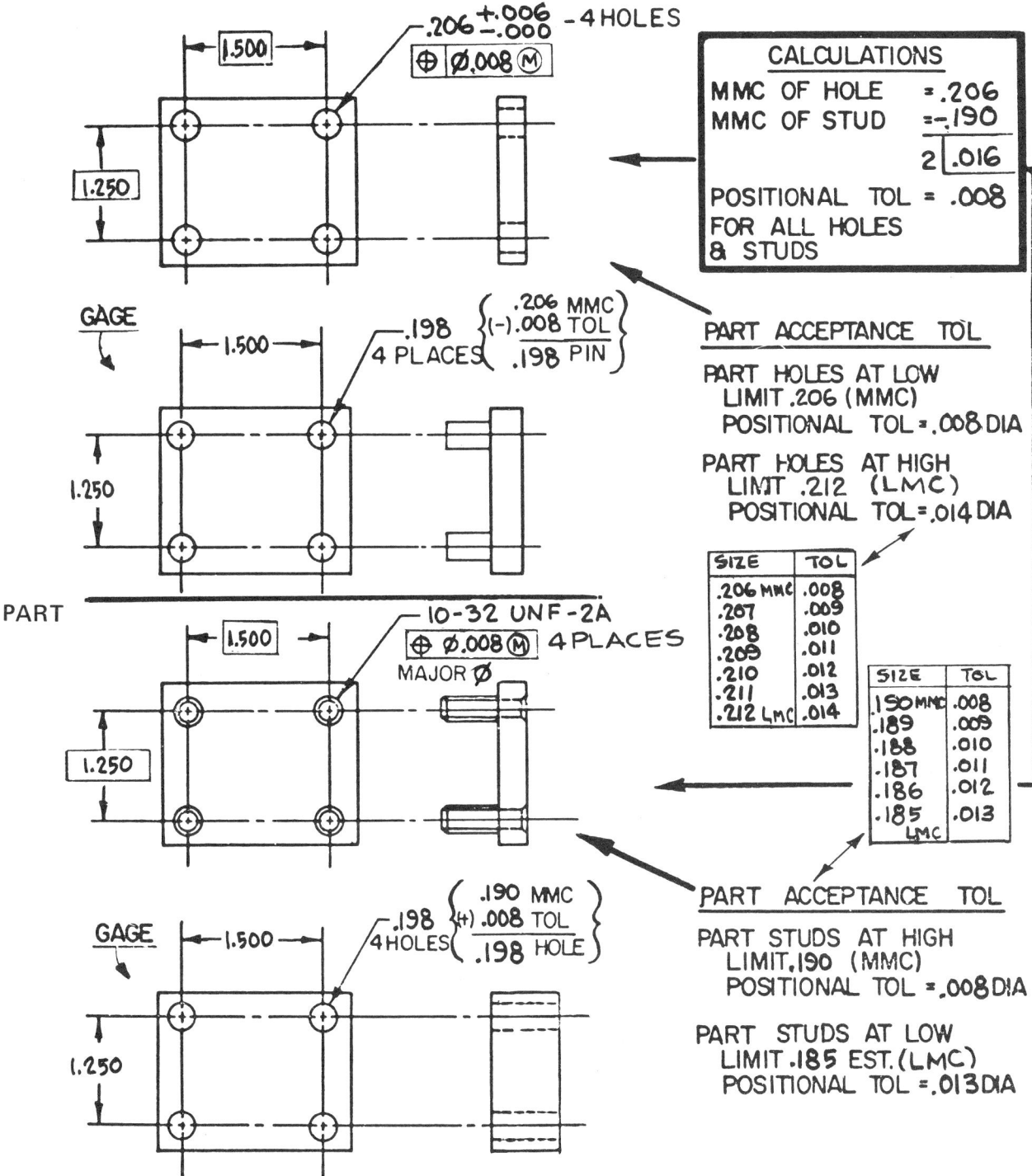

.206 +.006 -.000 - 4 HOLES
⊕ Ø.008 Ⓜ

1.500
1.250

CALCULATIONS

MMC OF HOLE = .206
MMC OF STUD = -.190
2 | .016

POSITIONAL TOL = .008
FOR ALL HOLES
& STUDS

GAGE

1.500
1.250

.198 { .206 MMC / (-).008 TOL / .198 PIN }
4 PLACES

PART ACCEPTANCE TOL

PART HOLES AT LOW
LIMIT .206 (MMC)
POSITIONAL TOL = .008 DIA

PART HOLES AT HIGH
LIMIT .212 (LMC)
POSITIONAL TOL = .014 DIA

SIZE	TOL
.206 MMC	.008
.207	.009
.208	.010
.209	.011
.210	.012
.211	.013
.212 LMC	.014

PART

10-32 UNF-2A
⊕ Ø.008 Ⓜ 4 PLACES
MAJOR Ø

1.500
1.250

SIZE	TOL
.190 MMC	.008
.189	.009
.188	.010
.187	.011
.186	.012
.185 LMC	.013

PART ACCEPTANCE TOL

PART STUDS AT HIGH
LIMIT .190 (MMC)
POSITIONAL TOL = .008 DIA

PART STUDS AT LOW
LIMIT .185 EST. (LMC)
POSITIONAL TOL = .013 DIA

GAGE

1.500
1.250

.198 { .190 MMC / (+).008 TOL / .198 HOLE }
4 HOLES

107

POSITION ⊕

Rule 4 would have invoked the tolerance on the basis of the pitch diameter of the threads. The major diameter (or O.D.) of the thread was the desired criterion in this example. See POSITION — EXTENDED PRINCIPLES section for additional examples of fixed fastener applications.

The calculations on these parts illustrate a balanced tolerance application in which the total permissible position tolerance of the two parts is equally divided, for example, .008 on each part. The total position tolerance can, however, be distributed as desired, as discussed earlier.

If more than two parts are assembled in a fixed fastener application, each part containing clearance holes must be calculated to mate with the part with the fixed features.

The position tolerance calculation method illustrated assumes the possibility of a zero interference-zero clearance condition of the mating part features at extreme tolerance limits. Additional compensation of the calculated tolerance values should be considered as necessary relative to the particular application.

Formulas used as a basis for the position fixed fastener (or locator) calculations are:

$$T = \frac{H - F}{2}$$

MMC hole	= H
MMC fastener	= F
(or pin, dowel, etc.)	
Tolerance	= T

Where the hole size or fastener (or pin, dowel, etc.) size is to be derived from an established tolerance, the formula is altered to:

$$H = F + 2T$$
$$F = H - 2T$$

MATING PARTS — FIXED FASTENER

This illustration shows position tolerancing applied to two mating parts with a circular hole pattern. The same reasoning applies here as in the preceding examples except that the basic dimensions are angular (45° angles, 8 places) and a diameter (the 1.500 diameter).

These two parts again are of the fixed fastening type, the studs of the lower part being the fixed elements. To determine the positional tolerances for each part, the MMC of the hole and the MMC of the stud are used to determine the total positional tolerance. This is divided by two to give the positional tolerance value for each part. The total value may be divided as desired, as previously described.

Note again how the positional tolerance *increases* as the holes in the upper part and the studs in the lower part depart from their MMC sizes, that is, when the holes get larger and the pins get smaller during the production process.

Functional gages are shown in the illustration for each of the parts. Note that the pins in the upper gage are calculated to the MMC or low limit of the holes in the part (which is .187 in this case) *minus* the positional tolerance (.0025), resulting in the .1845 gage pin size.

MATING PARTS, FIXED FASTENERS

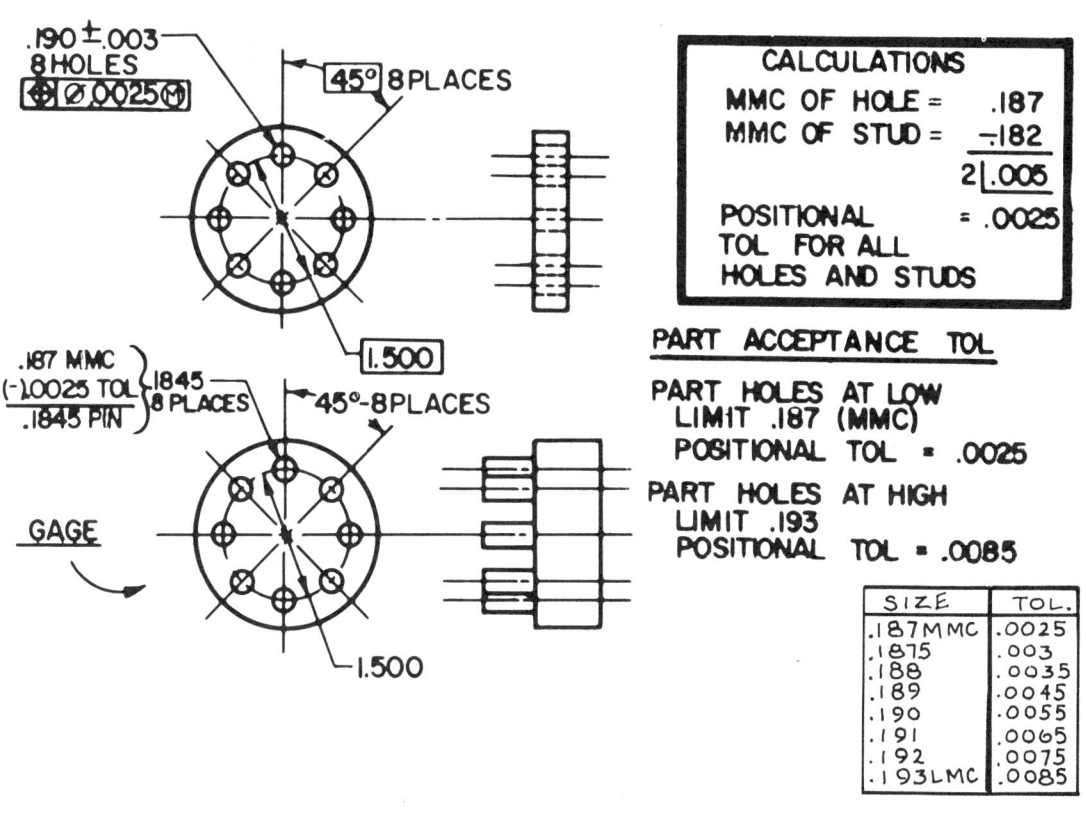

.190 ±.003
8 HOLES
⊕ Ø.0025 Ⓜ

45° 8 PLACES

1.500

.187 MMC
(-).0025 TOL }.1845
8 PLACES
.1845 PIN

45°-8 PLACES

GAGE

1.500

CALCULATIONS	
MMC OF HOLE =	.187
MMC OF STUD =	-.182
	2).005
POSITIONAL TOL FOR ALL HOLES AND STUDS	= .0025

PART ACCEPTANCE TOL

PART HOLES AT LOW
LIMIT .187 (MMC)
POSITIONAL TOL = .0025

PART HOLES AT HIGH
LIMIT .193
POSITIONAL TOL = .0085

SIZE	TOL.
.187 MMC	.0025
.1875	.003
.188	.0035
.189	.0045
.190	.0055
.191	.0065
.192	.0075
.193 LMC	.0085

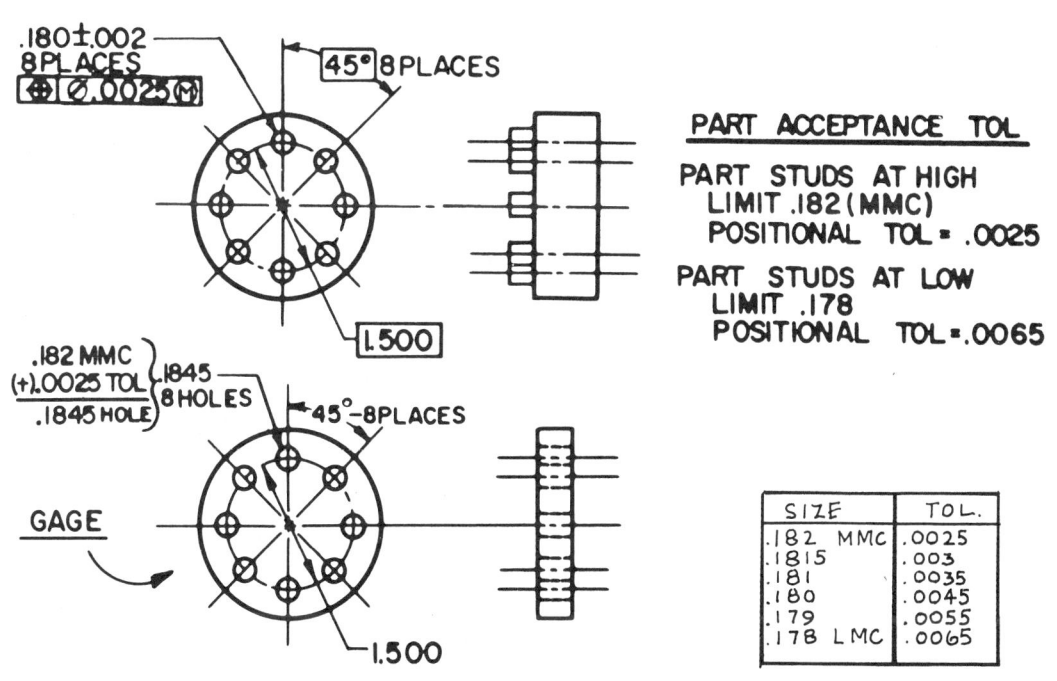

.180 ±.002
8 PLACES
⊕ Ø.0025 Ⓜ

45° 8 PLACES

1.500

.182 MMC
(+).0025 TOL }.1845
8 HOLES
.1845 HOLE

45°-8 PLACES

GAGE

1.500

PART ACCEPTANCE TOL

PART STUDS AT HIGH
LIMIT .182 (MMC)
POSITIONAL TOL = .0025

PART STUDS AT LOW
LIMIT .178
POSITIONAL TOL = .0065

SIZE	TOL.
.182 MMC	.0025
.1815	.003
.181	.0035
.180	.0045
.179	.0055
.178 LMC	.0065

The lower gage is calculated in reverse, using the MMC on high limit of the stud, .182, plus the positional tolerance (.0025), resulting in the .1845 gage hole size.

These calculations illustrate a balanced tolerance application in which the total permissible position tolerance of the two parts is equally divided, for example, .0025 on each part. The total position tolerance can, however, be distributed as desired, for example, .002 on one part, .003 on the other, etc., so long as it totals the tolerance calculated (in this case .005).

The position tolerance calculation method illustrated here and in preceding examples assumes the possibility of a zero interference-zero clearance condition of the mating part features at extreme tolerance limits. Additional compensation of the calculated tolerance value should be considered as necessary relative to the particular application.

RELATION TO IMPLIED DATUM SURFACES

Datum planes or surfaces as the basis for position relationships may be either *implied*, with no datum callouts on the drawing, or *specified* on the drawing. It should be made clear, however, that implied datums of any variety tend toward ambiguity. Caution should be exercised in their use. This discussion is intended to be an interpretation of principles applied on less critical requirements or where dimensional construction implies surface datums reasonably clearly. Datum precedence is, of course, left in doubt.

This illustration shows a part with three holes arranged in a position pattern. Datums are not specified on the drawing, because the datums are *implied* by the surfaces *from which* the position pattern dimensions are taken. In this case, the .750 dimensions obviously determine the surfaces which are to be used as datums and tie the surfaces to the position hole pattern.

The extremities or high points of the datum surfaces from which the .750 dimensions are taken establish the datum planes from which the position hole pattern is oriented. These datum planes are functional to the part requirement and are also used for tooling or fixturing reference and in establishing measuring planes to inspect the part.

The position tolerance zone is .010 when the holes are produced at MMC or the low limit of .245. The tolerance zone increases up to .020, if the holes are produced larger to the high limit of .255. Note that the position tolerance applies while the part is in contact with the datum surface; thus the position tolerance stated also controls the pattern location from the edges and provides the tolerance for the .750 pattern locating dimensions.

Where ambiguity exists in the use of implied surface datums or where datum precedence is required, datums should be specified. Implied datums, other than on outside surfaces of noncylindrical shapes, should normally not be permitted or used.

AS DRAWN

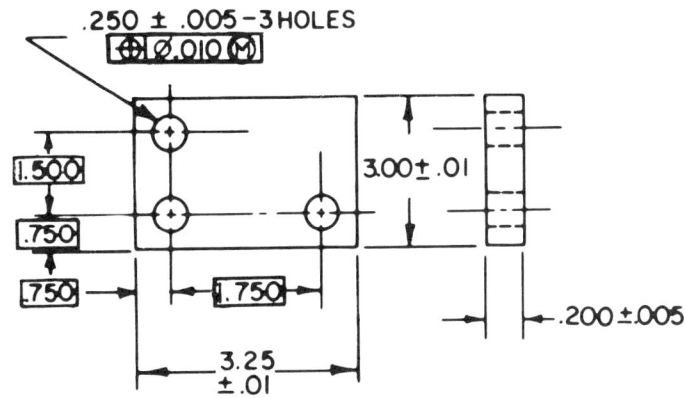

.250 ± .005 – 3 HOLES

⌖ ⌀ .010 Ⓜ

1.500

.750

.750

1.750

3.00 ± .01

3.25 ± .01

.200 ± .005

MEANING

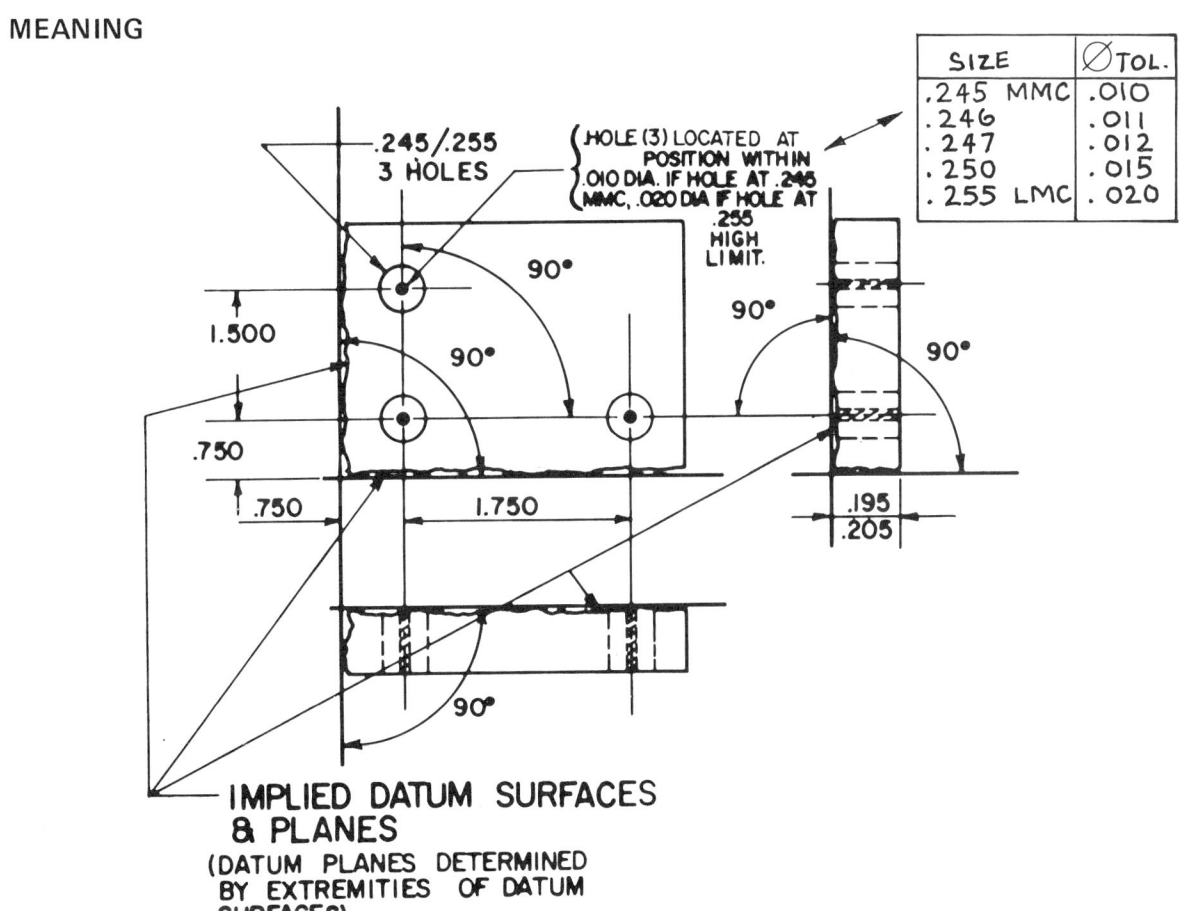

.245/.255
3 HOLES

HOLE (3) LOCATED AT POSITION WITHIN .010 DIA. IF HOLE AT .245 MMC, .020 DIA IF HOLE AT .255 HIGH LIMIT.

SIZE	⌀ TOL.
.245 MMC	.010
.246	.011
.247	.012
.250	.015
.255 LMC	.020

90°

90°

90°

90°

90°

1.500

.750

.750

1.750

.195
.205

IMPLIED DATUM SURFACES & PLANES
(DATUM PLANES DETERMINED BY EXTREMITIES OF DATUM SURFACES)

The illustration on the right shows the same part as that shown on the preceding page, but the datums are identified with the A, B, and C datum identification symbols and form tolerances are specified.

Where part function, and thus the stated drawing requirements, are more critical, specified datums and greater geometric control is necessary.

In this example, it was necessary to control the accuracy of the datum surfaces in their specific relationship to each other. To accomplish this, identification of the specific surfaces as datum references was required. Further, since the hole position pattern was critical in its orientation to the surfaces, datum identification was required for this purpose. With specification of the datums, precedence of the datum surfaces is established.

Datum surface A (top surface of the part) is to be held to a flatness of .001 total. Datum surface B is to be perpendicular to datum plane A within .001 total. Datum surface C is to be perpendicular to datum plane A within .001 total and also perpendicular to datum plane B within .002 total. Note that in this example we have also used combined datum identifying symbols and feature control symbols.

The extremities or high points of the datum surfaces from which the ⏐.750⏐ dimensions are taken establish the secondary and tertiary datum planes (B and C). The primary datum plane (A) is established by the extremities of datum surface A. The part orientation with respect to the position pattern is thus fixed. These datum planes are functional to the part requirement and are also used for tooling or fixturing reference.

The position tolerance zone is .010 when the holes are produced at MMC or the low limit of .245. The tolerance zone increases up to .020 if the holes are produced larger to the high limit of .255. Note that the position tolerance applies while the part is in contact with the datum surfaces according to their stated precedence or sequence. The position tolerance stated also controls the pattern location from the edges and provides the tolerance for the ⏐.750⏐ pattern locating dimensions.

In this example all geometric controls are specifically stated, removing all doubt as to design intent and follow-through manufacture and inspection requirements. The tools provided by specified datums and greater geometric control can be very effectively applied and will protect design integrity. However, caution should be exercised that these more complex requirements are introduced only where appropriate and as based on the specific needs of the application and part function.

AS DRAWN

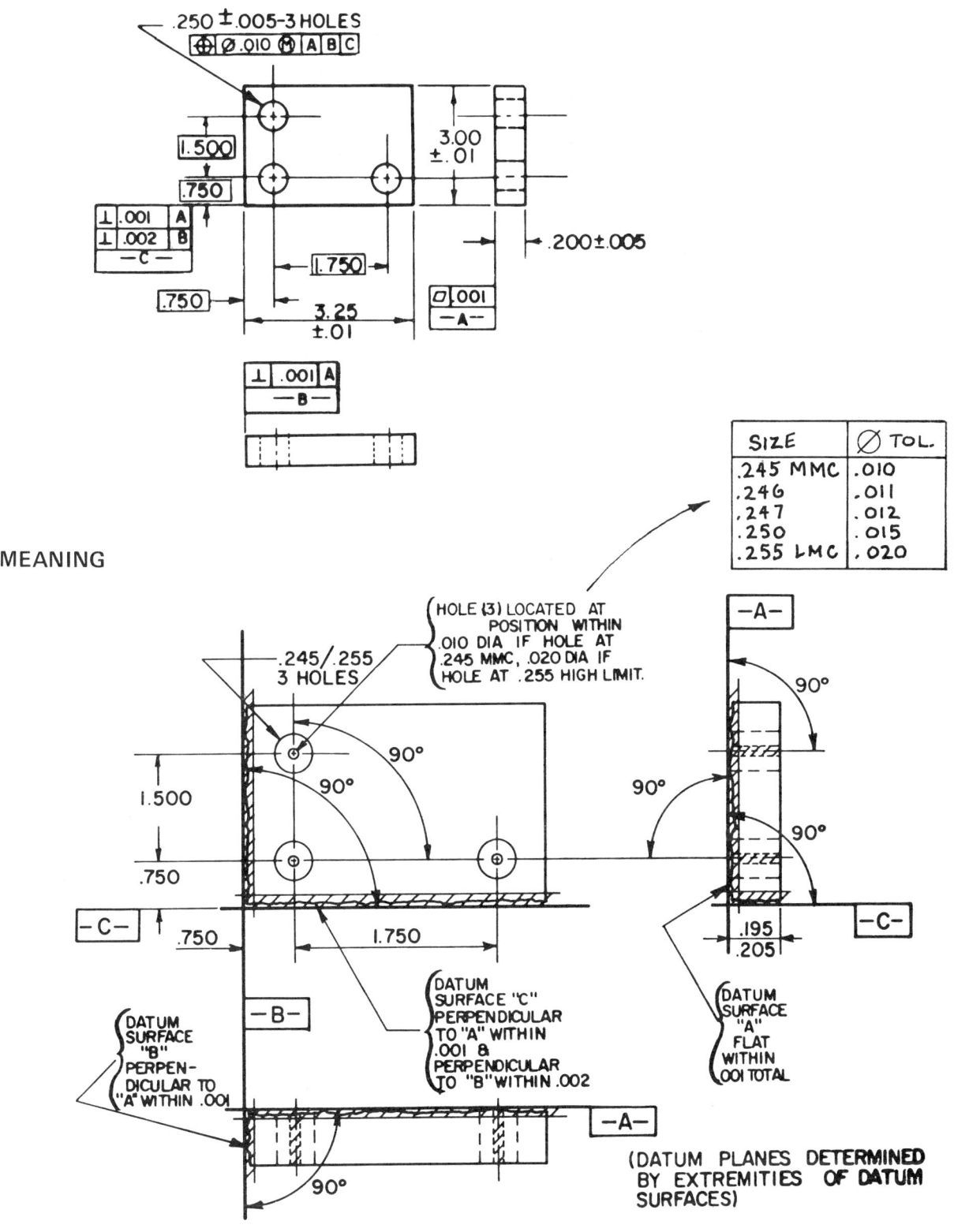

.250 ±.005-3 HOLES

⊕ | Ø.010 Ⓜ | A | B | C |

1.500

.750

3.00 ±.01

⊥ .001 | A
⊥ .002 | B
— C —

.200±.005

1.750

▱ .001
— A —

.750

3.25 ±.01

⊥ .001 | A
— B —

SIZE	Ø TOL.
.245 MMC	.010
.246	.011
.247	.012
.250	.015
.255 LMC	.020

MEANING

HOLE (3) LOCATED AT POSITION WITHIN .010 DIA IF HOLE AT .245 MMC, .020 DIA IF HOLE AT .255 HIGH LIMIT.

.245/.255 3 HOLES

—A—

90°

90°

1.500

90°

90°

90°

.750

—C—

.750

1.750

.195
.205

—C—

DATUM SURFACE "C" PERPENDICULAR TO "A" WITHIN .001 & PERPENDICULAR TO "B" WITHIN .002

DATUM SURFACE "A" FLAT WITHIN .001 TOTAL

—B—

DATUM SURFACE "B" PERPENDICULAR TO "A" WITHIN .001

90°

—A—

(DATUM PLANES DETERMINED BY EXTREMITIES OF DATUM SURFACES)

113

POSITION
RELATION TO DATUM SURFACES
LARGER TOLERANCES BETWEEN DATUM
AND PATTERN OF FEATURES—COMBINATION
COORDINATE AND POSITION TOLERANCING*

When the location of a pattern of features from datum surfaces is less important than the accuracy required within the pattern of features, both directly coordinated toleranced dimensions and position dimensions may be used. This approach may be used for mounting individual electrical or mechanical components, with the relationship of the mounting holes of the individual components as the most important consideration. When such combination dimensioning is used, the directly toleranced dimensions are to be interpreted as referring to theoretical centerlines (or center planes) in each pattern. Each individual feature is then permitted to deviate from its position in the pattern, within the position tolerance for that feature. In this case, a primary datum (A) is specified, with the less important outer edges implied as the secondary and tertiary datums.

In the upper right pattern of mounting holes in the example, the theoretical center of the pattern is located by toleranced dimensions from the datum surfaces. This center is the axis for the 1.990 basic diameter. The individual theoretical centers of the eight .188 holes are established by intersection of the 45° basic angles at the 1.990 basic diameter.

The actual centers of the individual holes are specified to be at position within a .005 diameter tolerance zone when the hole is at .188 MMC. The tolerance zone increases to .009 (the position tolerance of .005 plus the hole tolerance of .004) as the hole is produced at the high limit or largest size of .192 (LMC).

The lower rectangular pattern of holes shows the center planes located by toleranced dimensions from the datum surfaces and basic dimensions within the pattern. The meaning is shown in the lower portion of the page.

The position tolerance zone is .005 diameter total when the holes are at MMC of .250 and .009 diameter total when the holes are at .254, or the high limit (LMC).

Note that in each hole pattern, the pattern as a unit may vary within the more lenient pattern locating coordinate tolerance zone, while the hole-to-hole location within each pattern must be maintained within the position tolerance zone.

As is seen in the details of the lower illustration at right, the position tolerance zone (center or axis) must lie within, or on, the pattern location zone with the produced hole axis permitted to fall anywhere within the position tolerance zone.

Likewise seen is the restriction implied for the perpendicularity or attitude of each hole within (only) the position tolerance zone; the pattern location or rotation is, however, permitted the wider coordinate tolerance. Departure from MMC hole size then increases the tolerances as previously described.

---

* Before using this method, it is recommended that COMPOSITE POSITION TOLERANCE methods be considered.

AS DRAWN

MEANING

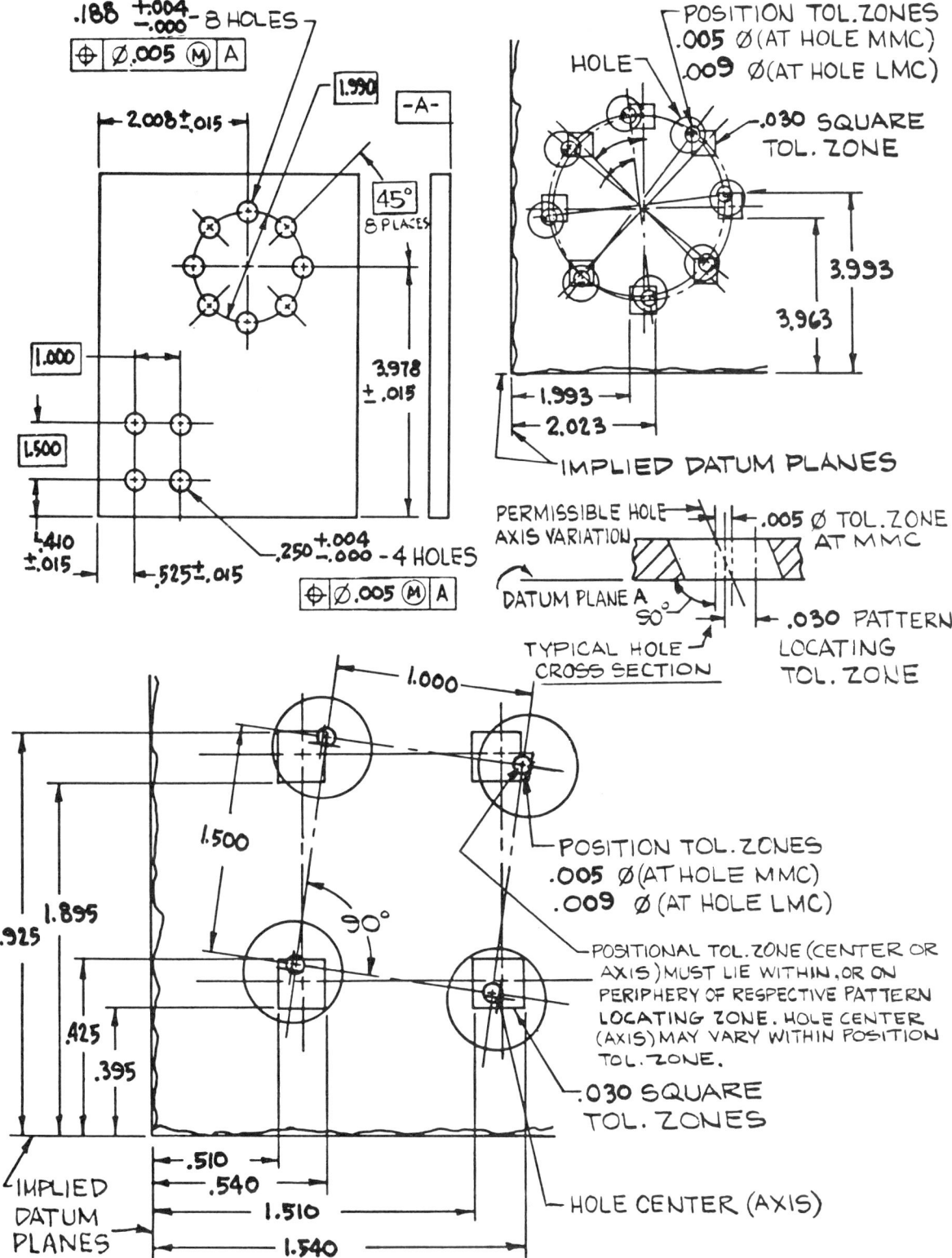

When the location of a pattern of features from datum surfaces is less important than the accuracy required *within* the pattern of features, composite position tolerancing may be used. Composite position tolerancing also extends the use of specified datum relationships and geometric tolerance control.

Composite position tolerancing incorporates a dual feature control symbol with two positional controls. One, the upper entry in the symbol, specifies the applicable datums and the pattern locating position tolerance. The lower entry specifies the applicable datum and the feature relating position tolerance. A single position tolerance symbol is used.

The composite position tolerance method utilizes the full advantages of MMC and extends the principles to control of patterns of features as well as of the individual feature interrelationship.

In the upper right pattern of eight mounting holes in the example, the theoretical center of the pattern is located by the 2.008 and 3.978 basic dimensions and is related to datums A, B, and C. This center is the axis for the 1.990 basic diameter. The individual theoretical centers of the eight .188 holes are established by the intersection of the 45° basic angles at the 1.990 basic diameter. The pattern as a unit, yet actually determined by the holes themselves, is to be located within a position tolerance of .03 when the holes are at MMC. As the individual holes in the pattern depart from MMC toward LMC, additional tolerance to that hole (and thus to the pattern) is acquired equal to that departure.

Within the hole pattern itself, the feature relating position tolerance is established at .005 diameter. Then, as previously described, each hole in the pattern may increase its position tolerance an amount equal to the departure from MMC as it is produced to a maximum of .009 at LMC. Note that the hole-to-hole interrelationship in the pattern, as well as the relationship to datum A, is maintained. The attitude (perpendicularity) of each individual hole must be within the .005 diameter tolerance zone as well as within position of .005 at MMC.

As can be seen in the typical hole cross-sectional view, the axes of both the large (pattern) and small (feature) tolerance zones are parallel. The axes of the holes must lie within *both* the larger and smaller tolerance zones. Portions of the smaller zones may fall outside the peripheries of the larger tolerance. However, this portion of the smaller zone is not usable since the axis of each hole must fall within both zones.

The lower left four hole pattern follows the same reasoning as described above and as seen in the illustration.

AS DRAWN

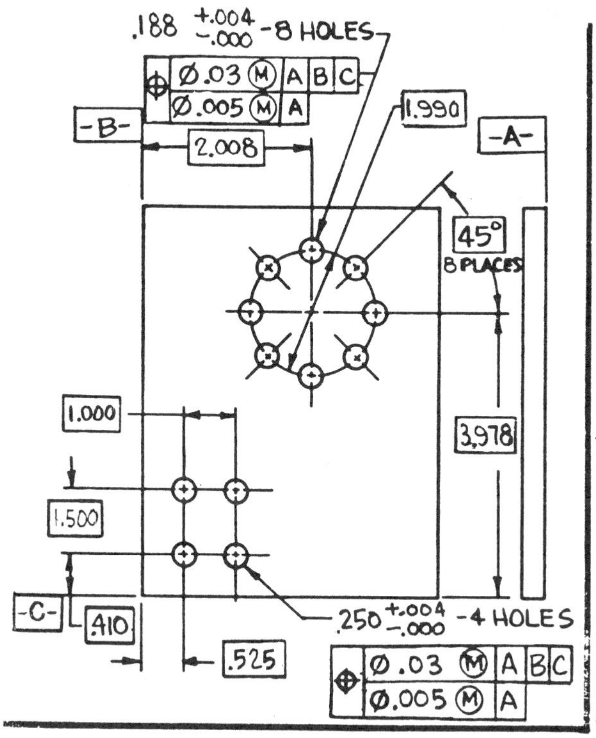

MEANING

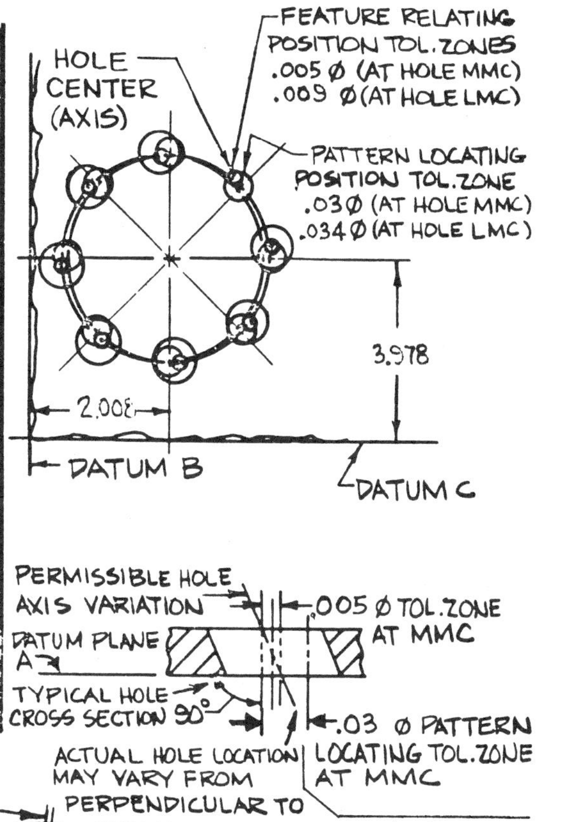

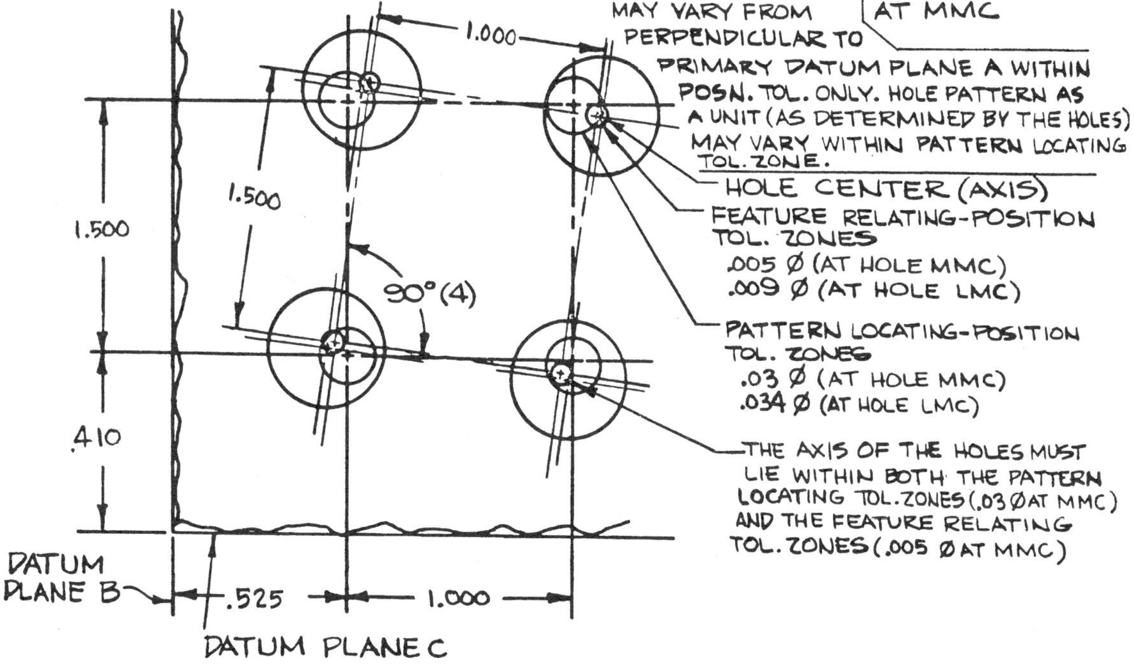

(See following pages for gaging.)

POSITION ⊕

RELATION TO DATUM SURFACES
COMPOSITE POSITION TOLERANCING — FUNCTIONAL GAGES

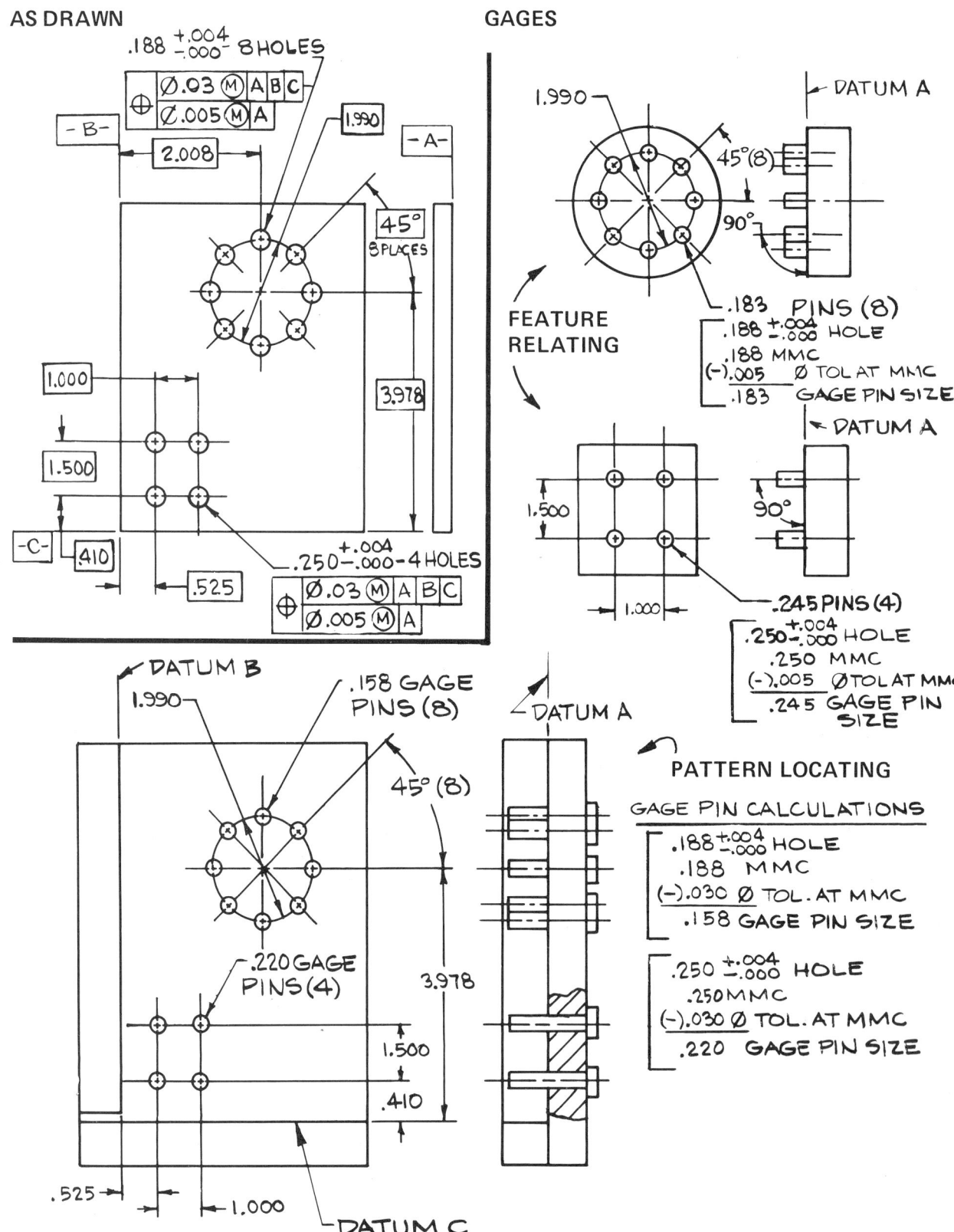

AS DRAWN

.188 +.004 -.000 8 HOLES

⊕	Ø.03 Ⓜ	A	B	C
	Ø.005 Ⓜ	A		

- B -

2.008

1.990

- A -

45° 8 PLACES

1.000

1.500

3.978

- C -   .410

.525

.250 +.004 -.000 - 4 HOLES

⊕	Ø.03 Ⓜ	A	B	C
	Ø.005 Ⓜ	A		

DATUM B

1.990

.158 GAGE PINS (8)

45° (8)

.220 GAGE PINS (4)

3.978

1.500

.410

.525   1.000

DATUM C

GAGES

1.990

DATUM A

45°(8)

90°

FEATURE RELATING

.183 PINS (8)
.188 +.004 -.000 HOLE
.188 MMC
(-).005 Ø TOL AT MMC
.183 GAGE PIN SIZE

DATUM A

90°

1.500

1.000

.245 PINS (4)
.250 +.004 -.000 HOLE
.250 MMC
(-).005 Ø TOL AT MMC
.245 GAGE PIN SIZE

DATUM A

PATTERN LOCATING

GAGE PIN CALCULATIONS

.188 +.004 -.000 HOLE
.188 MMC
(-).030 Ø TOL. AT MMC
.158 GAGE PIN SIZE

.250 +.004 -.000 HOLE
.250 MMC
(-).030 Ø TOL. AT MMC
.220 GAGE PIN SIZE

118

COMPOSITE POSITION TOLERANCING – FUNCTIONAL AND PAPER GAGING

The previous page illustrates functional gaging principles for the four hole pattern of the part on page 117. Another method, paper gaging, is represented by a sample situation on page 121.

Paper gaging methods are shown to demonstrate actual usable techniques which can also be used to quantify position tolerance principles pictorially.

Paper gaging is accomplished through plotting an enlarged scale of coordinately measured feature positions onto a piece of standard graph paper and then plotting the resulting differentials (actual position versus true position) to a selected scale (e.g., one square = .0005) with a dot on the graph. An overlay chart (gage) of tracing paper or other transparent material containing a series of graph-scale circles of desired increments is placed over the graph to depict the position tolerance zones. Note that the paper gaging method simulates part function and functional gaging. However, the individual tolerance zones are each assumed to be represented by the one exact (true) position on the graph. The exact (basic) dimensions of the pattern are assumed as 0 in the X and Y directions.

Step 1. The four .250 holes are located both as a pattern ($\varnothing$.03) and hole-to-hole within the pattern ($\varnothing$.005). In open set-up inspection methods, two steps are required to determine whether both requirements have been met. Step 1, using coordinate measuring and paper gaging, will determine whether the hole pattern as a unit (as based upon individual hole location) has met the $\varnothing$.03 tolerance requirement. The part is set up according to the datum surfaces and is measured to holes 1, 2, and 3 (see illustration on page 121); only three of the four holes are considered at this time.

The resulting X and Y measurements are compared to the specified coordinate dimensions on the drawing resulting in a differential (off position) value. This differential is plotted on the paper gage graph according to the above stated scale (i.e. one square = .0005).

When the actual location of the three holes are plotted, an overlay gage, with the circles, the scale of the graph plot, and a representation of the tolerances, is placed over the graph plot. The center of the overlay must be placed on the center of the plot. If the plotted centers fall within the position zone (in this case the $\varnothing$.030 zone), the pattern, via the holes, meets the requirement. Since hole #1 in our hypothetical example exceeds the $\varnothing$.030 tolerance zone, MMC principles may be invoked. The size of hole #1 is found to be .252 which is .002 departure from MMC, thus that hole has $\varnothing$.032 tolerance, and as seen in the illustration on the following page, is now acceptable. The hole pattern is also thus found acceptable within the $\varnothing$.030 tolerance.

Step 2. To evaluate the accuracy of the four holes in the pattern (part shown on page 117) relative to the individual .005 position tolerances, it is necessary to consider the hole-to-hole relationship in the pattern, exclusively. The four-hole grid pattern from which the individual hole positions can be compared is established by selecting two of the the holes as a basis. In our example, hole #1 is selected as the origin for the X- and Y-coordinate measurements. Since, one additional hole must be selected to give orientation or square-up to the pattern; hole #3 is selected for the X orientation.

POSITION ⊕

From the part orientation now established, each hole is measured in X and Y from this set-up. From the illustration on page 121, resultant differentials are derived from our hypothetical measurements. Note, of course, that hole #1 is zero in X and Y (as the origin for measurement), and hole #3 is zero in X as the square-up for the basic pattern.

The differentials are plotted on the graph paper to the desired scale (in this case every five squares = .001) using an origin on the graph as the hole #1 position. Each hole is plotted in the appropriate value and quadrant in X and Y from the hole #1 position.

With the holes plotted on the graph, the overlay gage with the circles representing the tolerances scaled to the graph plot is placed over the graph. The overlay is moved at random to try to encompass all the plotted hole centers simultaneously within the stated tolerance .005 circles. After trial and error, let us assume that the illustrated location of the overlay gage can successfully encompass holes #1, #2, and #3 but hole #4 is outside the .005 circle. MMC principles may now be invoked. By checking the size of hole #4, it is determined to be .2515, which is a departure of .0015 from MMC; thus hole #4 has .0065 position tolerance, and as is seen in the illustration, is now acceptable.

All holes in the pattern have met their position tolerance requirements. It is extremely important to note that hole #1, even though used as a zero location origin, is actually assumed as an equal partner in the pattern and must likewise be treated as imperfect in its relationship to its desired position.

Paper gaging simulates hard gaging and part function and thus is an effective technique. This method is further advantageous in that it visually detects error trends through periodic inspections, gives a permanent record, and requires no gaging tolerance.

STEP 1

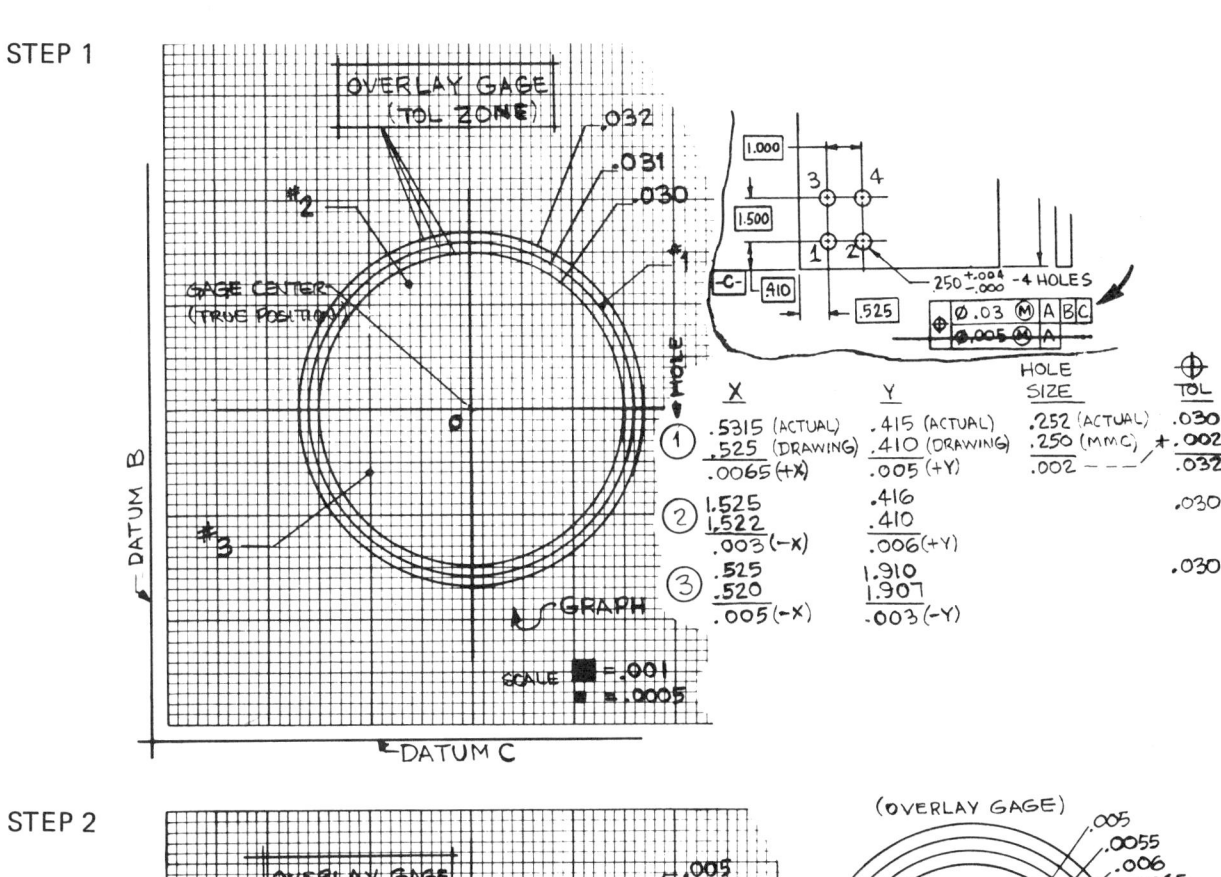

	X	Y	HOLE SIZE	TOL
①	.5315 (ACTUAL) / .525 (DRAWING) / .0065 (+X)	.415 (ACTUAL) / .410 (DRAWING) / .005 (+Y)	.252 (ACTUAL) / .250 (MMC) / .002 ---	.030 / +.002 / .032
②	1.525 / 1.522 / .003 (−X)	.416 / .410 / .006 (+Y)		.030
③	.525 / .520 / .005 (−X)	1.910 / 1.907 / .003 (−Y)		.030

STEP 2

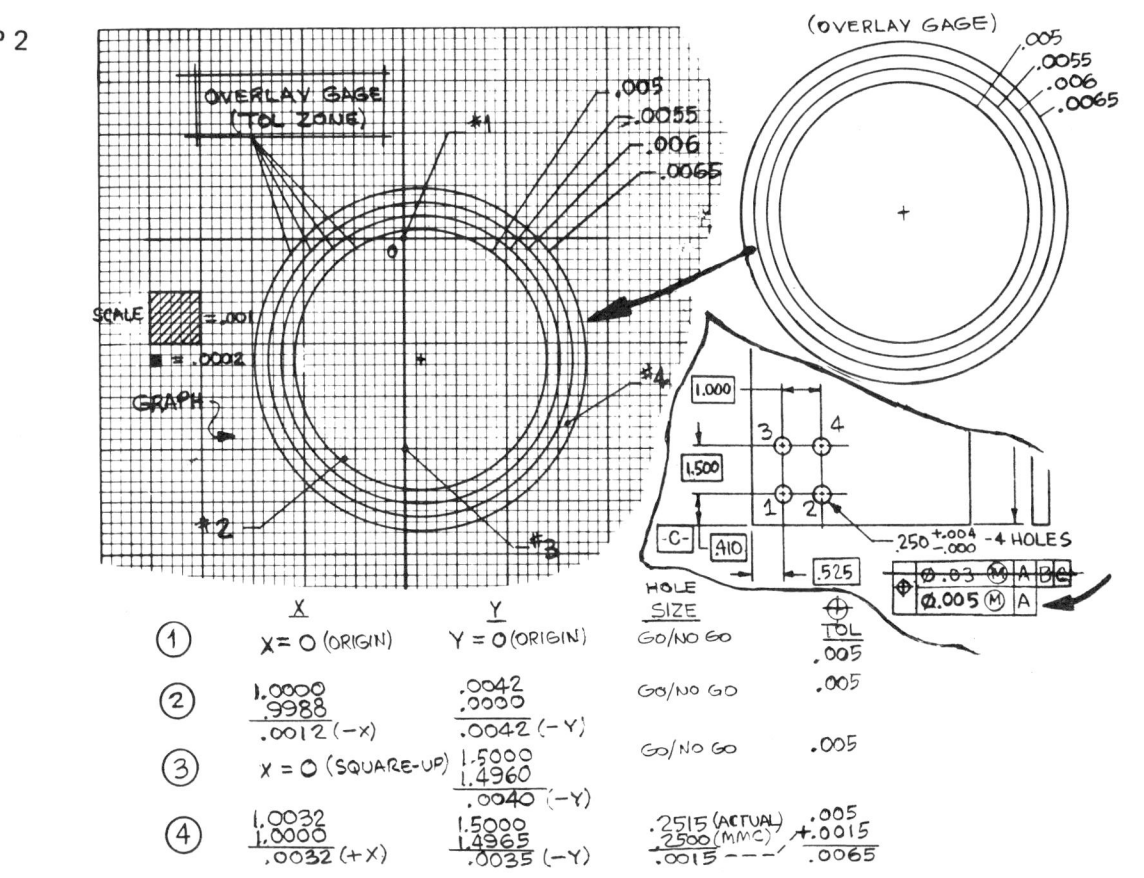

	X	Y	HOLE SIZE	TOL
①	X = 0 (ORIGIN)	Y = 0 (ORIGIN)	GO/NO GO	.005
②	1.0000 / .9988 / .0012 (−X)	.0042 / .0000 / .0042 (−Y)	GO/NO GO	.005
③	X = 0 (SQUARE-UP)	1.5000 / 1.4960 / .0040 (−Y)	GO/NO GO	.005
④	1.0032 / 1.0000 / .0032 (+X)	1.5000 / 1.4965 / .0035 (−Y)	.2515 (ACTUAL) / .2500 (MMC) / .0015 ---	.005 / +.0015 / .0065

121

POSITION ⊕
MMC RELATED TO MMC DATUM FEATURE

When a pattern of holes is dimensioned relative to the location of another hole, this hole is identified as a datum and the hole pattern is located dimensionally with respect to it.

In the example, the four holes are related to the center datum hole. As the position of the center datum hole shifts, the position of the four hole pattern itself must follow as dictated by the function of the part. Imagine that this part has a mating part with a shaft and four pins which must assemble with the five holes in the illustrated part.

Note that the .500 datum center hole is also located from datums (surfaces) A, B, and C. It is given a position tolerance of ⌀.010 at MMC and a refined attitude, or perpendicularity, tolerance of ⌀.003 at MMC. This properly controls the center hole relationship to the edge surfaces with a rather lenient position tolerance while the attitude to the primary datum A is maintained to closer tolerance. The center hole is identified as datum D so that the four .380 holes can be located with respect to it. Under MMC principles, the position and perpendicularity tolerances increase an amount equal to the hole size departure from MMC as shown in the illustration.

Wherever the datum hole D position varies in the design considerations, or in actual production, the four .380 hole pattern must follow. Note that the positional pattern dimensions originate at datum D to carry out this intent.

The four .380 holes are located by a position tolerance of ⌀.005 at MMC with respect to datum A (for attitude), datum D at MMC (virtual condition) (for location), and datum B (for orientation). Reference to the lower portion of the illustration under Meaning will assist in the understanding of the effect of the three datums and the importance of datum precedence. Although not illustrated, imagine the four .380 hole position tolerances have been calculated relative to a mating part using the "fixed fastener" method previously explained.

The four holes individually with respect to their own true positions can vary in location up to ⌀.005 at MMC. While at any size in a departure from MMC a hole can vary an additional amount (enlarges positional tolerance) equal to that departure. For example (as shown), if the hole size is produced at .381, the position tolerance becomes ⌀.006, at .382 it becomes ⌀.007, etc., up to ⌀.008 if the hole is produced at .383 LMC.

Since the datum feature is a size feature, its variation in size (.500 to .502) has an effect on the four hole pattern relationship which locates from it. That is, as the size of the datum feature increases, its relationship to the mating part corresponding feature (e.g. a shaft) changes; if the imagined mating part shaft and corresponding pins (at MMC) will insert into holes larger than those we used to calculate the position tolerances, greater latitude (off position) of the pattern as a unit can be realized. This latitude is, however, to the hole pattern as a unit and not relative to the four holes individually or hole-to-hole in the four hole pattern.

Please see the following gaging illustrations and text describing evaluation of this part with representative techniques.

AS DRAWN

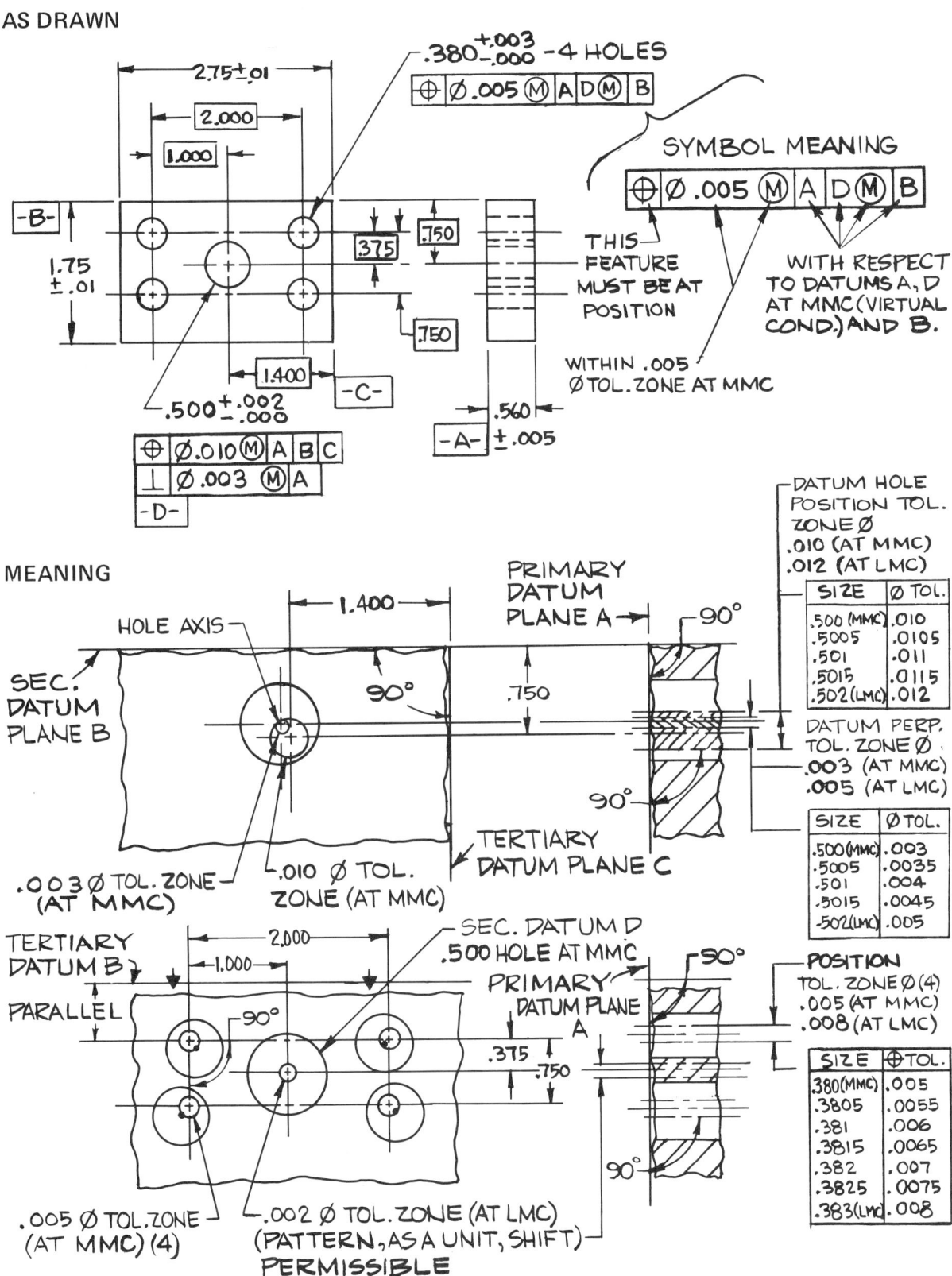

SYMBOL MEANING

THIS FEATURE MUST BE AT POSITION

WITH RESPECT TO DATUMS A, D AT MMC (VIRTUAL COND.) AND B.

WITHIN .005 Ø TOL. ZONE AT MMC

MEANING

HOLE AXIS

SEC. DATUM PLANE B

PRIMARY DATUM PLANE A

TERTIARY DATUM PLANE C

.003 Ø TOL. ZONE (AT MMC)

.010 Ø TOL. ZONE (AT MMC)

DATUM HOLE POSITION TOL. ZONE Ø .010 (AT MMC) .012 (AT LMC)

SIZE	Ø TOL.
.500 (MMC)	.010
.5005	.0105
.501	.011
.5015	.0115
.502 (LMC)	.012

DATUM PERP. TOL. ZONE Ø .003 (AT MMC) .005 (AT LMC)

SIZE	Ø TOL.
.500 (MMC)	.003
.5005	.0035
.501	.004
.5015	.0045
.502 (LMC)	.005

TERTIARY DATUM B

PARALLEL

SEC. DATUM D .500 HOLE AT MMC

PRIMARY DATUM PLANE A

POSITION TOL. ZONE Ø (4) .005 (AT MMC) .008 (AT LMC)

SIZE	⊕ TOL.
.380 (MMC)	.005
.3805	.0055
.381	.006
.3815	.0065
.382	.007
.3825	.0075
.383 (LMC)	.008

.005 Ø TOL. ZONE (AT MMC) (4)

.002 Ø TOL. ZONE (AT LMC) (PATTERN, AS A UNIT, SHIFT) PERMISSIBLE

Functional gaging principles which can be applied to the preceding example (duplicated at the upper left of the opposite page) are shown in the illustration.

It must first be clarified that functional gaging is not required when position tolerancing is used. Other techniques (i.e., open set-up coordinate methods, optical methods, etc.) can also be used. For example, the paper gaging method illustrated earlier in this section could effectively be applied using the datum D feature as zero and attitude and orientation from Datums A and B.

The purpose of this illustration is to depict representative functional gages and show actual methods as well as to demonstrate principle.

The illustration is intended to be self descriptive and to explain the necessary details. Note that three gages (or operations) are required to fulfill the requirements.

Note specifically the clarity afforded by the clearly specified design requirements via the datums. Also, note the manner in which the gages pattern after the part function and simulate mating part relationships.

Also worthy of note is the manner in which the datum D pick-up pin size virtual condition) is established based upon Rule 5. The virtual condition applicable (.497) is derived from the smallest form or position tolerance controlling the hole (i.e., ∅ .003) as subtracted from MMC size .500.

AS DRAWN

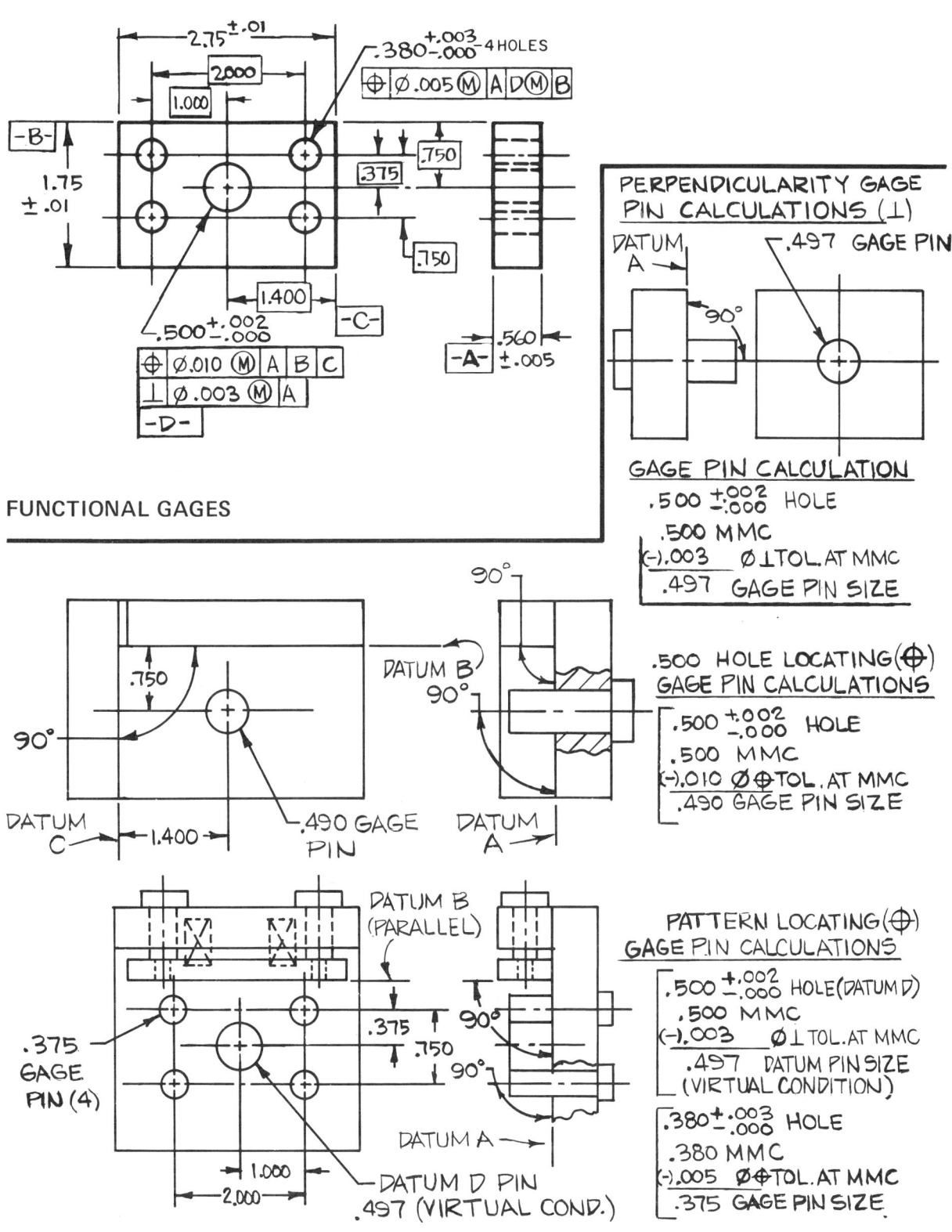

FUNCTIONAL GAGES

PERPENDICULARITY GAGE PIN CALCULATIONS (⊥)

DATUM A →

90°

.497 GAGE PIN

GAGE PIN CALCULATION

.500 $\pm$.002/-.000 HOLE

.500 MMC

(-).003 Ø⊥TOL. AT MMC

.497 GAGE PIN SIZE

.500 HOLE LOCATING (⊕) GAGE PIN CALCULATIONS

.500 $\pm$.002/-.000 HOLE

.500 MMC

(-).010 Ø⊕TOL. AT MMC

.490 GAGE PIN SIZE

PATTERN LOCATING (⊕) GAGE PIN CALCULATIONS

.500 $\pm$.002/-.000 HOLE (DATUM D)

.500 MMC

(-).003 Ø⊥TOL. AT MMC

.497 DATUM PIN SIZE (VIRTUAL CONDITION)

.380 $\pm$.003/-.000 HOLE

.380 MMC

(-).005 Ø⊕TOL. AT MMC

.375 GAGE PIN SIZE

When a pattern of holes is dimensioned relative to the location of another hole, this hole is identified as a datum and the hole pattern is located dimensionally with respect to it.

In the illustrated example, the four holes are related to the center datum hole. As the position of the center datum hole shifts, the position of the four hole pattern itself must follow as dictated by the function of the part. Imagine that this part has a mating part with a shaft and four pins which must assemble with the five holes in the illustrated part.

Differing from the preceding illustration, assume there is a precision fit required between the shaft of the imagined part as it fits closely into the .500 hole. The four pins of the imagined part, however, are to relate to the four .380 holes on an MMC basis as in the preceding illustration. Therefore, since in this case the relationship between the four holes and their datum is critical or more precise, the datum D is referenced regardless of feature size (RFS).

Note that the .500 datum center hole is located from surface datums A, B, and C. It is given a position tolerance relative to these edges of $\varnothing$ .010 at MMC since the relationship to the edges on this basis can have a rather lenient position tolerance. The attitude of the datum hole relative to the primary datum A is, however, to be maintained to a closer tolerance. Since datum D position and attitude is controlled on an MMC basis, the tolerances (position and perpendicularity) increase an amount equal to the produced size departure from MMC shown in the illustration. It should be noted here, however, that the reference to datum D in the relationship of the four .380 holes is on an RFS basis. This means the four hole pattern takes reference from the exact center of the datum hole at whatever size it is produced (RFS) within .500 to .502.

Wherever the datum hole D position varies in the design considerations, or in actual production, the four .380 hole pattern must follow. Note that the positional pattern dimensions originate at datum D to carry out this intent.

The four .380 holes are located by a position tolerance of $\varnothing$ .005 at MMC with respect to datum A (for attitide), datum D at RFS (for location), and datum B (for orientation). Reference to the lower portion of the illustration under Meaning will assist in understanding the effect of the three datums and the importance of datum precedence. Although not illustrated, imagine that the four .380 hole position tolerances have been calculated relative to a mating part using the "fixed fastener" method previously explained.

The four holes individually, with respect to their own true or exact positions, can vary in location up to $\varnothing$ .005 at MMC. While at any size in a departure from MMC, a hole can vary an additional amount (enlarge positional tolerance) equal to that departure. For example (as shown), if the hole size is produced at .381, the position tolerance becomes $\varnothing$ .006, at .382 it becomes $\varnothing$ .007, etc., up to $\varnothing$ .008, if the hole is produced at .383 (LMC).

Although the datum feature is a size feature as in the preceding illustration, its variation in size (.500 to .502) will have *no* effect on the four hole pattern relationship. This is because the pattern relationsip is to the center (axis) of the datum D, .500 hole, no matter to which size it is produced in its size tolerance range (i.e., RFS).

In this instance, it is seen that a more critical or precise relationship is maintained between the four holes and their datum. The design requirements and mating part functional interface determined this approach in our example.

Please see the following gaging illustrations and text describing evaluation of this part with representative techniques.

AS DRAWN

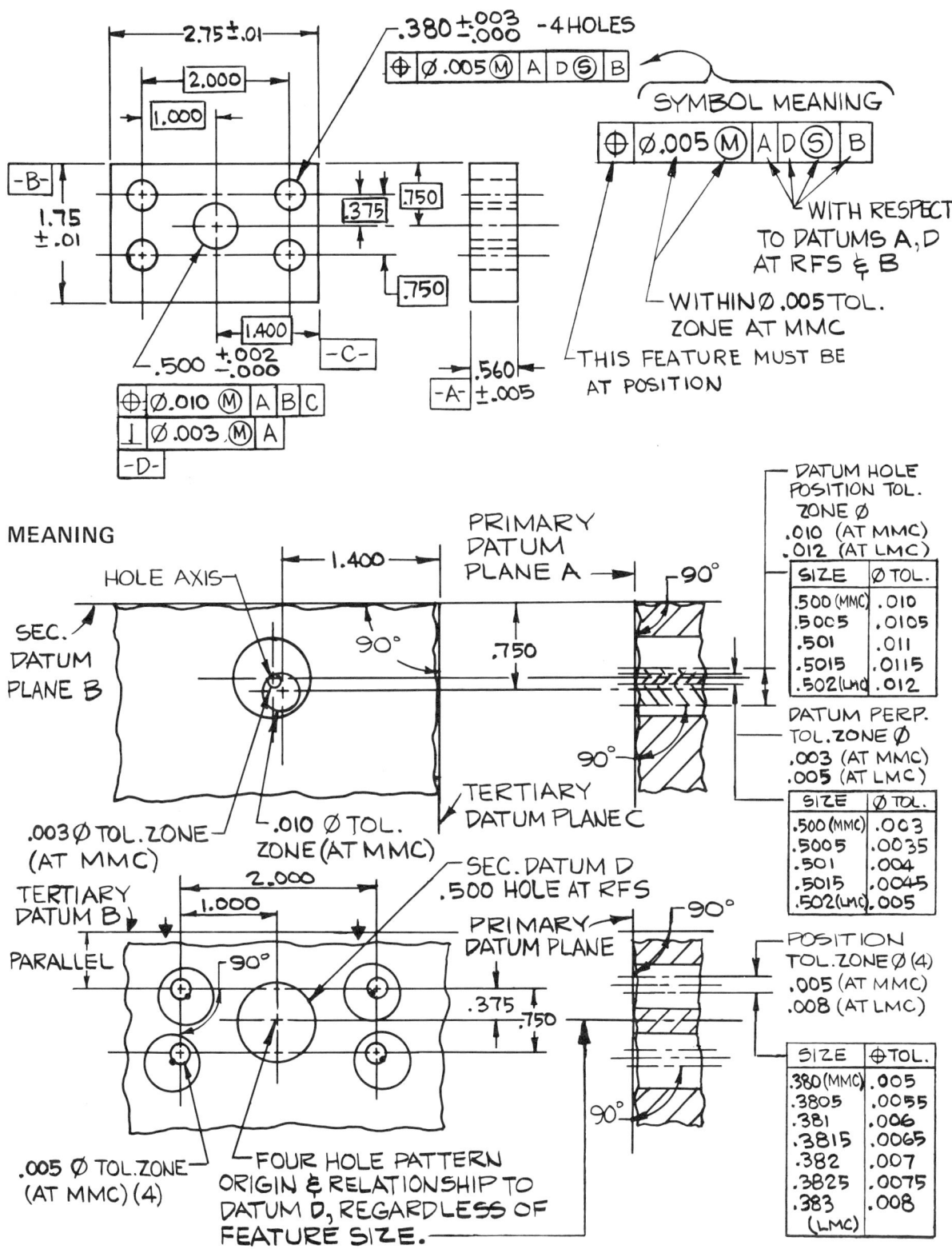

MEANING

Functional gaging principles which can be applied to the preceding example (duplicated at the upper left of the opposite page) are shown in the illustration.

It must first be clarified that functional gaging is not required when position tolerancing is used. Other techniques (i.e. open setup coordinate methods, optical methods, etc.) can also be used. For example, the paper gaging methods illustrated earlier in this section, could effectively be applied using the datum D feature as zero and attitude and orientation from datums A and B.

The purpose in this illustration is to depict representative functional gages to show actual methods as well as to illustrate principle. Note that functional principles, identical to the preceding part, can be used except for the RFS pick-up of the datum feature.

The illustration is intended to be self-descriptive and to explain the necessary details. Note that three gages (or operations) are required to fulfill the requirements.

Note specifically the clarity afforded by the clearly specified design requirements via the datums. Also, note the manner in which the gages pattern after the part function and simulate mating part relationships.

In this example, the use of the RFS datum has tightened requirements. Costs in manufacturing will probably be higher than with total use of MMC as in the preceding example; yet, functional principles can be applied as shown.

AS DRAWN

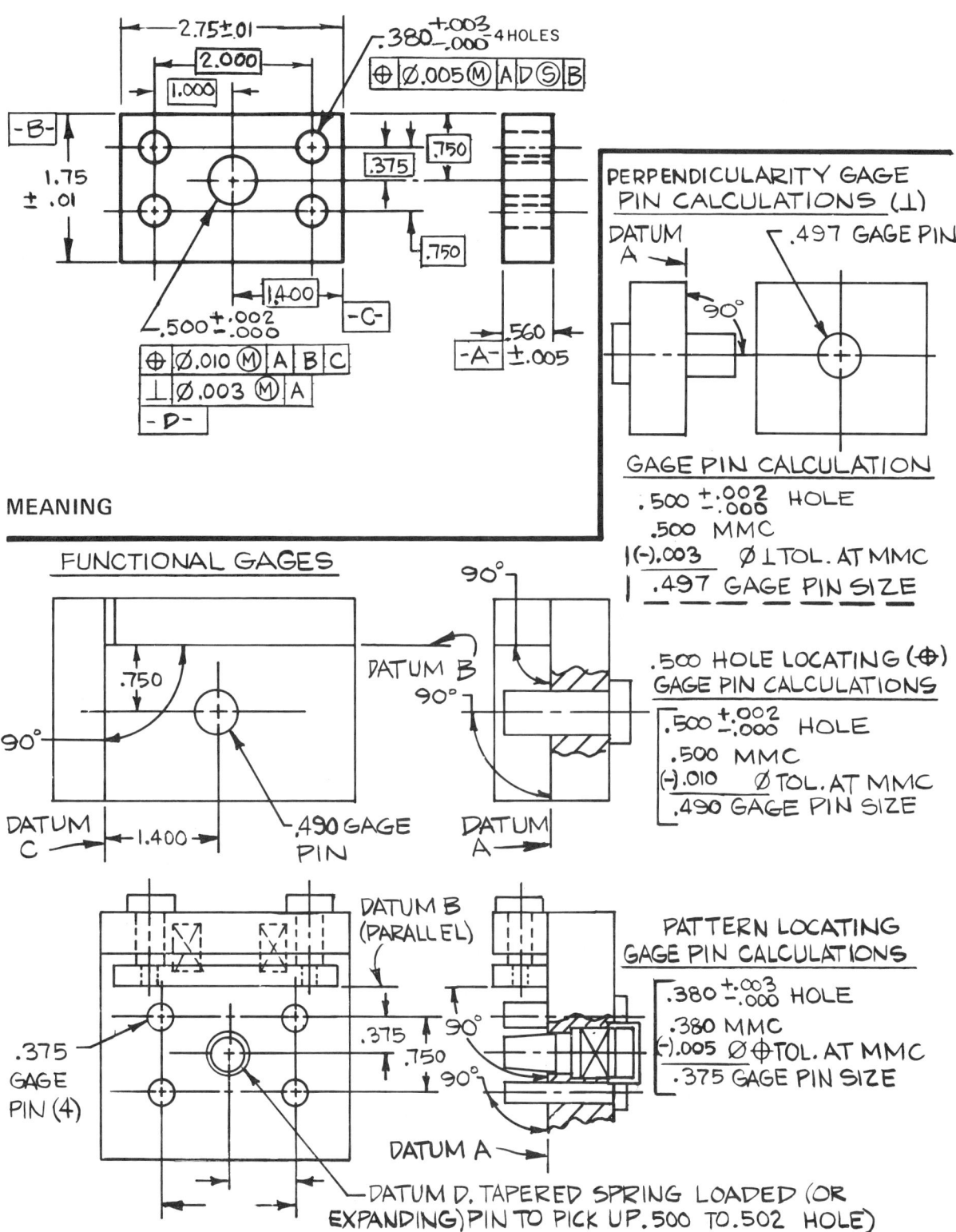

.380 $^{+.003}_{-.000}$ 4 HOLES

⊕ | Ø.005 Ⓜ | A | D Ⓢ | B

2.75 ± .01

2.000

1.000

-B-

1.75 ± .01

.750

.375

.750

1.400

-C-

.500 $^{+.002}_{-.000}$

⊕ | Ø.010 Ⓜ | A | B | C

⊥ | Ø.003 Ⓜ | A

-D-

.560 ± .005

-A-

PERPENDICULARITY GAGE PIN CALCULATIONS (⊥)

DATUM A

90°

.497 GAGE PIN

GAGE PIN CALCULATION

.500 $^{+.002}_{-.000}$ HOLE

.500 MMC

(-).003 Ø ⊥ TOL. AT MMC

.497 GAGE PIN SIZE

MEANING

FUNCTIONAL GAGES

.750

90°

DATUM C

1.400

.490 GAGE PIN

90°

DATUM B

90°

DATUM A

.500 HOLE LOCATING (⊕) GAGE PIN CALCULATIONS

.500 $^{+.002}_{-.000}$ HOLE

.500 MMC

(-).010 Ø TOL. AT MMC

.490 GAGE PIN SIZE

PATTERN LOCATING GAGE PIN CALCULATIONS

.380 $^{+.003}_{-.000}$ HOLE

.380 MMC

(-).005 Ø ⊕ TOL. AT MMC

.375 GAGE PIN SIZE

DATUM B (PARALLEL)

.375

.750

90°

90°

.375 GAGE PIN (4)

DATUM A

DATUM D. TAPERED SPRING LOADED (OR EXPANDING) PIN TO PICK UP .500 TO .502 HOLE)

POSITION $\oplus$

RFS RELATED TO RFS DATUM FEATURE

When a pattern of holes is dimensioned relative to the location of another hole, this hole is identified as a datum and the hole pattern is located dimensionally with respect to it.

In the illustrated example, the four holes are related to the center datum hole. As the position of the center datum hole shifts, the position of the four hole pattern itself must follow as dictated by the function of the part.

In this instance, note that the four holes of the pattern are assigned an RFS position tolerance and that they are related to the datum hole D also on an RFS basis. Imagine that this part has precision requirements between the holes either to provide accurate relationship with a mating part or to maintain accuracy for a mating situation; such as a semi-critical gear plate mounting.

Note that the .500 datum center hole is located from surface datums A, B, and C. It is given a position tolerance relative to these edges of $\varnothing$ .010 at MMC since the relationship to the edges can be on this basis with a rather lenient position tolerance. The attitude of the datum hole relative to the primary datum A is, however, to be maintained to a closer tolerance. Since datum D position and attitude is controlled on an MMC basis, the tolerances (position and perpendicularity) increase an amount equal to the produced size departure from MMC as shown in the illustration. Note, however, that the reference to datum D in the relationship of the four .380 holes is on an RFS basis. The four hole pattern, therefore, takes its positional reference from the exact center of the datum hole at whatever size it is produced (RFS) within .500 to .502.

Wherever the datum hole D position varies in the design considerations or in actual production, the four .380 hole pattern must follow. Note that the positional pattern dimensions originate at datum D to carry out this intent.

The four .380 holes are located by a position tolerance of $\varnothing$ .005 at RFS with respect to datum A (for attitude), datum D at RFS (for location), and datum B (for orientation). Reference to the explanation portion of the illustration (under Meaning) will assist in understanding the effect of the three datums and the importance of datum precedence. Although not illustrated, imagine that, as previously stated, the four .380 hole position tolerances have been determined to relate to a mating part or to maintain accuracy in a mating situation where other features or components must relate with precision regardless of the produced sizes (RFS) of the .380 holes (.380 to .383).

The four holes individually, with respect to their own true or exact positions, can vary in location up to $\varnothing$ .005. Under the RFS method, however, this tolerance applies to each hole regardless of the size to which it is produced. That is to say (as shown in the lower right corner of the illustration), the $\varnothing$ .005 position tolerance is the maximum allowable to each hole no matter to which size it is produced (.380 to .383). If, as was shown in previous illustrations, the MMC method had been applied, the position tolerance would increase to the extent of departure of the .380 holes from MMC. Not so, however, in this example since the RFS method has been invoked. The choice of proper approach, be it MMC or RFS, is of course decided by the design requirements. MMC methods are obviously recommended wherever possible due the added functional (interchangeability) and tolerance advantages.

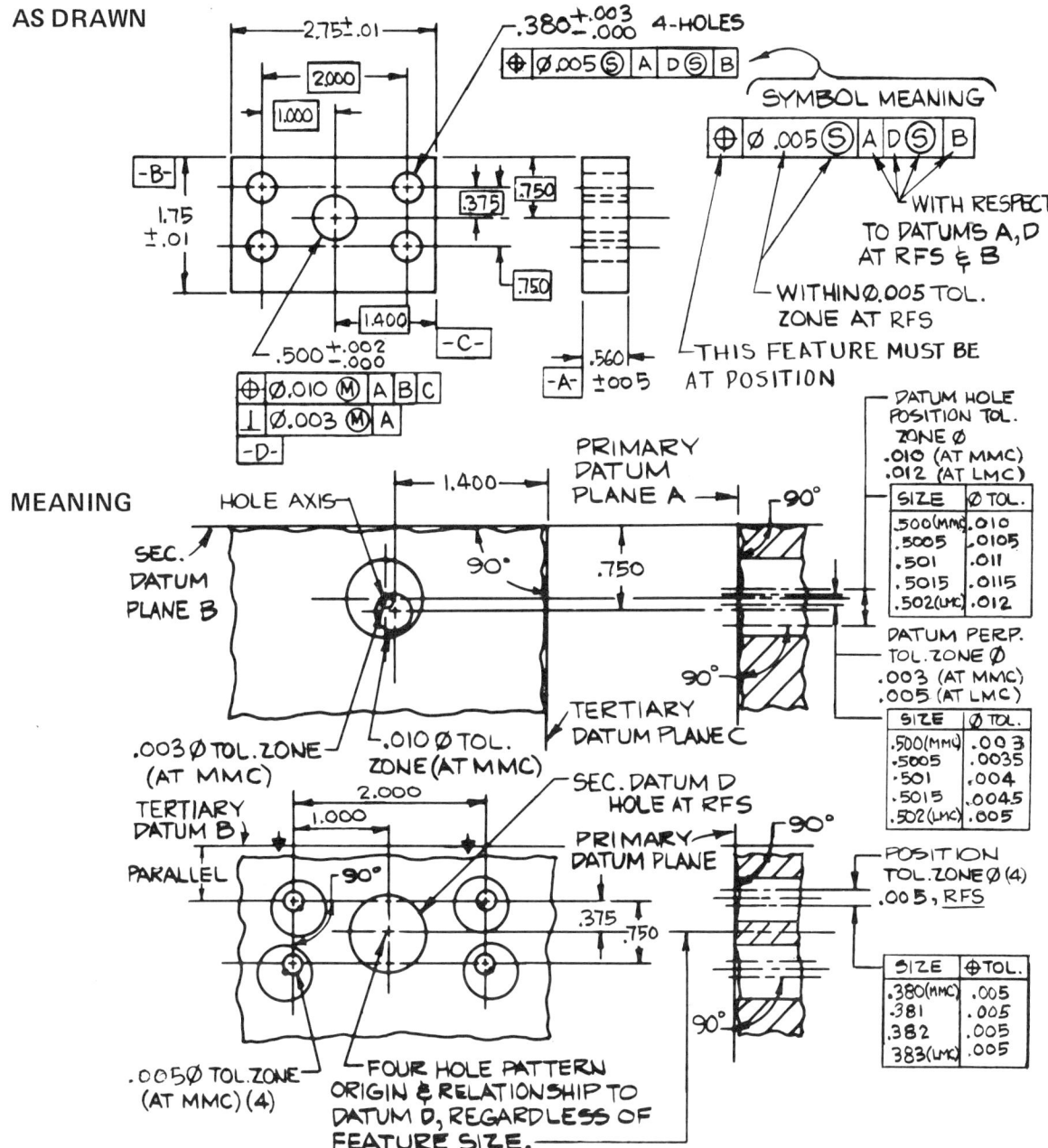

AS DRAWN

SYMBOL MEANING

WITH RESPECT TO DATUMS A,D AT RFS & B

WITHIN Ø.005 TOL. ZONE AT RFS

THIS FEATURE MUST BE AT POSITION

MEANING

In this instance, it is seen that a more critical or precise relationship is maintained between the four holes and their datum. The design requirements based on part function determined this approach in our example.

As may be seen in the previous examples in this section, functional gaging principles can be used to evaluate the datum hole location and relationships. However, the four hole (.380) pattern relative to datum D in this example cannot utilize such methods because it is an RFS application; open set-up principles would be necessary. The techniques of "paper gaging," described earlier in the text could, of course, be used as desired.

POSITION $\oplus$
MMC RELATED TO MMC DATUM FEATURE

PROJECTED TOLERANCE ZONE *

This illustration shows a part with four tapped mounting holes. A cover plate mechanism is assumed as a mating part (not shown). Position tolerancing is used to assure assembly of the mating parts.

We wish to establish a relationship between the mounting surface, the mounting hole pattern, and the 1.506 counterbored seat diameter (identified as datum B). The mating part has a seat mechanism which must fit within this counter-bore and attach and locate with screws to the four mounting holes on the flange surface.

The flange surface itself is established as the primary datum and is identified as datum A to ensure clarity of the hole pattern positional relationship with the top surface.

We calculate the position tolerances using the "fixed-fastener" method as based on the .250–28 screw and the clearance hole in the mating part cover plate. Assuming that the cover plate has .256 clearance holes and .250 maximum thickness, we calculate the position tolerance and assign .003 to each part (.006/2 = .003).

The four .250–28 holes are thus designated as shown. The feature control symbol specifies "at position within $\varnothing$.003 at MMC size of the holes with respect to datums A and B (at MMC) and to datum C."

Conventional position tolerancing discussed previously reveals that a position (MMC) application is affected by size departure of the involved features from MMC. However, in this instance, the peculiarity of thread assembly requires further consideration and may cause an exception to the usually inferred interpretation.

The centering effect of a screw as it tightens in the tapped hole tends to negate or diminish any *added* position tolerance based on greater clearance between the pitch diameters of the screw and tapped hole. The screw seeks center as, in tightening, the flank of the mating thread forms (screw and hole) come into contact or bottom out. Thus additional position tolerance relative to the tapped hole increase in pitch diameter size (departure from MMC) may not be fully realized. There may be some additional tolerance derived in actual assembly but it should not normally be counted upon.

The same reasoning as in the foregoing paragraph applies to attempts to assign position tolerance to countersinks. Position tolerance would be appropriate to the through hole, but, as a rule, is not practical for the countersink itself.

Added tolerance may be acquired, however, for the tapped hole pattern as a unit relative to datum B (counterbore) as it departs from MMC size (gets larger). Datum B is a straight sided feature and size deviation from MMC will have the effect previously discussed in other examples.

*See also page 175 for further explanation of principle.

AS DRAWN

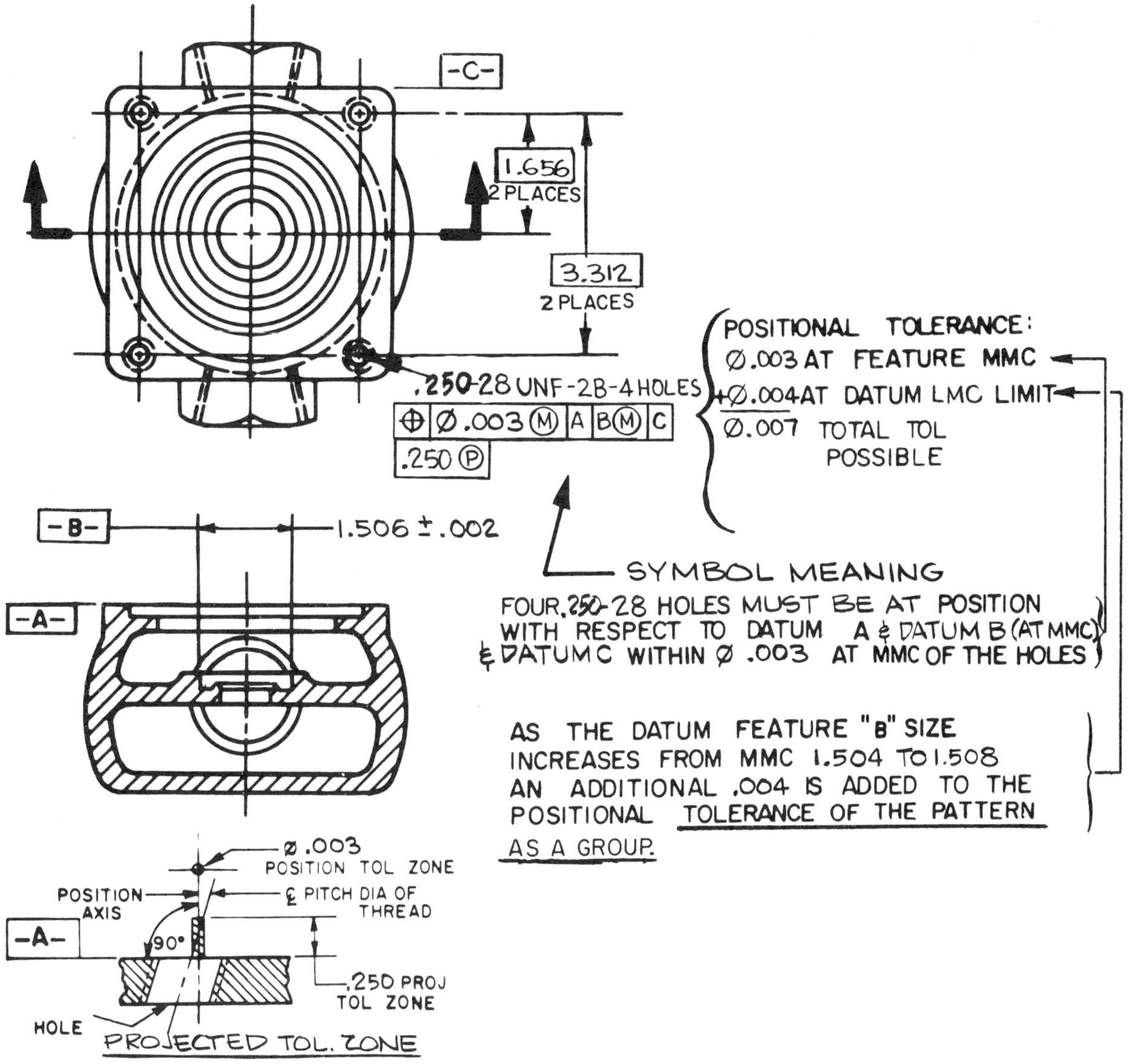

-C-

1.656
2 PLACES

3.312
2 PLACES

.250-28 UNF-2B-4 HOLES

⊕ | Ø.003 Ⓜ | A | B Ⓜ | C
.250 Ⓟ

POSITIONAL TOLERANCE:
Ø.003 AT FEATURE MMC ◄
+ Ø.004 AT DATUM LMC LIMIT ◄
Ø.007 TOTAL TOL
POSSIBLE

-B- 1.506 ± .002

-A-

SYMBOL MEANING
FOUR .250-28 HOLES MUST BE AT POSITION
WITH RESPECT TO DATUM A & DATUM B (AT MMC)
& DATUM C WITHIN Ø .003 AT MMC OF THE HOLES

AS THE DATUM FEATURE "B" SIZE
INCREASES FROM MMC 1.504 TO 1.508
AN ADDITIONAL .004 IS ADDED TO THE
POSITIONAL TOLERANCE OF THE PATTERN
AS A GROUP.

Ø .003
POSITION TOL ZONE

POSITION
AXIS

₵ PITCH DIA OF
THREAD

90°

-A-

.250 PROJ
TOL ZONE

HOLE

PROJECTED TOL. ZONE

GAGE

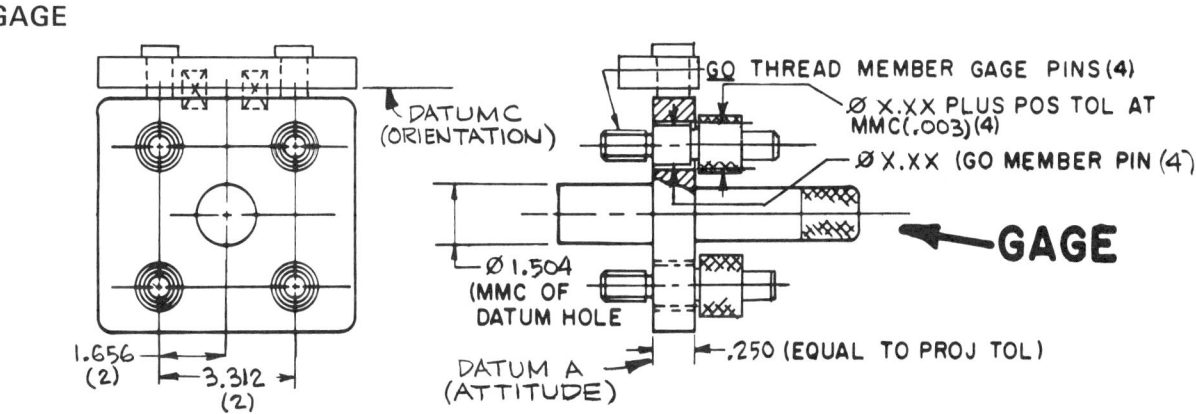

DATUM C
(ORIENTATION)

GO THREAD MEMBER GAGE PINS (4)

Ø X.XX PLUS POS TOL AT
MMC (.003)(4)

Ø X.XX (GO MEMBER PIN (4)

GAGE

Ø 1.504
(MMC OF
DATUM HOLE

1.656
(2)

3.312
(2)

DATUM A
(ATTITUDE)

.250 (EQUAL TO PROJ TOL)

133

POSITION

Since the threaded holes are to attach a mating part, we see that in assembly the critical position location of these holes will actually concern the inserted screws or the actual projection from these holes which must accommodate the mating part. Therefore, the location of the threaded holes is controlled by a *projected* position tolerance zone to represent the projected screw locations. As is seen in the sectioned projected tolerance zone view, the tolerance zone is cylindrical and extends above, and perpendicular to, the datum plane A.

The height of the projected position tolerance zone is determined by the application. It may be established by the thickness of the mating part through which the screws are to extend, or it may be determined by the thread hole depth when a thin mating part is involved, or any desired height which fulfills the part design requirements may be selected. When the projected tolerance zone is intended, it should be specified as shown with the symbol immediately below the feature control symbol box (see illustration). This system is also used on parts in which pins, studs, or other features are to be inserted and the critical assembled location is *above* the surface of the part.

The representative production gage shown on the preceding page also illustrates this principle. The gage is made up of the main gage body containing the datum B hole pin, the plate which contacts the datum surface A, a spring plate to provide orientation to datum C, four GO thread members per standard tolerances and, if desired, torquing flats to represent appropriate screw tightening and setting. The plate contains four holes which are .003 (the positional tolerance) larger than the four GO thread pins which are inserted into the threaded holes of the part through the gage plate.

The relationship of the GO thread member projections and the holes in the gage plate simulate the functional assembly requirement of the two mating parts and the four screws.

Gage-maker's tolerances would, of course, also apply but are not shown.

MMC RELATED TO MMC DATUM FEATURE

The illustration on the opposite page shows a circular pattern of (8) holes located by position tolerancing with respect to the center hole.

The requirement states "the eight holes are to be located at position within ∅ .002 at .403 MMC size of the holes with respect to datum A and datum hole B (at 1.025 MMC)."

As the size of the holes increases in production from MMC .403 to the high limit of .405, .002 is added to the positional tolerance. As the datum hole A increases from MMC of 1.025 to 1.027, .002 is added to the hole pattern's positional tolerance as a group. Depending on the size increase of the holes, the total tolerance may be .006 instead of .002.

A production gage for this part is also shown. The pins in the pattern are spaced at 45° intervals on a 3.000 circle, with the pin size at .401 or at MMC size of the part hole .403 less than the positional tolerance. The center pin is 1.025 (MMC size of the datum hole).

Gage-maker's tolerances also apply.

AS DRAWN

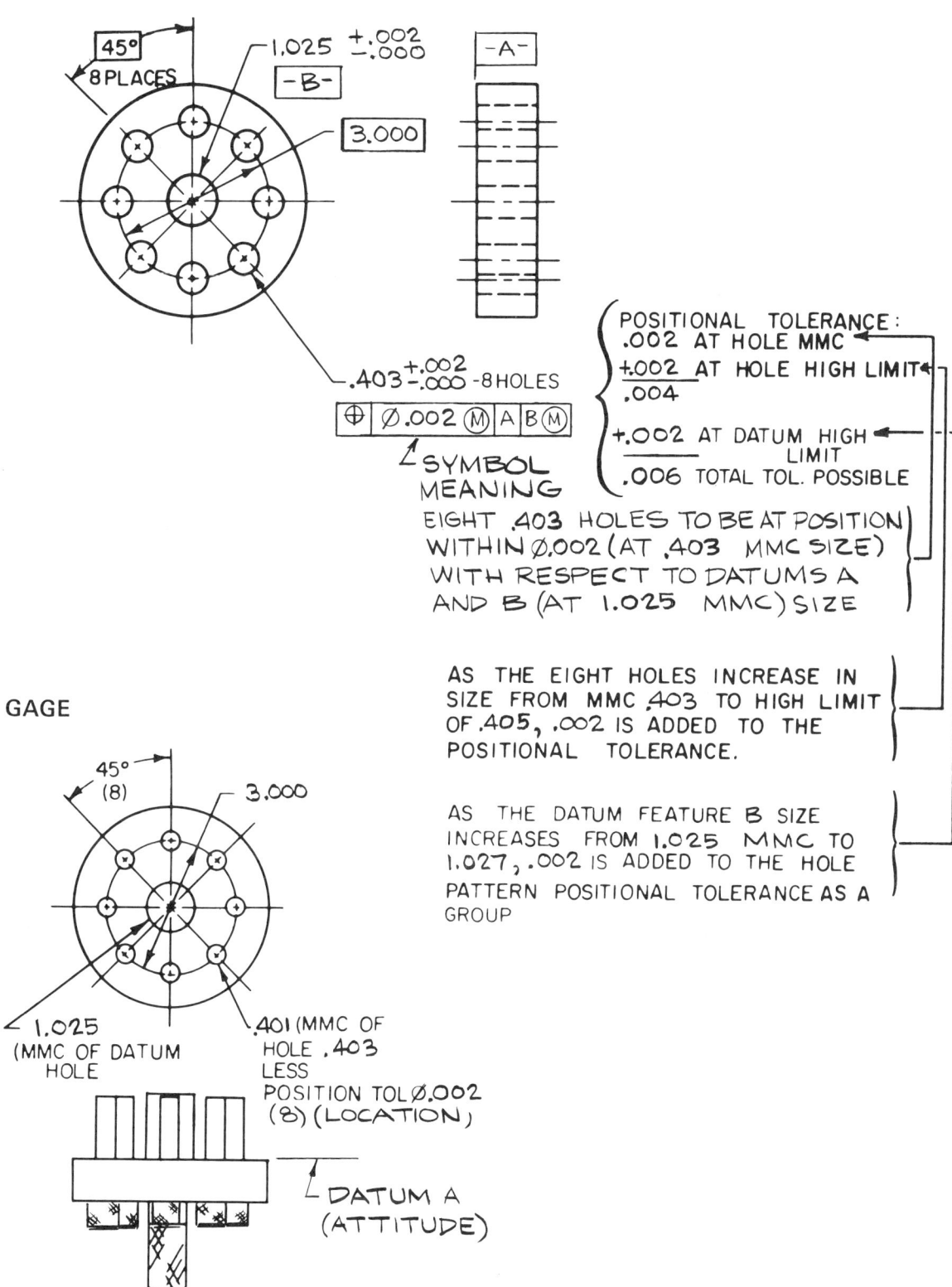

45°
8 PLACES

1.025 +.002 -.000
-B-

3.000

-A-

.403 +.002 -.000 -8 HOLES

⊕ Ø.002 Ⓜ A B Ⓜ

SYMBOL MEANING

EIGHT .403 HOLES TO BE AT POSITION
WITHIN Ø.002 (AT .403 MMC SIZE)
WITH RESPECT TO DATUMS A
AND B (AT 1.025 MMC) SIZE

POSITIONAL TOLERANCE:
.002 AT HOLE MMC
+.002 AT HOLE HIGH LIMIT
.004

+.002 AT DATUM HIGH
LIMIT
.006 TOTAL TOL. POSSIBLE

AS THE EIGHT HOLES INCREASE IN
SIZE FROM MMC .403 TO HIGH LIMIT
OF .405, .002 IS ADDED TO THE
POSITIONAL TOLERANCE.

AS THE DATUM FEATURE B SIZE
INCREASES FROM 1.025 MMC TO
1.027, .002 IS ADDED TO THE HOLE
PATTERN POSITIONAL TOLERANCE AS A
GROUP

GAGE

45°
(8)

3.000

1.025
(MMC OF DATUM
HOLE

.401 (MMC OF
HOLE .403
LESS
POSITION TOL Ø.002
(8) (LOCATION)

DATUM A
(ATTITUDE)

135

POSITION $\oplus$

MMC RELATED TO MMC DATUM FEATURE

This example shows the use of positional tolerancing on two locating pins on a part. The two pins are to take reference from the center hole which accommodates a shaft mechanism of the mating part (not shown). The center hole is selected as the datum reference. This example reminds us that positional tolerancing is not restricted to holes. It can also be applied to pins, or to any feature on which a center (axis) is the basis for location.

The meaning of the position requirement is, "the two pins are to be located at position within $\emptyset$ .002 tolerance zone at .125 MMC size of the pins with respect to datums A, B (at MMC .500), and C."

Note that MMC of the pins is the high limit of size. As the two pins reduce in size from MMC of .125 to the low limit (LMC) of .122, .003 is added to the positional tolerance. As the datum feature A (the center hole) increases from MMC of .500 to .502 high limit, .002 is added to the two pins' positional tolerance as a group.

Depending on the production sizes of the bosses and datum hole, the total positional tolerance on the boss location may vary from $\emptyset$ .002 to .007.

A representative gage to check the position location and the relationship with the datum hole is also shown. Note that the holes to check the position location of the pins are to the high limit of MMC of the pins *plus* the position diameter tolerance. The datum hole pin is to the MMC size of the datum hole of .500.

Gage-maker's tolerances also apply.

AS DRAWN

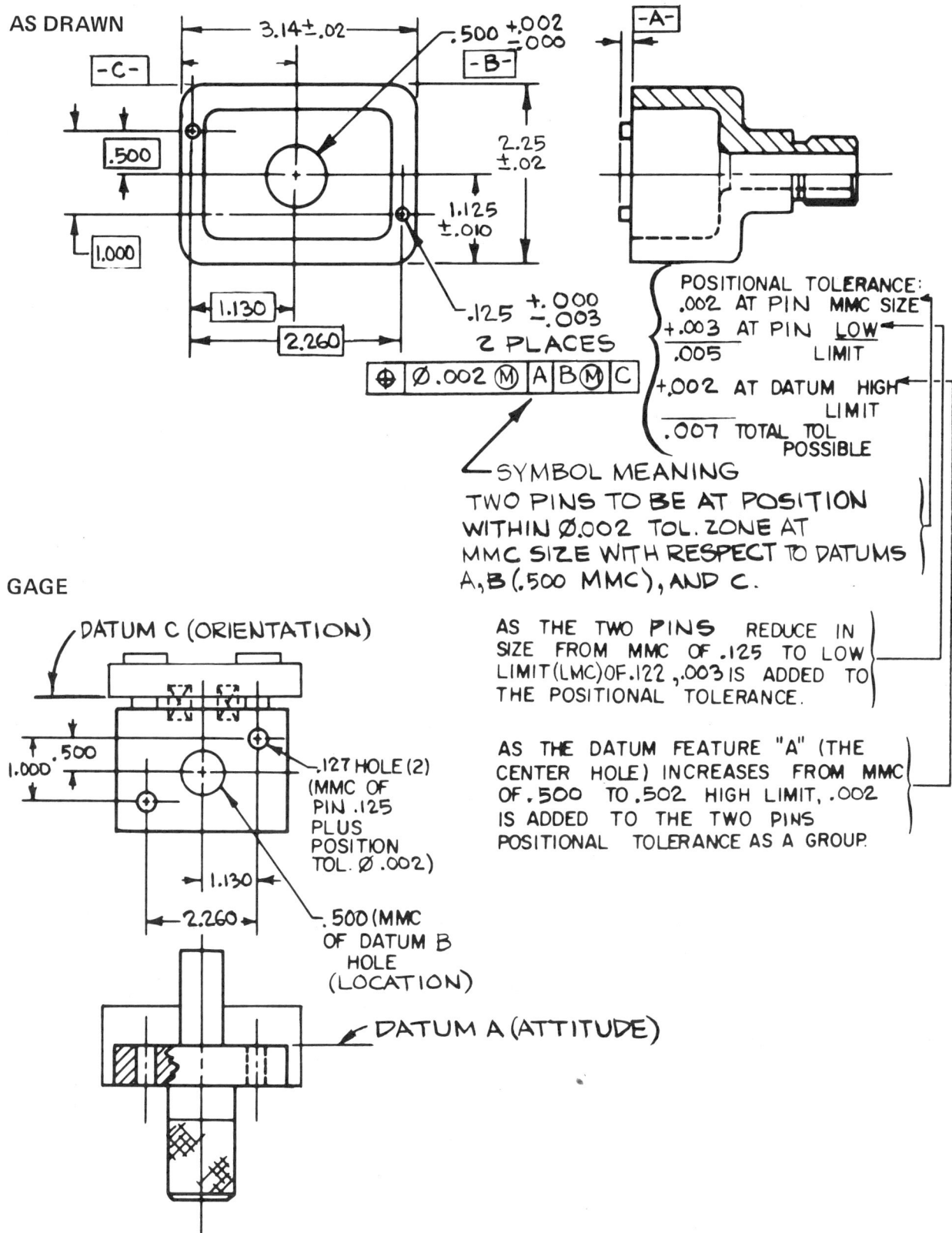

3.14 ± .02

-C-

.500 +.002 -.000

-B-

2.25 ± .02

.500

1.125 ± .010

1.000

1.130

2.260

.125 +.000 -.003

2 PLACES

⊕ | Ø.002 Ⓜ | A | B Ⓜ | C

-A-

POSITIONAL TOLERANCE:
.002 AT PIN MMC SIZE
+.003 AT PIN LOW LIMIT
.005
+.002 AT DATUM HIGH LIMIT
.007 TOTAL TOL POSSIBLE

SYMBOL MEANING
TWO PINS TO BE AT POSITION
WITHIN Ø.002 TOL. ZONE AT
MMC SIZE WITH RESPECT TO DATUMS
A, B (.500 MMC), AND C.

AS THE TWO PINS REDUCE IN
SIZE FROM MMC OF .125 TO LOW
LIMIT (LMC) OF .122 , .003 IS ADDED TO
THE POSITIONAL TOLERANCE.

AS THE DATUM FEATURE "A" (THE
CENTER HOLE) INCREASES FROM MMC
OF .500 TO .502 HIGH LIMIT, .002
IS ADDED TO THE TWO PINS
POSITIONAL TOLERANCE AS A GROUP.

GAGE

DATUM C (ORIENTATION)

.500

1.000

1.130

2.260

.127 HOLE (2)
(MMC OF
PIN .125
PLUS
POSITION
TOL. Ø .002)

.500 (MMC
OF DATUM B
HOLE
(LOCATION)

DATUM A (ATTITUDE)

137

POSITION
OF
NONCYLINDRICAL
FEATURES

POSITION $\oplus$

MMC WITH RESPECT TO A CENTER PLANE AND RELATED TO A DATUM FEATURE

Position tolerance relationships are more often associated with round holes or features and establish a cylindrical tolerance zone around theoretically exact axes. The cylindrical tolerance zone is not applicable to slots, dial markings, tabs, etc., for which non-cumulative tolerance and MMC aspects of positional tolerancing may also be desired.

Such features may be allowed to vary with respect to a center plane rather than an axis. The position tolerance zone is a total wide zone with one-half the total tolerance assigned to each side of the center plane.

In this example, we present two mating parts in order to illustrate the calculations and relationships. The top part could be either a thin metal part or a type of drive shaft with three tab projections. The mating part below might be a sleeve or collar which must fit the upper part. To simplify initial explanation, side views and primary (attitude) datums are not shown on either part.

Both parts have corresponding datum reference diameters which are related, in turn, to the controlled features of each individual part. The datums are identified by the letter A in the datum identification symbol. The feature control symbol for the top part (Example 1) reads, "these features (3 tabs) must be at position within .006 total wide zone with the feature at MMC size and with respect to datum A at MMC." Although the symbol used is the same as that for cylindrical zones, there is no confusion, since the drawing always clearly shows the feature being dimensioned and the $\emptyset$ symbol is not used, thus designating the tolerance zone as a noncylindrical width zone.

The feature control symbol for the bottom part (Example 2) reads, "these features (3 slots) must be at position within .006 total wide zone with the features at MMC size and with respect to datum A at MMC."

Note that the tolerance zones are *not* cylindrical but are total widths (parallelepiped) equally disposed about, and parallel to, the center plane as established by the 120° basic angles and extending the full depth and length of the produced feature.

When designated as shown, the width of the tolerance zone is always total and is equally disposed on either side of the true position center plane. In this case, the total wide zone is .006, with .003 on each side of the basic center plane.

The calculations of the positional tolerance zones for mating parts of this type are shown at the upper right. They are based on the same reasoning previously discussed for "fixed fasteners" using cylindrical features. The tolerance zones in this case are however not cylindrical but total width.

As in any positional calculation, the MMC sizes of the two corresponding mating part features are used to determine their individual positional tolerances. The MMC width of the tab, .250, is subtracted from the MMC width of the slot, .262, giving a combined clearance of .012. This is divided by the fixed factor 2 to give the total tolerance zone for each mating part feature at MMC. As previously discussed on round feature positional tolerance calculations, the total combined tolerance (in this case .012) may be divided as desired in other combinations (e.g., .008 and .004, .009 and .003, etc.).

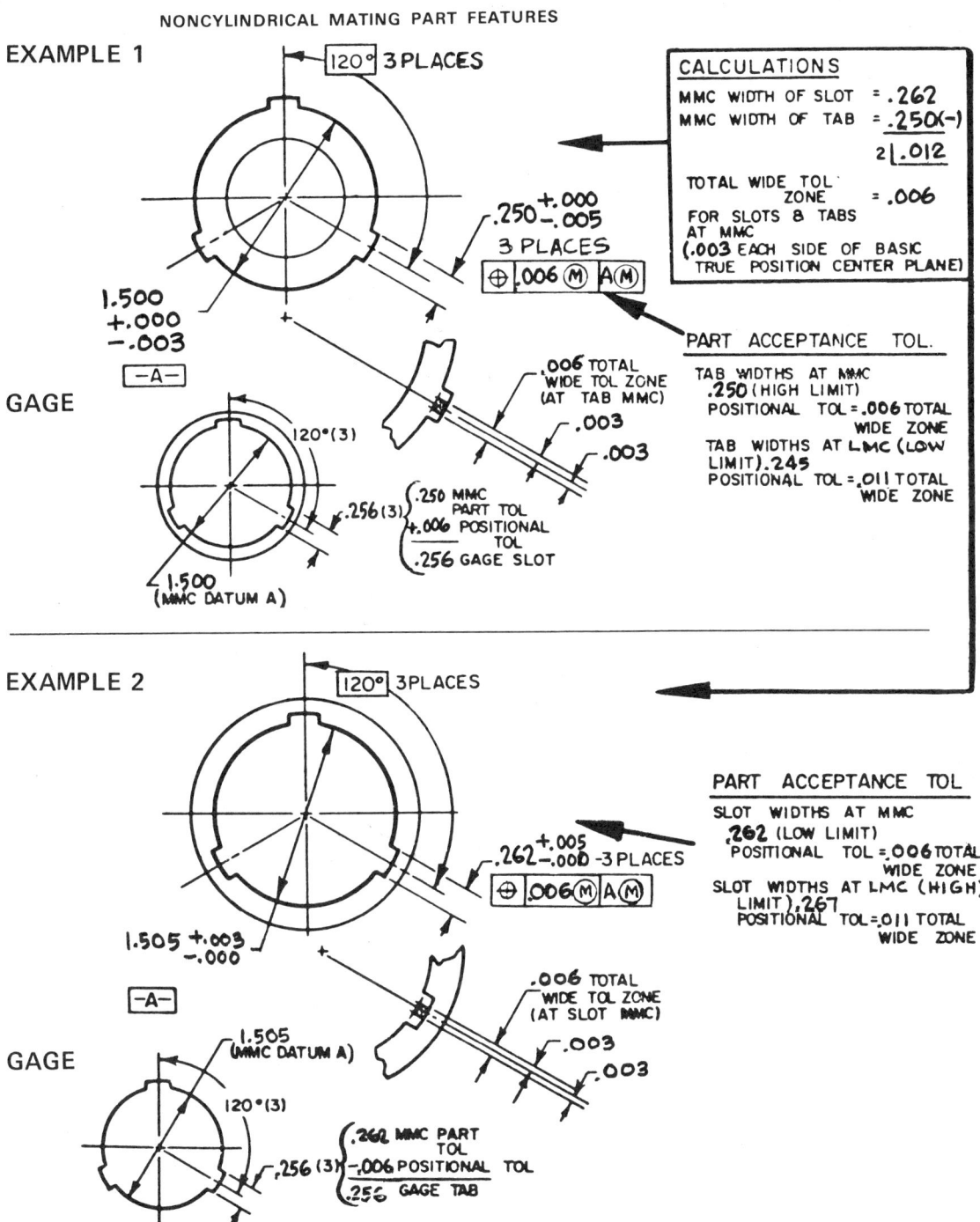

EXAMPLE 1

120° 3 PLACES

.250 +.000 −.005
3 PLACES

⊕ |.006 Ⓜ |A Ⓜ

1.500
+.000
−.003

−A−

GAGE

120°(3)

.256(3) { .250 MMC PART TOL
+.006 POSITIONAL TOL
_____
.256 GAGE SLOT }

1.500
(MMC DATUM A)

.006 TOTAL WIDE TOL ZONE (AT TAB MMC)

.003

.003

CALCULATIONS

MMC WIDTH OF SLOT = .262
MMC WIDTH OF TAB = .250(−)
2 | .012

TOTAL WIDE TOL ZONE = .006
FOR SLOTS & TABS AT MMC
(.003 EACH SIDE OF BASIC TRUE POSITION CENTER PLANE)

PART ACCEPTANCE TOL.

TAB WIDTHS AT MMC
.250 (HIGH LIMIT)
POSITIONAL TOL = .006 TOTAL WIDE ZONE
TAB WIDTHS AT LMC (LOW LIMIT).245
POSITIONAL TOL =.011 TOTAL WIDE ZONE

EXAMPLE 2

120° 3PLACES

.262 +.005 −.000 -3 PLACES

⊕ |.006 Ⓜ |A Ⓜ

1.505 +.003 −.000

−A−

GAGE

1.505 (MMC DATUM A)

120°(3)

.256 (3)

.006 TOTAL WIDE TOL ZONE (AT SLOT MMC)

.003

.003

.262 MMC PART TOL
−.006 POSITIONAL TOL
_____
.256 GAGE TAB

PART ACCEPTANCE TOL

SLOT WIDTHS AT MMC
.262 (LOW LIMIT)
POSITIONAL TOL =.006 TOTAL WIDE ZONE
SLOT WIDTHS AT LMC (HIGH) LIMIT),.267
POSITIONAL TOL=.011 TOTAL WIDE ZONE

In Example 1, the notation "Part Acceptance Tolerance" indicates that the total positional tolerance zone increases from .006 to .011 as the actually produced tab width reduces from MMC of .250 to .245.

The same is true for Example 2. The slot width positional tolerance increases from .006 to .011 as the slot is produced to the high limit size of .267. Simulated gages are also shown.

Note that the .250 MMC tab width in Example 1 is accommodated by a .256 gage slot determined by adding the .006 positional tolerance to the MMC size of the part tab. The gage for Example 2 shows the reverse, with the positional tolerance of .006 being subtracted from the .262 MMC size of the part slot to establish the gage tab size of .256.

Functionally gaging these parts will permit additional positional tolerance for the tabs and slots as a group as the datum feature sizes depart from MMC. However, because of the unique geometry, the actual amount of this effect (due to the datum diameter relationship to the slot and tab flats) may not be conveniently predicted or calculated.

NONCYLINDRICAL MATING PART FEATURES

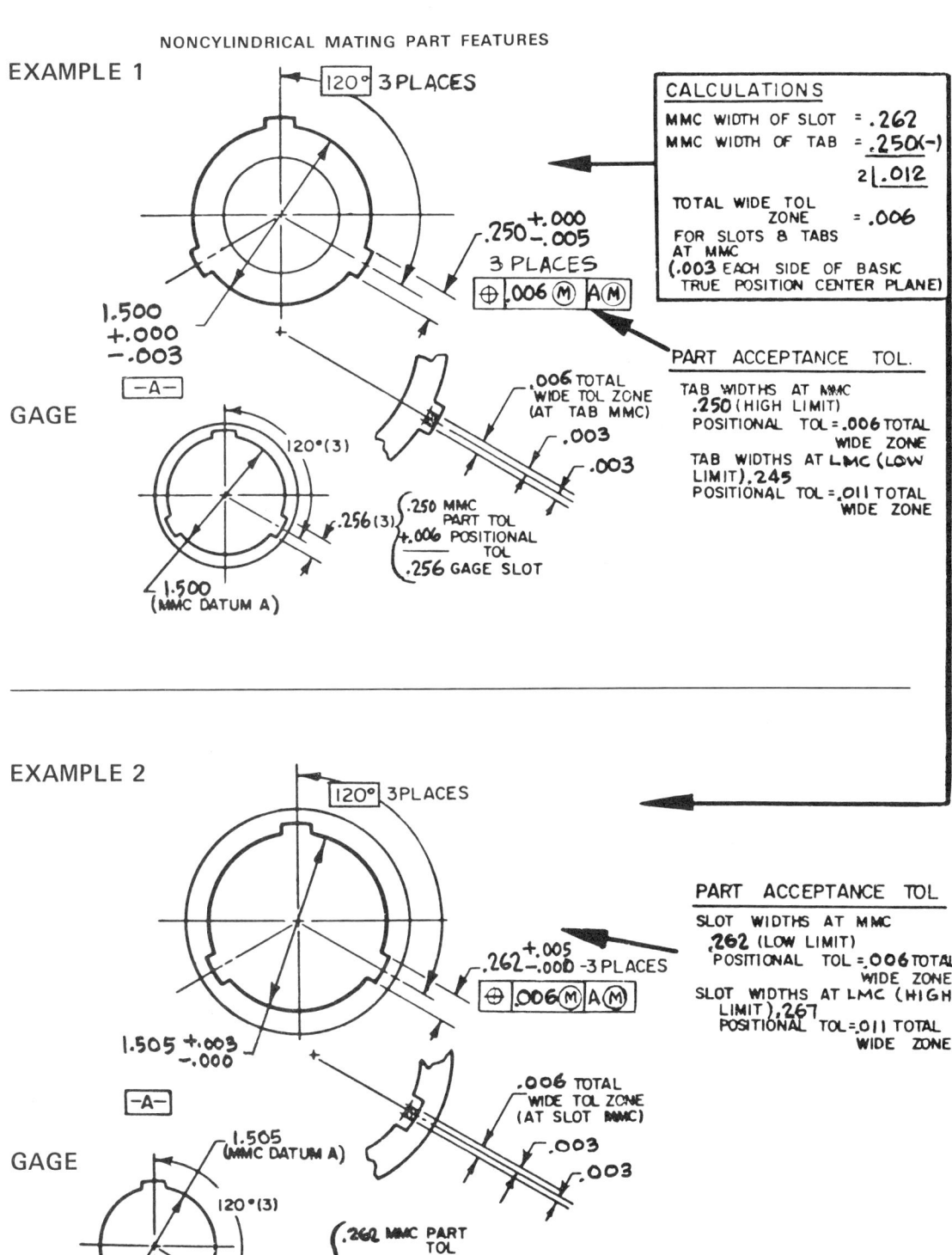

EXAMPLE 1

120° 3 PLACES

.250 +.000 -.005
3 PLACES
⊕ .006 Ⓜ A Ⓜ

1.500
+.000
-.003

─A─

GAGE

120°(3)

.256 (3)

1.500
(MMC DATUM A)

.006 TOTAL
WIDE TOL ZONE
(AT TAB MMC)

.003

.003

.250 MMC
PART TOL
+.006 POSITIONAL
TOL
.256 GAGE SLOT

CALCULATIONS
MMC WIDTH OF SLOT = .262
MMC WIDTH OF TAB = .250(-)
2|.012
TOTAL WIDE TOL
ZONE = .006
FOR SLOTS & TABS
AT MMC
(.003 EACH SIDE OF BASIC
TRUE POSITION CENTER PLANE)

PART ACCEPTANCE TOL.

TAB WIDTHS AT MMC
.250 (HIGH LIMIT)
POSITIONAL TOL = .006 TOTAL
WIDE ZONE
TAB WIDTHS AT LMC (LOW
LIMIT) .245
POSITIONAL TOL = .011 TOTAL
WIDE ZONE

EXAMPLE 2

120° 3 PLACES

.262 +.005 -.000 3 PLACES
⊕ .006 Ⓜ A Ⓜ

1.505 +.003 -.000

─A─

GAGE

1.505
(MMC DATUM A)

120°(3)

.256 (3)

.262 MMC PART
TOL
-.006 POSITIONAL TOL
.256 GAGE TAB

.006 TOTAL
WIDE TOL ZONE
(AT SLOT MMC)

.003

.003

PART ACCEPTANCE TOL

SLOT WIDTHS AT MMC
.262 (LOW LIMIT)
POSITIONAL TOL = .006 TOTAL
WIDE ZONE
SLOT WIDTHS AT LMC (HIGH)
LIMIT) .267
POSITIONAL TOL = .011 TOTAL
WIDE ZONE

143

POSITION ⊕

MMC WITH RESPECT TO A LINE (DIAL) AND RELATED DATUM FEATURE

Positional tolerancing may also be applied effectively to parts which do not mate. Since position tolerancing is noncumulative, it is ideally suited to dial markings, etc.

The example illustrates a dial with graduated markings which have critical angular location with respect to one another, and with the center hole identified as the datum reference. The length of the markings are relatively unimportant. In cases such as this, it is usually important that the angular distances between the markings not accumulate tolerance. The use of basic angle locations with position tolerance zones straddling these locations is ideal.

Referring to Meaning, note that the tolerance zones for the groove center lines are parallel to the 15° basic center planes which are taken from the actual center of the .190 datum hole A. The tolerance zones are total *wide* tolerance zones with half the zone on each side of these basic center planes. For each marking, the tolerance zone is .010 total when the marking grooves are at MMC size of .030. The tolerance zones can increase to as much as .015 if the dial markings are produced to their least material condition (high limit) of .035.

The center hole datum feature is specified at RFS with the RFS symbol included with the datum reference letter A. The RFS datum is used to ensure that the size variation of the center datum hole will not affect the position pattern location with respect to the center hole axis.

USE OF "LEAST MATERIAL CONDITION" APPLICATION*

When greatest accuracy of the slot angular position is required while the slot marking is at its largest width .035 (high limit size), the principles of "least material condition" (LMC) may be desired.

EXAMPLE

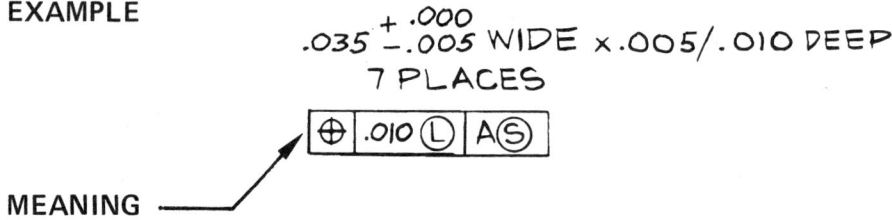

.035 +.000/-.005 WIDE x.005/.010 DEEP
7 PLACES

⊕ | .010 Ⓛ | AⓈ

MEANING

These markings are to be located at position within .010 total wide zone at marking LMC size of .035 with respect to datum A (RFS). As the marking groove gets smaller, or departs from LMC (approaches MMC), the position tolerance is increased an equal amount; e.g., at the marking groove size of .030 width, the position tolerance is .015 total wide zone.

---

*. LEAST MATERIAL CONDITION, a term used to describe the opposite size of MMC, is contained in ANSI-Y14.5-1973. However, use of the LMC concept and symbol Ⓛ is *not* included in the preceding document. It is presented here as advisory information to be used only at the discretion of the reader. Future standards may contain this principle.

AS DRAWN

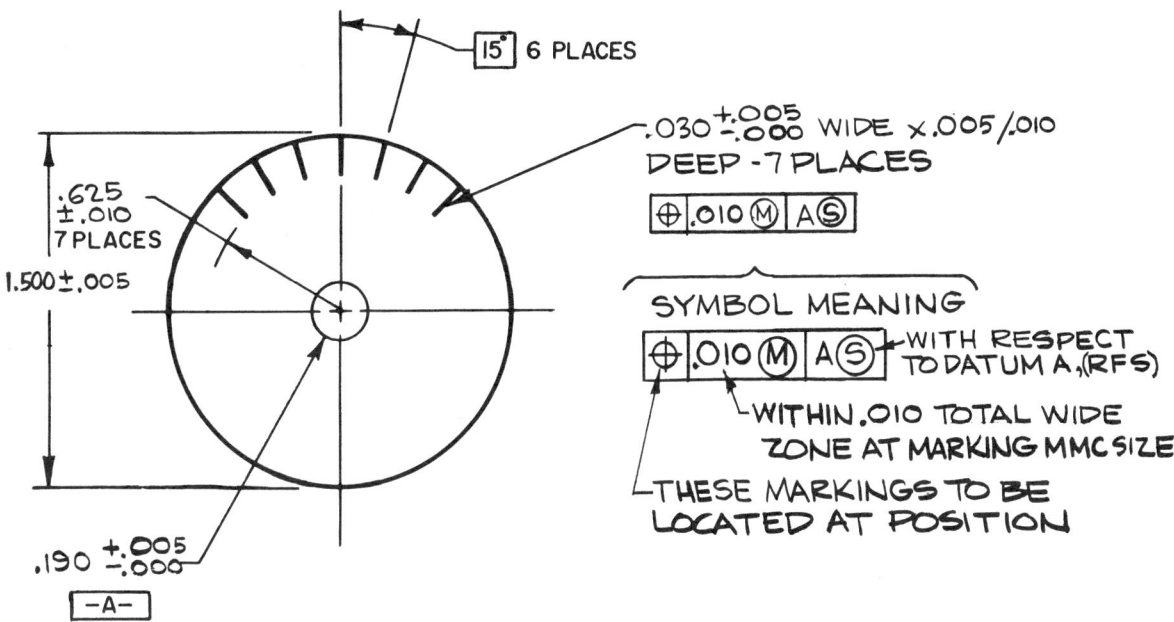

15° 6 PLACES

.625 ±.010 7 PLACES

1.500 ±.005

.190 +.005/-.000

-A-

.030 +.005/-.000 WIDE x .005/.010 DEEP - 7 PLACES

⊕ .010 Ⓜ A Ⓢ

SYMBOL MEANING

⊕ .010 Ⓜ A Ⓢ ← WITH RESPECT TO DATUM A, (RFS)

WITHIN .010 TOTAL WIDE ZONE AT MARKING MMC SIZE

THESE MARKINGS TO BE LOCATED AT POSITION

MEANING

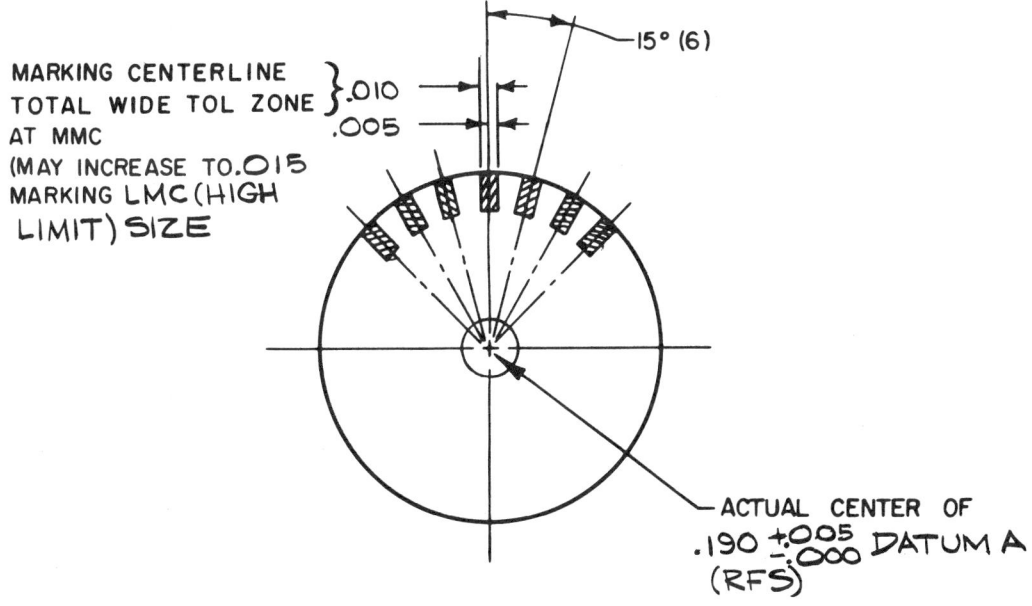

MARKING CENTERLINE TOTAL WIDE TOL ZONE AT MMC (MAY INCREASE TO .015 MARKING LMC (HIGH LIMIT) SIZE

.010
.005

15° (6)

ACTUAL CENTER OF .190 +.005/-.000 DATUM A (RFS)

This illustration presents a unique mating part relationship. Position tolerancing principles are extended to utilize a combination of total wide position requirements. The horizontal slots of the upper part are to mate with screw clearance holes of the lower part.

The illustration clearly shows that assembly requirements demand closer control of the hole location in the vertical direction than in the horizontal direction since the slots can compensate for considerable horizontal variation. Therefore the conventional cylindrical zone position method does not seem appropriate but total wide zone methods appear to satisfy the requirement. Using this approach, we can specify separate positional tolerance zones of different values for vertical and horizontal displacements of the slot and holes.

In Example 1, it is quite apparent from the elongated slots that a width type positional tolerancing rather than diametral positional tolerancing is desired. However, in Example 1 as well as in Example 2, the absence of the diameter symbol ⌀ and the direction of the arrows is the key indicator that the total wide tolerance zone is to apply. The vertical and horizontal extension lines from the features indicate the directions of the tolerance zones, and the datums and tolerance of the feature control symbol state the orientation and total width of the tolerance zone. Note that the tolerance zone consists of the total width equally disposed over the center planes; it extends the full depth of the feature perpendicular to the primary datum.

One may ask why the positional tolerancing system was used here, since the resulting tolerance zone seems to be the same as that which would have been obtained from coordinate dimensioning and tolerancing. The answer is that the positional MMC system recognizes that the *size* to which these features are produced affects their position relationships. As the feature sizes depart from MMC, that is, as the slots or holes get larger in their size tolerance range, additional positional tolerance is permitted. For instance, in Example 1, the .012 total wide tolerance zone on the slots could be increased to as much as .017, and the .012 and .030 total wide zones on the holes in Example 2 could be increased to as much as .017 and .035, respectively, depending on the sizes to which these features are actually produced.

Functional gaging techniques could be used on this part.

See also EXTENDED USES OF POSITION TOLERANCING for similar applications.

EXAMPLE 1

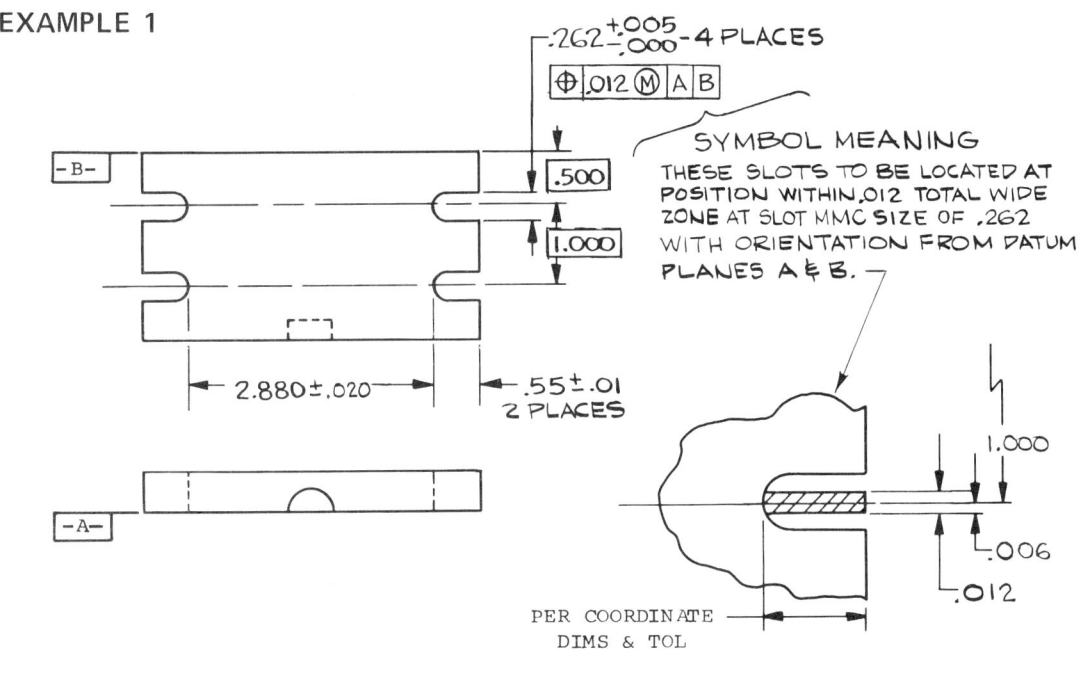

$.262 {}^{+.005}_{-.000}$ - 4 PLACES

⌖ .012 Ⓜ A B

SYMBOL MEANING
THESE SLOTS TO BE LOCATED AT
POSITION WITHIN .012 TOTAL WIDE
ZONE AT SLOT MMC SIZE OF .262
WITH ORIENTATION FROM DATUM
PLANES A & B.

-B-

.500

1.000

2.880 ±.020

.55 ±.01
2 PLACES

-A-

1.000

.006

.012

PER COORDINATE
DIMS & TOL

EXAMPLE 2

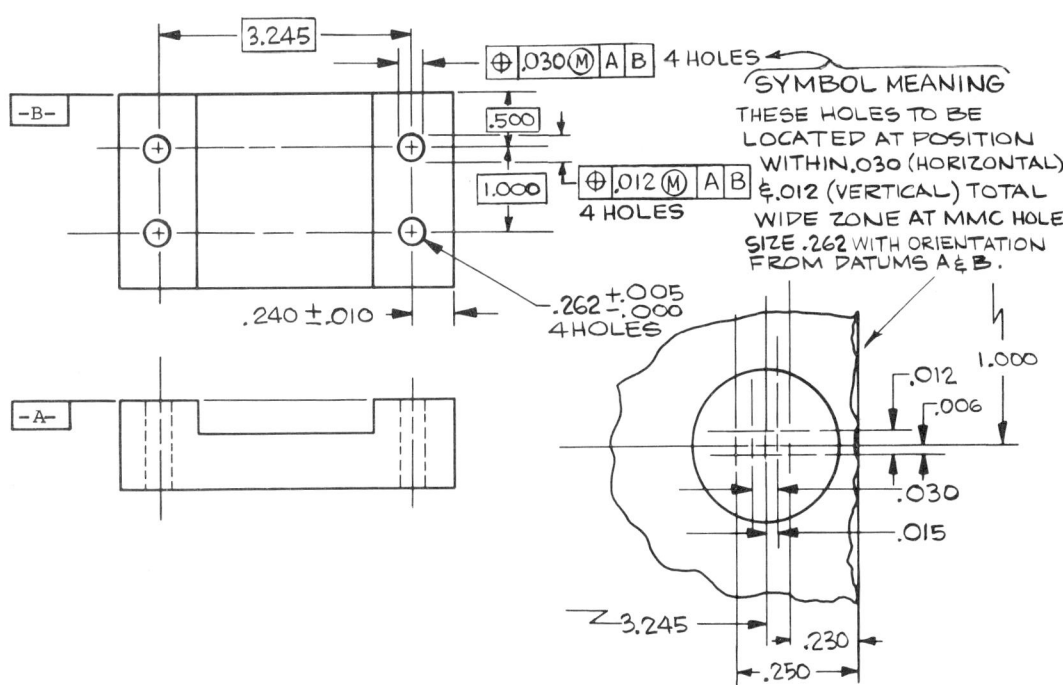

3.245

⌖ .030 Ⓜ A B 4 HOLES

SYMBOL MEANING
THESE HOLES TO BE
LOCATED AT POSITION
WITHIN .030 (HORIZONTAL)
& .012 (VERTICAL) TOTAL
WIDE ZONE AT MMC HOLE
SIZE .262 WITH ORIENTATION
FROM DATUMS A & B.

-B-

.500

1.000

⌖ .012 Ⓜ A B
4 HOLES

.240 ±.010

$.262 {}^{+.005}_{-.000}$
4 HOLES

-A-

.012   1.000
.006

.030

.015

3.245

.230

.250

147

POSITION $\oplus$

This illustration shows a pair of mating parts involving noncylindrical features. Part 1 is to fit within the opening of part 2.

Part 1 has a width of $1.000 \, ^{+.000}_{-.006}$ which is to fit within the $1.005 \, ^{+.003}_{-.000}$ opening width on part 2. Simultaneously, the $.500 \, ^{+.004}_{-.000}$ slot on part 1 is to fit onto the $.495 \, ^{+.000}_{-.003}$ projection on part 2.

The $.500 \, ^{+.004}_{-.000}$ slot on part 1 has a position feature control symbol which states, "this feature is to be at position within .005 at MMC size of the feature with respect to datum A at MMC size." The width of the part is established as datum A.

Part 2 has an identical position feature control symbol on the $.495 \, ^{+.000}_{-.003}$ dimension, and the $1.005 \, ^{+.003}_{-.000}$ opening is established as datum A.

Figure 1(a) shows the relationship of these two parts as they would appear if both parts were produced perfectly at the feature MMC sizes. Note the common center or median planes established on both parts. The parts are assembled in Fig. 1(b).

Figure 2(a) illustrates the slot feature on part 1 offset the maximum permissible amount (.0025) at the extreme of the .005 total tolerance zone when the part is at MMC size. Also, the mating projection of part 2(b) is shown offset in the opposite direction the maximum permissible amount (.0025) at the extreme of the .005 total tolerance zone when the part is at MMC size.

Figure 2(b) shows the assembly of the two parts. They still assemble satisfactorily. Figure 2 also emphasizes that the .005 total tolerance zone, as stated in the symbol boxes on parts 1 and 2, applies at the MMC size of the features and is the maximum tolerance permissible under this condition.

Figure 3 illustrates the increase in the permissible total position tolerance zone as the feature sizes *depart* from MMC to the opposite extreme of LEAST MATERIAL CONDITION. For part 1 (Fig. 3a), with the slot at its *high* limit size of .504 and the datum width at its *low* limit of .994, the permissible position tolerance zone becomes .015 total or a .0075 offset off the median plane of the slot with respect to the datum median plane.

For part 2 in Fig 3(b), with the projection at its low limit of .492 and the datum opening width at its high limit of 1.008, the position tolerance zone becomes 0.11 total or a .0055 offset off the median plane of the projection with respect to datum median plane.

Figure 3(b) shows the assembly of the two parts under these conditions. They still assemble satisfactorily with considerably more clearance as a result of the feature size variation to size limits opposite MMC, or their LEAST MATERIAL CONDITION.

From these illustrations it is evident that positional MMC applications permit greater tolerance and ensure a satisfactory fit of mating parts. For example, the possible tolerance on part 1 has been increased from .005 to .015, and on part 2 from .005 to .011. The actual tolerance to be realized is, of course, dependent upon the sizes to which the concerned features are actually produced.

MAXIMUM MATERIAL CONDITION (MMC)

EXAMPLES

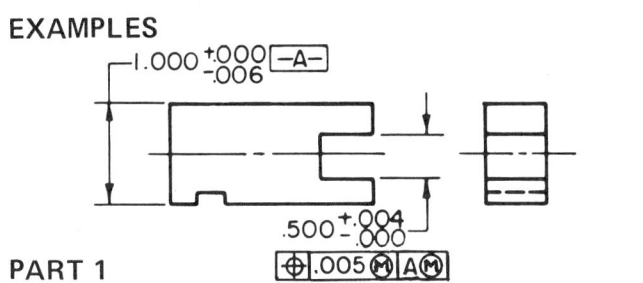

PART 1

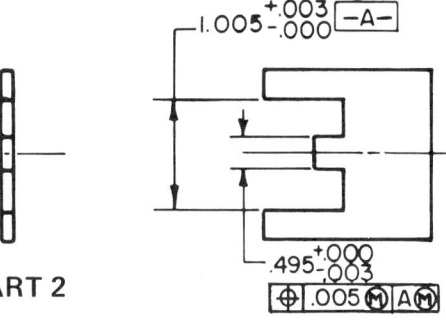

PART 2

PERFECT POSITION AT MMC

MEANING

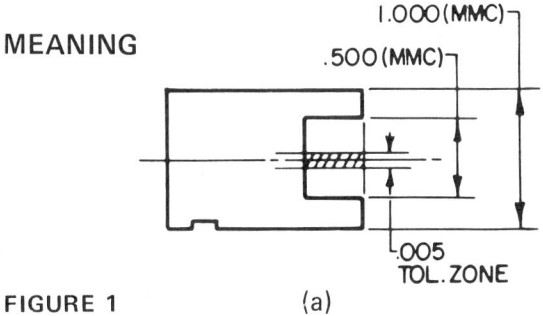

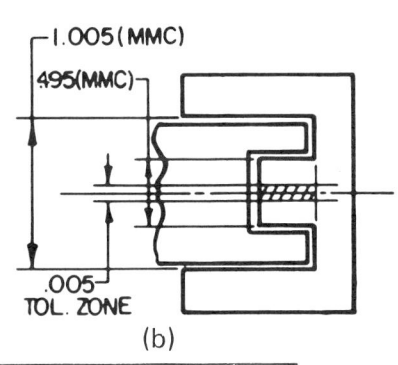

FIGURE 1   (a)   (b)

POSITION TOLERANCE ZONES AT MMC

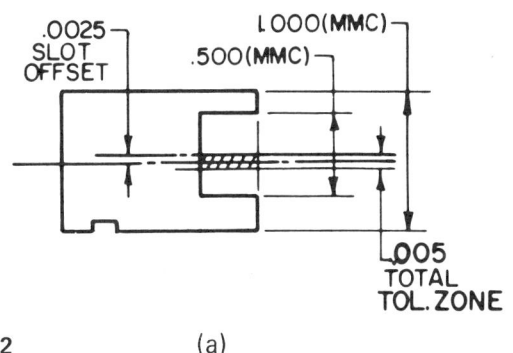

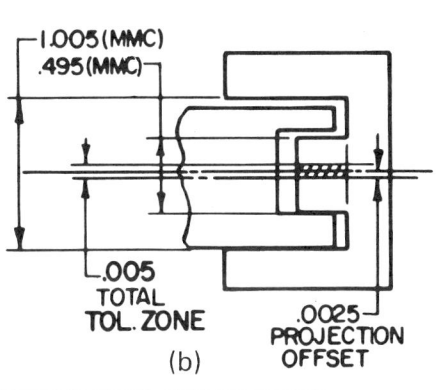

FIGURE 2   (a)   (b)

POSITION TOLERANCE ZONES AT LEAST MATERIAL CONDITION

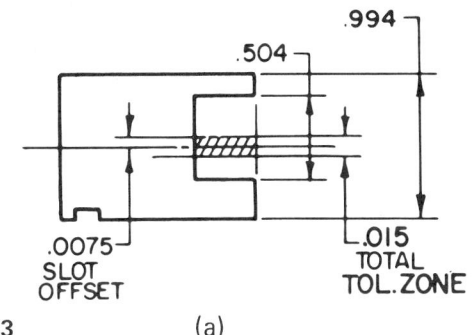

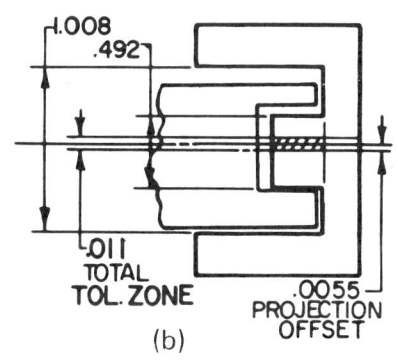

FIGURE 3   (a)   (b)

POSITION ⊕
NONCYLINDRICAL MATING PART FEATURES

MMC CALCULATIONS TO DETERMINE TOLERANCE

In this example we present the calculations required to determine the positional tolerance for the mating parts shown in the previous example.

Since one part is to fit within the other, the first step is to determine the clearance of the features and which feature is to receive the position tolerance. In this case, it seems more functional to control the position of the slot in part 1 and the position of the projection in part 2. The clearance of the two mating part features is to be .005 minimum. The projection on part 2 is .495 and the slot on part 1 is larger at .500. These are MMC sizes, or the largest projection possible on part 2 and the smallest slot possible on part 1.

The width features on both parts are given .005 clearance at MMC size of the features and are selected as the datum features for each part.

Under the subheading, POSITION TOLERANCE CALCULATIONS, the .495 MMC size of the projection on part 2 is subtracted from the .500 MMC size of the slot on part 1. This results in a difference of .005. Next the 1.000 MMC datum projection feature of part 1 is subtracted from the 1.005 MMC datum slot of part 2, resulting in a difference of .005.

The .005 result of the first calculation and the .005 result of the second calculation are added to give the .010 total combined positional tolerance for both parts and their interrelated features. This total tolerance is then divided to establish the required position tolerance on *each* individual part. How we allocate the total tolerance is optional, so long as it totals the calculated combined tolerance, in this case .010.

For the purposes of this example, the .010 total tolerance was divided evenly, with .005 selected as the position tolerance for both the .500 slot on part 1 and the .495 projection on part 2. These two figures, .005 plus .005, total .010 and comply with the .010 allowable total combined positional tolerance calculated.

Once the position tolerance is established for both mating part features based on their relationship to each other and to common datum axes, possible extra position tolerance for each part may be determined as shown in the lower half of the figure.

To do these calculations, we must first determine the relationship of one mating part feature to another and, then, we must consider each of these part features individually with respect to the size variations which could occur within their size tolerances. As has been shown, the size of features affects their positional tolerances, and it is this fact that makes positional tolerancing advantageous, since it permits economical production with greater tolerances and ensures assembly of the mating parts.

On part 1, the permissible tolerance may be increased from .005 up to .015 and on part 2 the permissible tolerance may be increased from .005 up to .011. The *actual* tolerance permissible in each case is, of course, dependent on the *actual* sizes of the features as produced.

This method of calculating position tolerances assumes the possibility of zero clearance – zero interference fits of mating part features if all features are at extreme tolerance limits. It also assumes parallel orientation or permissible float of one part to the other at assembly. Additional compensation of the calculated tolerance values should be considered as necessary for any particular application or where additional datum orientation may restrict this float.

EXAMPLES

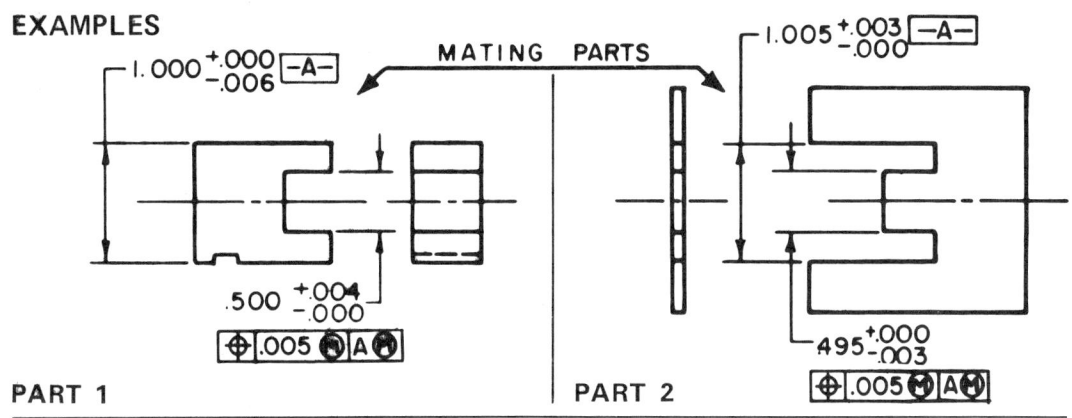

PART 1   PART 2

POSITION TOLERANCE CALCULATIONS

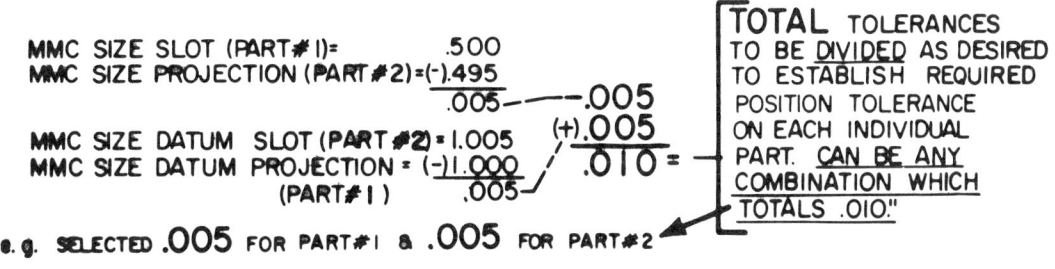

MMC SIZE SLOT (PART#1)= .500
MMC SIZE PROJECTION (PART#2)=(-).495
.005 — — .005
MMC SIZE DATUM SLOT (PART#2)= 1.005   (+).005
MMC SIZE DATUM PROJECTION = (-)1.000   .010 =
(PART#1)   .005

e.g. SELECTED .005 FOR PART#1 & .005 FOR PART#2

TOTAL TOLERANCES TO BE <u>DIVIDED</u> AS DESIRED TO ESTABLISH REQUIRED POSITION TOLERANCE ON EACH INDIVIDUAL PART. <u>CAN BE ANY COMBINATION WHICH TOTALS .010.</u>"

EXTRA TOLERANCE FOR EACH PART

<u>PERMISSIBLE **SLOT** POSITION TOL.</u> AS FEATURE SIZES DEPART FROM

	MMC:
STATED POSITION TOL WITH SLOT AT .500 MMC	= .005
PLUS TOTAL .500 SLOT SIZE TOL. _ _ _ _ _ _ _ _ _ _ _ _	+.004
POSITION TOL WITH DATUM WIDTH AT 1.000 MMC	= .009
PLUS TOTAL 1.000 DATUM WIDTH SIZE TOL. _ _ _ _ _ _ _ _	+.006
TOTAL POSITION TOL WITH SLOT & DATUM WIDTH AT LEAST MAT'L CONDITION (LARGEST SLOT, SMALLEST DATUM WIDTH)	=.015

PART 1

<u>PERMISSIBLE **PROJECTION** POSITION TOL</u> AS FEATURE SIZES DEPART FROM MMC:

STATED POSITION TOL WITH PROJECTION AT .495 MMC	= .005
PLUS TOTAL .495 PROJECTION SIZE TOL. _ _ _ _ _ _ _ _ _	+.003
POSITION TOL WITH DATUM OPENING AT 1.005 MMC	= .008
PLUS TOTAL 1.005 DATUM SLOT SIZE TOL. _ _ _ _ _ _ _ _	+.003
TOTAL POSITION TOL WITH PROJECTION & DATUM OPENING AT LEAST MAT'L CONDITION (SMALLEST PROJ., LARGEST DATUM OPENING)	=.011

PART 2

POSITION ⊕

Functional gages may be utilized on noncylindrical parts when position tolerancing is used.

Part 1 shown at the right is the same part used in the preceding examples. Immediately below part 1 is a representative functional gage used to check position on this part. Gage-makers' tolerances would, of course, be included in the actual contruction of this gage.

From the previous explanations and this illustration, we see that as the part feature sizes depart from MMC within their size tolerances, they become equally acceptable to the gage while permitting greater position tolerance. The gage essentially simulates the mating part situation, and therefore a part which passes the gage will assemble with its mating part.

Note, however, that the high and low limits of *size* are determined by other gages or measurements. The illustrated gage checks only the position requirement.

Part 2 is the mating part used in the previous examples. Below it is a representative functional gage used to check the position on this part. Gage-makers' tolerances would, of course, also be included in the actual construction of this gage.

As in the mating part above, we see that as the part feature sizes of part 2 depart from MMC within their size tolerances, they become equally acceptable to the gage while permitting greater position tolerance. The gage essentially simulates the mating part situation, and therefore a part which passes the gage will assemble with its mating part.

As in part 1, the high and low limits of *size* are determined by other gages or measurements. The illustrated gage checks the position requirement only.

PART 1

$1.000 \, {}^{+.000}_{-.006}$ [—A—]

$.500 \, {}^{+.004}_{-.000}$

⊕ .005 Ⓜ A Ⓜ

MMC FUNCTIONAL GAGE

1.000*
.500
.2475
.495*

*GAGE MAKERS TOL AS REQUIRED

GAGE CALCULATIONS
MMC SIZE PART SLOT = .500
MINUS POSN
TOL = (-).005
GAGE SIZE = .495

MMC SIZE OF PART
DATUM WIDTH (1.000)
ESTABLISHES GAGE
WIDTH SIZE = 1.000

PART 2

$1.005 \, {}^{+.003}_{-.000}$ [—A—]

$.495 \, {}^{+.000}_{-.003}$

⊕ .005 Ⓜ A Ⓜ

MMC FUNCTIONAL GAGE

1.005*
.250
.500*
.5025

*GAGE MAKERS TOL AS REQUIRED

GAGE CALCULATIONS
MMC SIZE PART PROJ. = .495
PLUS POSN TOL = (+).005
GAGE SIZE = .500

MMC SIZE OF PART
DATUM WIDTH (1.005)
ESTABLISHES GAGE
WIDTH SIZE = 1.005

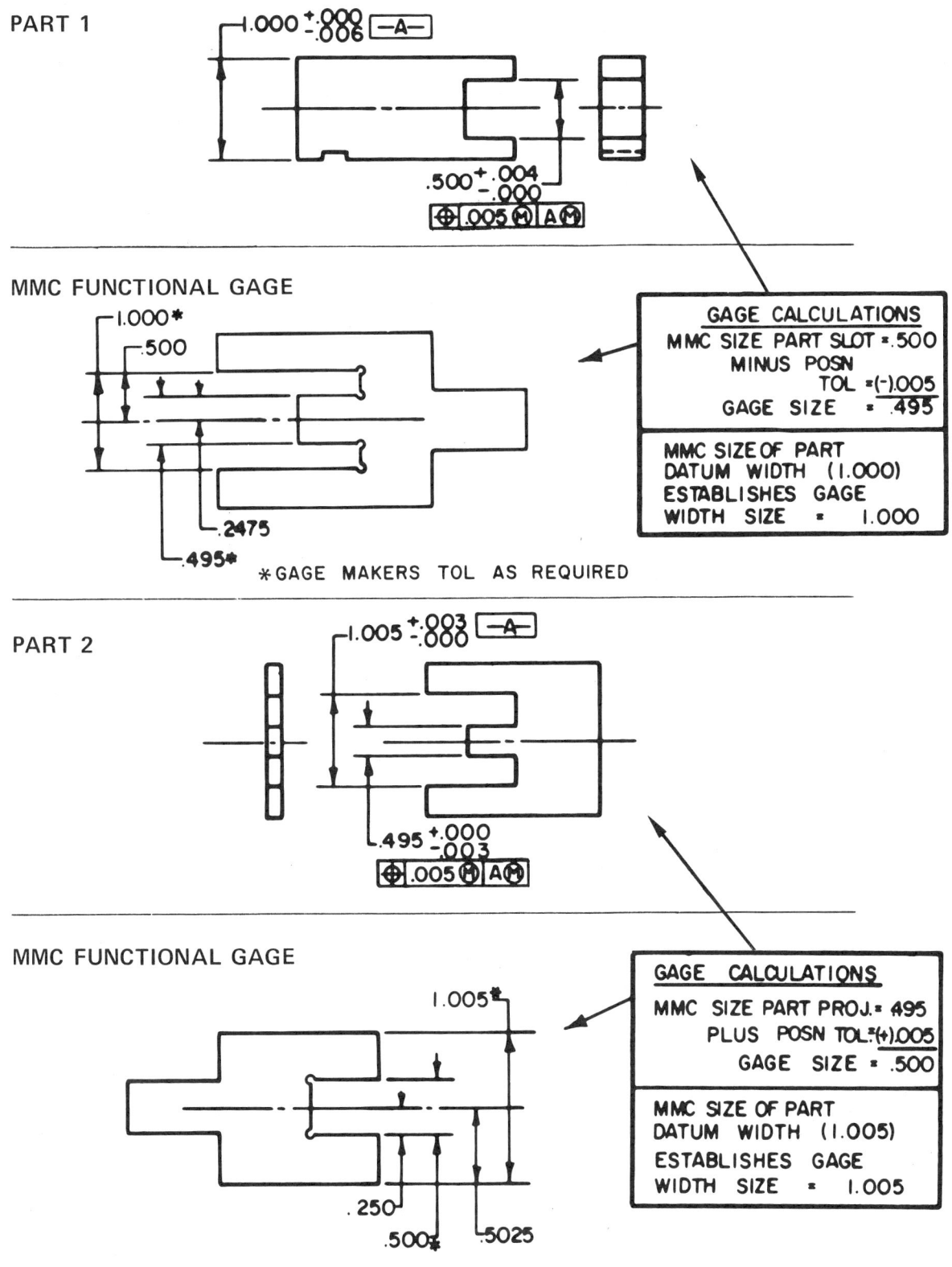

The example below illustrates control of a flat feature in a position relationship to a cylindrical datum. This situation is typical of many parts for which position functional techniques will assure agreement with design intent and also permit functional or receiver gaging.

The part is shown with a representative functional gage. Note that the possible position tolerance increases from .010 to .016 maximum as the feature sizes depart from MMC toward the opposite or least material limits. The actual position tolerance depends on the sizes to which the features are produced with the amount of increase equal to the departure from MMC size.

EXAMPLE

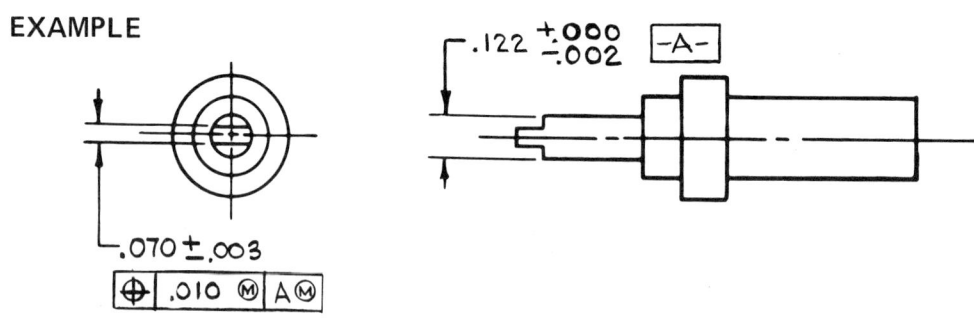

GAGE

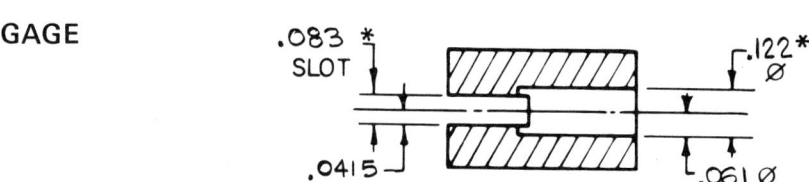

```
        GAGE CALCULATIONS
   MMC SIZE PART =   .073
   PLUS POSN TOL. = (+).010
   GAGE              .083 *

   MMC SIZE PART
   DATUM DIA      = .122
   ESTABLISHES
   GAGE DIA.SIZE =   .122 *
```

∗ GAGE MAKERS TOL. AS REQUIRED

POSITION
OF
COAXIAL
FEATURES

COAXIAL FEATURES — SELECTION OF PROPER CONTROL

There are three methods of controlling interrelated coaxial features:

1. RUNOUT TOLERANCE (circular or total) (RFS)
2. POSITION TOLERANCE   (MMC)
3. CONCENTRICITY TOLERANCE   (RFS)

Any of the above methods provide effective control. However it is important to select the *most appropriate* one to both meet the design requirements and provide the most economical manufacturing conditions. (See also details of above sections).

Below are recommendations to assist in selecting the proper control:

If the need is to control only CIRCULAR cross sectional elements in a composite relationship to the datum axis, RFS, e.g., multi-diameters on a shaft, use

CIRCULAR RUNOUT   **EXAMPLE :**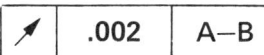

(This method controls any composite error effect of roundness and circular cross-sectional profile variations)

If the need is to control the TOTAL cylindrical or profile surface in composite relative to the datum axis' RFS, e.g., multi-diameters on a shaft, bearing mounting diameters, etc., use

EXAMPLE: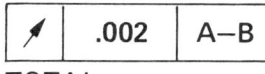

TOTAL RUNOUT   **TOTAL**   TOTAL

(This method controls any composite error effect of roundness, cylindricity, straightness, angularity, and parallelism.)

Note: Runout is always implied as an RFS application. It cannot be applied on an MMC basis, since an MMC situation involves functional interchangeability or assemblability (probably of mating parts), in which case POSITION tolerance would be used. See below.

If the need is to control the total cylindrical or profile surface and its axis in composite location relative to the datum axis on an MMC basis, e.g., on mating parts to assure interchangeability or assemblability, use

EXAMPLE:

POSITION   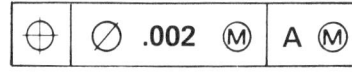

If the need is to control the *axis* of one or more features in composite relative to a *datum axis*, RFS, e.g., to control balance of a rotating part, use

EXAMPLE:

CONCENTRICITY   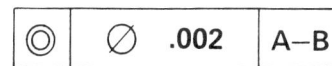

Note: Concentricity is always implied as an RFS application. Variations in size (departure from MMC size, out-of-roundness, out-of-cylindricity, etc.) do not in themselves conclude *axis* error.

POSITION TOLERANCING (MMC) OF COAXIAL FEATURES

This illustration shows a common application of position tolerancing of coaxial features. A functional MMC relationship is desired.

A functional datum is selected and the feature is specified to be "at position (coaxial) within $\oslash$.001 tolerance zone with feature at MMC with respect to datum A at MMC."

A functional gage (or equivalent technique) will always be used on this part. The interpretation and the illustration gage show how the part remains functionally acceptable as the concerned features depart from their MMC (worst condition) size within their size tolerance range. Further, the advantages of position tolerancing are realized since greater position tolerance is permitted as the feature and datum sizes depart from MMC. Functionally good parts are always accepted on this basis; as in all types of MMC position tolerance, size tolerance *and* position tolerance are considered together.

Positional tolerance controls axis or center plane displacement. This part illustrates the manner in which position of coaxial features incorporates the same considerations as conventional positional tolerancing of hole patterns. The axes provide the common denominator needed to relate displacement of one feature to another and to allow calculations and gage determinations based on functional design requirements. The functional gage will accomodate permissible axis displacement in terms of the surface configuration, including both form and position errors. The functional gage checks *position* only. The hole high size limit and low size limit (if necessary) are checked separately.

Zero tolerancing on position of coaxial features may occasionally be useful. See EXTENDED USES OF POSITION TOLERANCING in this section.

EXAMPLE

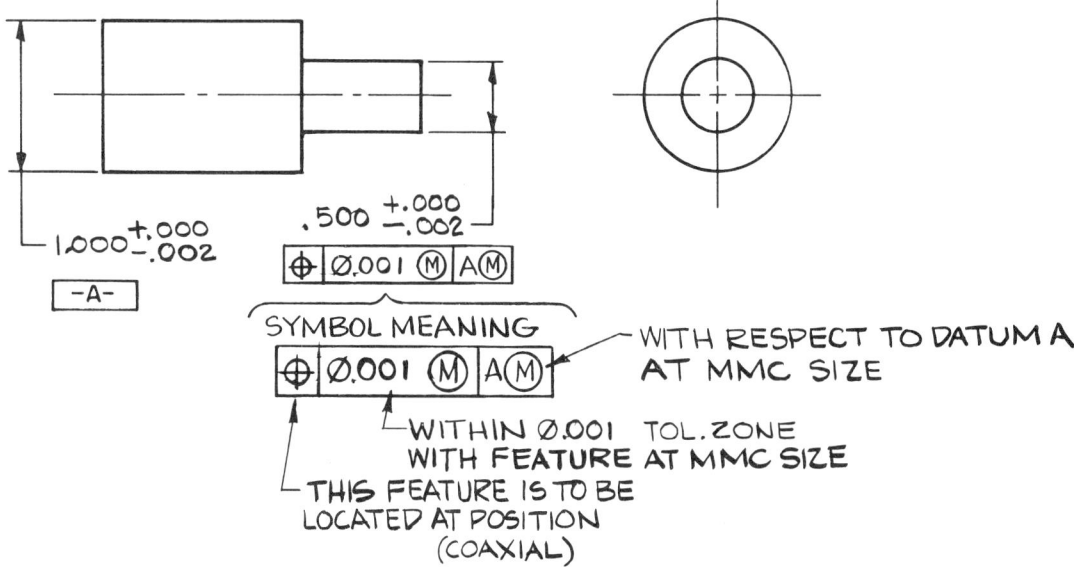

SYMBOL MEANING

⊕ | Ø.001 Ⓜ | AⓂ ← WITH RESPECT TO DATUM A
AT MMC SIZE

WITHIN Ø.001 TOL. ZONE
WITH FEATURE AT MMC SIZE

THIS FEATURE IS TO BE
LOCATED AT POSITION
(COAXIAL)

MEANING

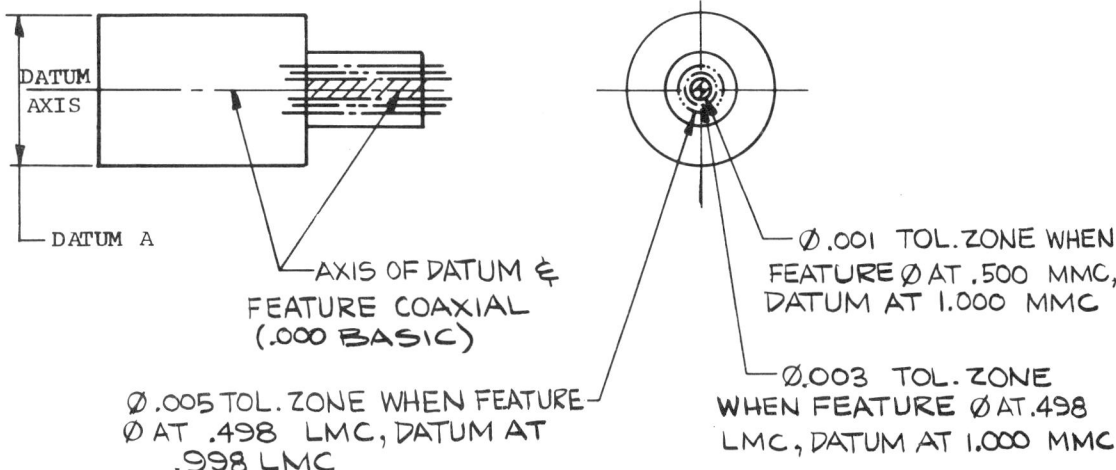

AXIS OF DATUM &
FEATURE COAXIAL
(.000 BASIC)

Ø.005 TOL. ZONE WHEN FEATURE
Ø AT .498 LMC, DATUM AT
.998 LMC

Ø.001 TOL. ZONE WHEN
FEATURE Ø AT .500 MMC,
DATUM AT 1.000 MMC

Ø.003 TOL. ZONE
WHEN FEATURE Ø AT .498
LMC, DATUM AT 1.000 MMC

GAGE

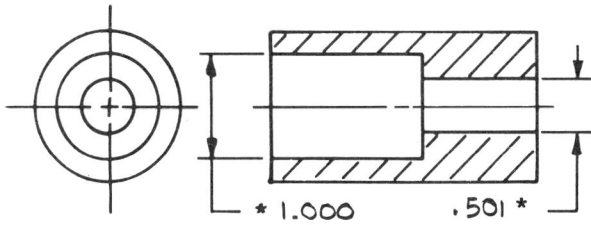

* 1.000   .501 *

* GAGE MAKERS TOL APPLY

POSITION ⊕

COMPARISON BETWEEN COAXIAL FEATURES SPECIFIED RFS (RUNOUT) AND COAXIAL FEATURES SPECIFIED MMC (POSITION)

In a mating part situation, positional tolerancing of coaxial features can be used to good advantage. In this illustration we compare a part with coaxial features dimensioned and toleranced using total runout (RFS) with the same part dimensioned and toleranced using position (MMC).

The example at the top shows a part similar to that of the previous example except that total runout (RFS) tolerancing is used. Note that the $.500 \, {}^{+.000}_{-.002}$ diameter in the top example has a tolerance box which states "this feature is to be within .001 total runout when mounted on datum A." This is an RFS application where the stated runout tolerance of .001 is the total allowable tolerance FIM, (TIR, FIR) regardless of the size to which the .500 diameter and 1.000 datum diameter are produced. The meaning is shown at the right; the probable method of checking is also illustrated. It is seen that the dial indicator reading of the rotated .500 diameter with respect to the 1.000 datum diameter and its axis will entirely determine the runout reading and thus the acceptance or rejection of the part. The dial indicator reading must be .001 or less for part acceptance, and this tolerance is maximum.

A closer study of this part and its function may reveal that it has a *mating part* into which it must fit. Knowing this, we can determine the control of the dimensions and tolerances of the *two* parts as they are related to each other, to ensure proper fit and economic production. Where this problem of assemblability or functional requirement exists, position tolerancing and its associated advantages should be considered.

Note that the same part is illustrated under the MMC (Specified as Position) subheading. The dimensions are identical. The position characteristic symbol is, however, indicated with a ∅.001 tolerance requirement and is specified as an MMC application. At the right is the mating part containing the holes which must fit the shaft at the left. It also utilizes position control with a tolerance requirement of ∅.002 at MMC.

Figures 1, 2, and 3 illustrate the mating part positional relationships and the manner in which additional tolerance is achieved dependent on part size variations.

In Fig. 1(a), the .001 tolerance zone permits a .0005 axis displacement of the .500 diameter on the shaft with respect to the 1.000 datum diameter A axis. On the mating part (Fig. 1b), the .002 tolerance zone on the hole permits a .001 axis displacement of the .501 hole with respect to the 1.002 datum hole A axis. This displacement includes all the errors of form and position of the feature which have an effect. Actually, in the use of MMC position, the reference to axes is primarily theoretical as a common denominator for comparison and calculation. In reality, the functional gage (or technique) which represents each mating part and its fit determines acceptance of the part on the basis of surface contact as related to the axis displacement.

The sectional view in Fig. 1(c), showing the shaft inserted into the hole, illustrates the fit of the two parts at MMC size with the .500 shaft diameter and the .501 hole

RFS (SPECIFIED AS TOTAL RUNOUT)

MEANING

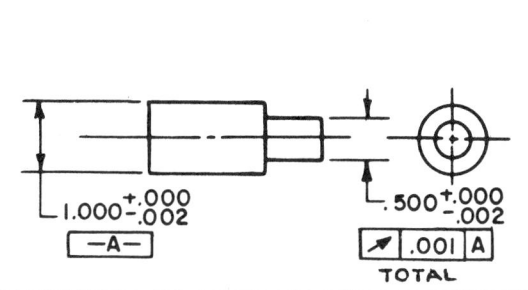

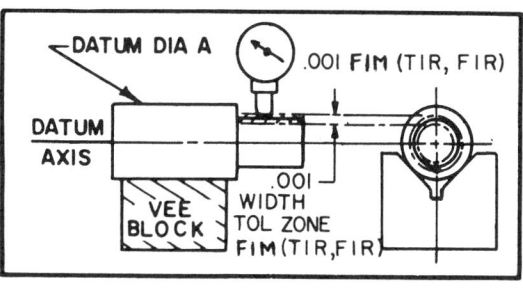

MMC (SPECIFIED AS POSITION)

MATING PART

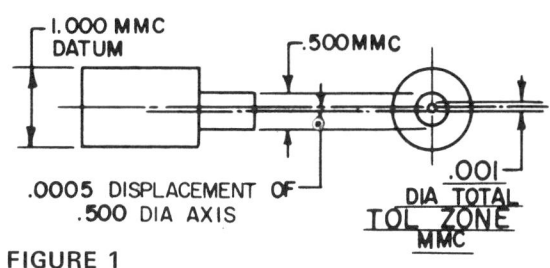

POSITION TOLERANCE ZONES AT MMC

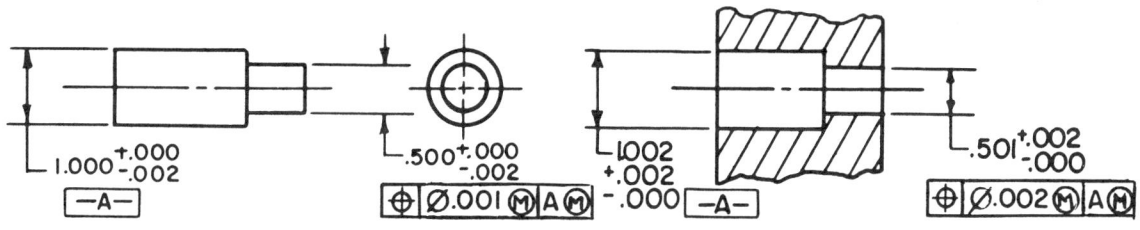

FIGURE 1

POSITION TOLERANCE ZONES AT LEAST MATERIAL CONDITION

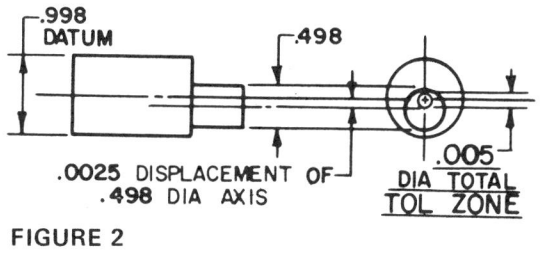

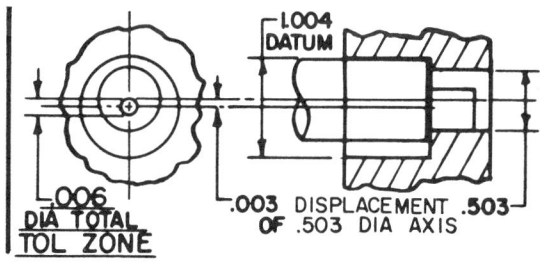

FIGURE 2

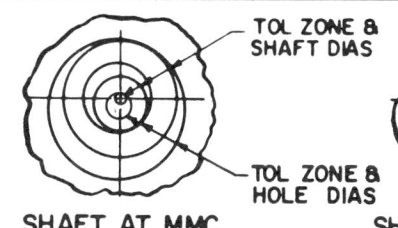

FIGURE 3

161

diameter displaced in opposite directions to the full extent of the position tolerance. The parts assemble satisfactorily.

In this situation the part shaft and hole sizes are at MMC: that is, we have the largest permissible shaft size and the smallest permissible hole size. If the parts assemble under these conditions, it is reasonable to assume that as the shaft gets smaller and the hole larger, the tolerance could be increased while continuing to permit assembly.

Figure 2(a) illustrates how the position tolerance zone increases as the part and hole sizes depart from MMC to the opposite size extreme (LEAST MATERIAL CONDITION). Note that the .500 shaft diameter is now .498 and the 1.000 datum is now .998, or the low limit of the shaft diameter sizes. In Fig. 2(b) the .501 hole diameter is now .503 and the 1.002 datum diameter is now 1.004, or the high limit of the hole sizes.

It may now be seen that the reduction in shaft sizes and increase in hole sizes permit a greater displacement of the mating features, resulting in a permissible ⌀ .005 total position tolerance zone on the stepshaft at the left and a ⌀ .006 total position tolerance zone on the mating part at the right.

The sectional view at the far right showing the shaft inserted into the hole illustrates the fit. The parts still assemble satisfactorily.

With the use of position, the permissible tolerance has been increased from ⌀ .001 to ⌀ .005 on the shaft at the left, and from ⌀ .002 to ⌀ .006 on the mating part at the right.

Referring to the part examples under the MMC (Position) subheading, note that a rule-of-thumb method of determining the maximum possible position tolerance is to *add* the stated tolerance found in the symbol box and the total size tolerance of the features involved. For example, on the shaft at left, the position tolerance of ⌀ .001 found in the symbol box, plus the .002 total size tolerance of the .500 shaft diameter, plus the .002 total size tolerance of the shaft 1.000 datum diameter (or .001, plus .002, plus .002) equals .005 total. On the mating part at the right, the position tolerance of ⌀ .002 found in the symbol box, plus the .002 total size tolerance of the .500 hole, plus the .002 total size tolerance of the 1.002 datum hole diameter (or .002, plus .002, plus .002) equals .006 total.

Remember, however, that the resultant tolerance is dependent on the actual size to which the features are produced. Thus the tolerance under an MMC application is usually somewhere between the stated tolerance and the opposite limit as determined by actual feature sizes.

Figure 3 illustrates the assembly relationship of the shaft and holes at various extremes. Figure 3(a) shows the shaft at MMC or its largest size, with the mating part holes at LEAST MATERIAL CONDITION, or their *largest* size. Figure 3(b) shows both parts at LEAST MATERIAL CONDITION or at the size limits opposite to the calculated MMC sizes; that is, the shaft is at its smallest allowable size and the holes are at their largest allowable size. This is the condition illustrated in Fig. 2; it allows the greatest amount of tolerance on both parts. Figure 3(c) shows the holes at MMC or their smallest size and the shaft at LEAST MATERIAL CONDITION or its smallest size. Note that the parts assemble in each case.

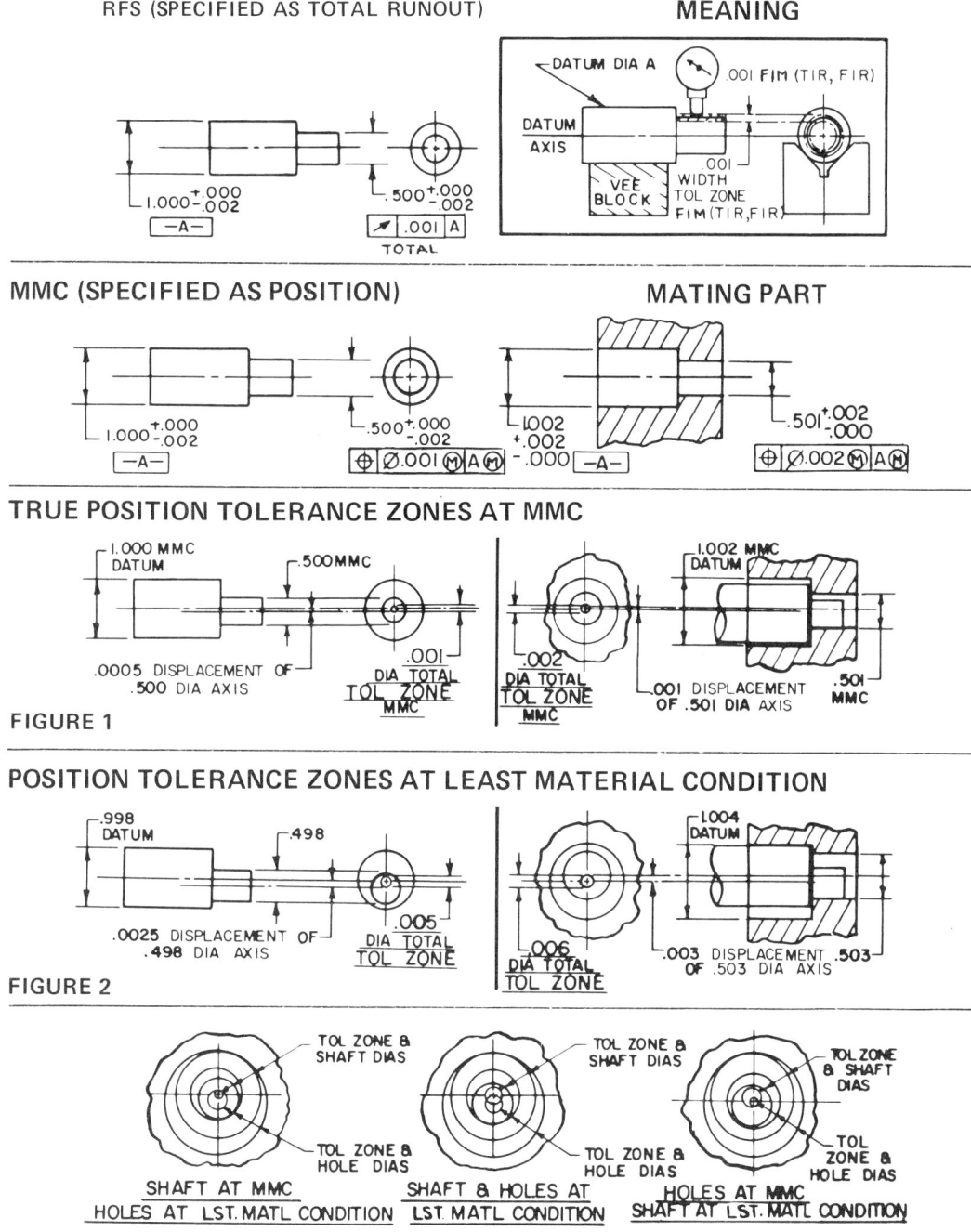

RFS (SPECIFIED AS TOTAL RUNOUT)

MEANING

MMC (SPECIFIED AS POSITION)

MATING PART

TRUE POSITION TOLERANCE ZONES AT MMC

FIGURE 1

POSITION TOLERANCE ZONES AT LEAST MATERIAL CONDITION

FIGURE 2

FIGURE 3

Numerous other combinations are, of course, also possible and acceptable, so long as the established size and location tolerances are held.

These examples again illustrate that in position tolerancing, size tolerances affect position tolerances. It is to advantage to consider this fact whenever possible so that maximum tolerance yield may be achieved.

163

POSITION OF COAXIAL FEATURES
MMC CALCULATIONS TO DETERMINE TOLERANCE

The actual calculations necessary to determine position tolerances of coaxial features of mating parts are illustrated in this example. The mating parts are the same as those in the preceding example.

Note that we have selected functional datums: the diameters of mating features, the 1.000 shaft diameter, and the 1.002 hole diameter. This establishes a common datum axis and the relationship between the diameters of the two parts. This information is necessary for our calculation.

Selected *clearances* of the holes to the mating shaft diameters are established by the design requirements. These clearances in turn establish the position tolerances to be stated for these features.

Note that the first step illustrated under the subheading Position Tolerance Calculations shows the .500 MMC shaft size of part 2 subtracted from the .501 MMC hole size of part 1, resulting in a difference of .001. Also, the 1.000 MMC datum shaft size of part 2 is subtracted from the 1.002 MMC datum hole size of part 1, resulting in a difference of .002.

The .001 result of the first calculation and the .002 result of the second calculation are added to give the .003 total combined position tolerance for both parts and for their interrelated features. This total tolerance is then divided to establish the required position tolerance on each individual part. The division can be made in any combination, so long as the individual part tolerances total the calculated combined tolerance, in this case, .003.

For this example, .002 has been selected as the position tolerance for the .501 hole in part 1, and .001 as the position tolerance for the .500 diameter shaft of part 2. These two figures total .003 and comply with the allowable total combined position tolerance.

Dependent on feature size variation, we may realize an extra or bonus tolerance.

This possible additional position tolerance as the feature sizes depart from MMC size, is calculated as shown on the lower portion of the illustration.

Remember, however, that this increase in position tolerance is dependent upon the *actual* sizes to which the features are produced within their size tolerances.

Since most parts are produced somewhere between the high and low size tolerance extremes, the actual position tolerance permissible on part number 1 would be somewhere between .002 and .006, say approximately .004. It could, however, be .006 and still provide the proper fit to the mating part. The actual position tolerance permissible on part 2 would probably be somewhere between .001 and .005, say approximately .003. It could, however, be .005 and still provide the proper fit to the mating part.

The advantages of using position tolerances on coaxial features are now apparent.

1. Greater tolerance is permitted.

2. The fit of the mating part is guaranteed when the dimension and the tolerance requirement are calculated simultaneously and consideration is given to the possible variables which can occur in the assembly relationship.

The method of calculating position tolerances described above assumes the possibility of a zero interference – zero clearance condition of the mating part features at extreme tolerance limits. Additional compensation of the calculated tolerance values should be considered as necessaary relative to a particular application.

EXAMPLES

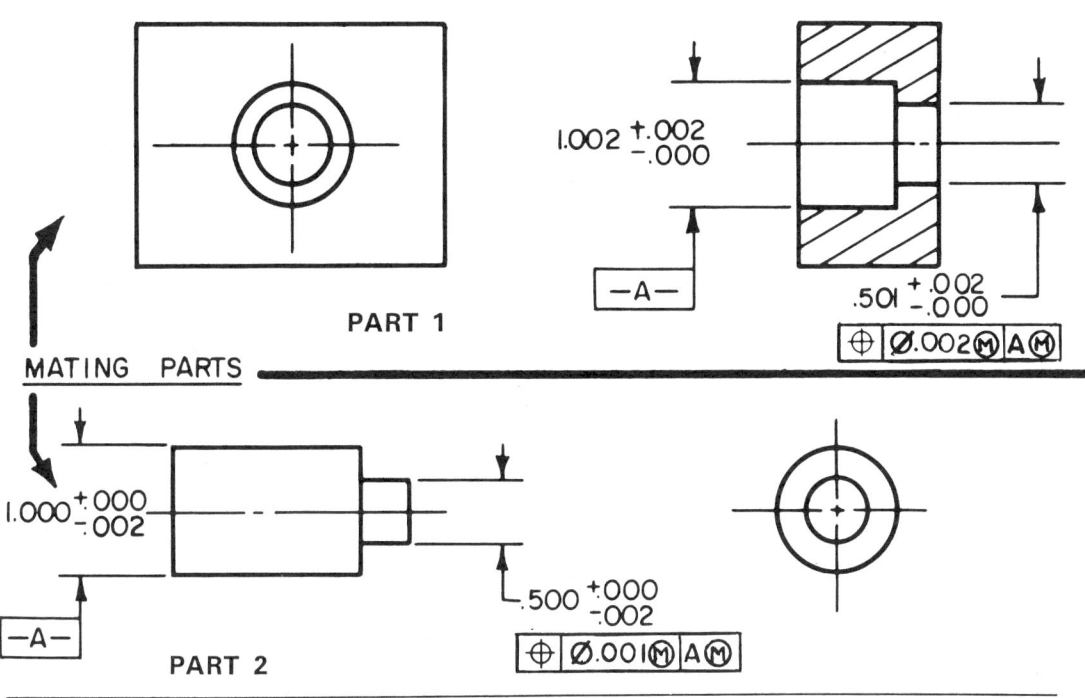

MATING PARTS

PART 1

PART 2

POSITION TOLERANCE CALCULATIONS

MMC SIZE HOLE (PART #1) = .501
MMC SIZE SHAFT (PART #2) =(-).500
.001

MMC SIZE DATUM HOLE (PART #1) = 1.002
MMC SIZE DATUM SHAFT (PART #2)=(-)1.000
.002

.001
(+).002
.003 =

TOTAL TOLERANCE TO BE DIVIDED AS DESIRED TO ESTABLISH REQUIRED POSITION TOLERANCE ON EACH INDIVIDUAL PART. CAN BE ANY COMBINATION WHICH TOTALS .003".

e.g. SELECTED .002 FOR PART#1 & .001 FOR PART#2

EXTRA TOLERANCE FOR EACH PART

PERMISSIBLE **HOLE** POSITION TOL AS HOLE SIZES DEPART (GET LARGER) FROM MMC:

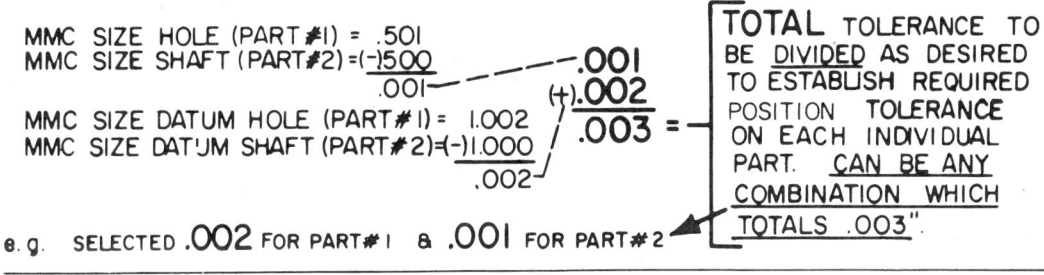

STATED POSN TOL. WITH HOLE AT .501 MMC	=.002
PLUS TOTAL .501 HOLE TOL	+.002
POSN TOL WITH DATUM HOLE 1.002 AT MMC	.004
PLUS TOTAL 1.002 DATUM HOLE TOL	+.002
PART 1 TOTAL POSN TOL WITH BOTH HOLES AT **LEAST** MAT'L CONDITION.	=.006

PERMISSIBLE **SHAFT** POSITION TOL AS SHAFT SIZES DEPART (GET SMALLER) FROM MMC:

STATED POSN TOL WITH SHAFT AT .500 MMC	=.001
PLUS TOTAL .500 DIA TOL	+.002
POSN TOL WITH DATUM SHAFT DIA 1.000 AT MMC	=.003
PLUS TOTAL 1.000 DATUM DIA TOL	+.002
PART 2 TOTAL POSN TOL WITH BOTH SHAFT DIAS AT **LEAST** MAT'L CONDITION	.005

POSITION OF COAXIAL FEATURES
MMC GAGES

Positional tolerances may, of course, be used on any part, even though there is no direct assembly relationship between mating part features. The tolerance advantages and the possibility of using functional or receiver gages for effective and economic inspection often makes the position technique very desirable under such conditions. However, where the assemblability of mating part features is involved, position tolerancing is always recommended. Position tolerancing of coaxial features of mating parts is an ideal example of this use.

Functional or receiver gages for coaxial features of mating parts simulate the fit of the actual part features in a manner similar to the hole pattern functional gages. As in the hole pattern functional gage, part feature size departure from MMC permits greater part acceptance based on the functional interrelationship of size and position variations.

The illustration shows the two mating parts previously discussed and examples of functional gages to check the position requirements of each part. The upper illustration, part 1, is shown with a gage for functionally checking the position tolerance of the holes. The simple calculations to determine the gage dimensions are shown at the right. Gage-maker's tolerances also apply as required.

The lower illustration shows part 2 and a functional gage for checking the position of the shaft diameters of this part. The calculations are shown at right. Gage-maker's tolerances also apply as required.

The illustrated gages will check only the position tolerances of the part. Sizes of the associated features require a separate size check.

PART 1

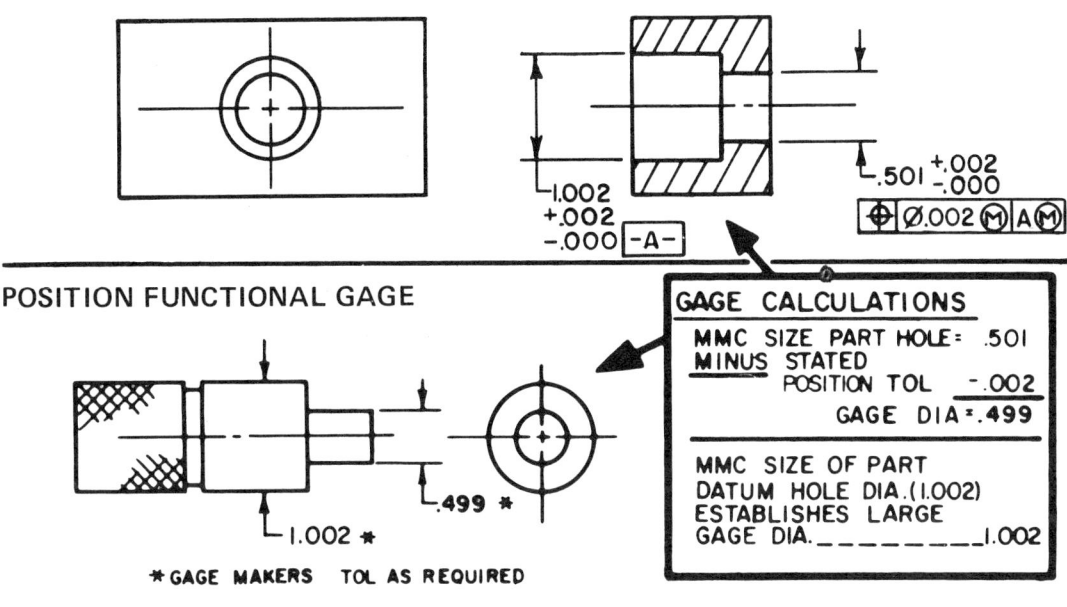

POSITION FUNCTIONAL GAGE

1.002 +.002 -.000 -A-

.501 +.002 -.000

⊕ Ø.002 Ⓜ A Ⓜ

.499 *

1.002 *

*GAGE MAKERS TOL AS REQUIRED

GAGE CALCULATIONS

MMC SIZE PART HOLE= .501
MINUS STATED
     POSITION TOL   - .002
        GAGE DIA =.499

MMC SIZE OF PART
DATUM HOLE DIA.(1.002)
ESTABLISHES LARGE
GAGE DIA._____1.002

PART 2

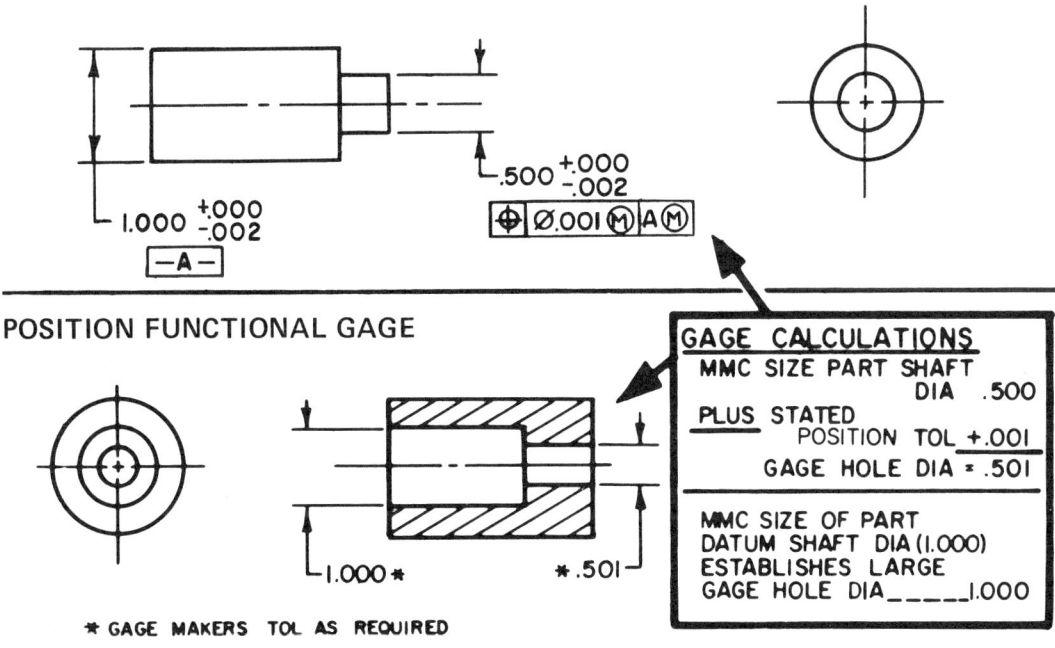

1.000 +.000 -.002
-A-

.500 +.000 -.002

⊕ Ø.001 Ⓜ A Ⓜ

POSITION FUNCTIONAL GAGE

1.000 *

* .501

*GAGE MAKERS TOL AS REQUIRED

GAGE CALCULATIONS

MMC SIZE PART SHAFT
             DIA .500
PLUS STATED
     POSITION TOL +.001
   GAGE HOLE DIA =.501

MMC SIZE OF PART
DATUM SHAFT DIA(1.000)
ESTABLISHES LARGE
GAGE HOLE DIA_____1.000

POSITION ⊕
RELATION TO DATUM SURFACES USING A RADIAL HOLE PATTERN

Radial hole patterns may be controlled by position tolerancing in a manner similar to that used on more conventional hole patterns.

Since the four hole pattern in the example at the right requires location as well as feature-to-feature relationship (in-line), composite positional tolerancing may be used to advantage.

Datums are established relative to the part function, the face surface A and the .900 diameter, datum B. Imagine this part mating with another at assembly on the surface A and located and centered on B so pins will pass into the holes. MMC is desired so the hole and mating part pin sizes are used to determine the hole-to-hole position tolerances.

In this example, the holes are desired to be within one plane (coplanar), in-line (colinear), and spaced 90° apart. Therefore, as a pattern they may vary from that basic configuration only to the extent of permissible tolerances. In this case, the pattern as an entity (established from the holes themselves) may vary from position (and parallel to A) within .030 diameter at MMC. In the pattern, the holes may vary from their basic colinear alignment within .005 diameter at MMC. The axes of the four holes must lie within both the pattern locating tolerance zones and the feature relating tolerance zones.

In terms of gaging this part, two separate gages or gaging operations would be required: one to orient to datums A and B and evaluate the .030 diameter tolerance, and another to evaluate the .005 diameter tolerance. The latter could be simply accomplished by passing a gage pin of .120 diameter through two holes in-line simultaneously. The hole size requirements must, of course, also be met.

As the four holes depart from MMC, more position tolerance is acquired (as has been previously described in other examples). In this case, although the example is more complex, the principles are the same.

AS DRAWN

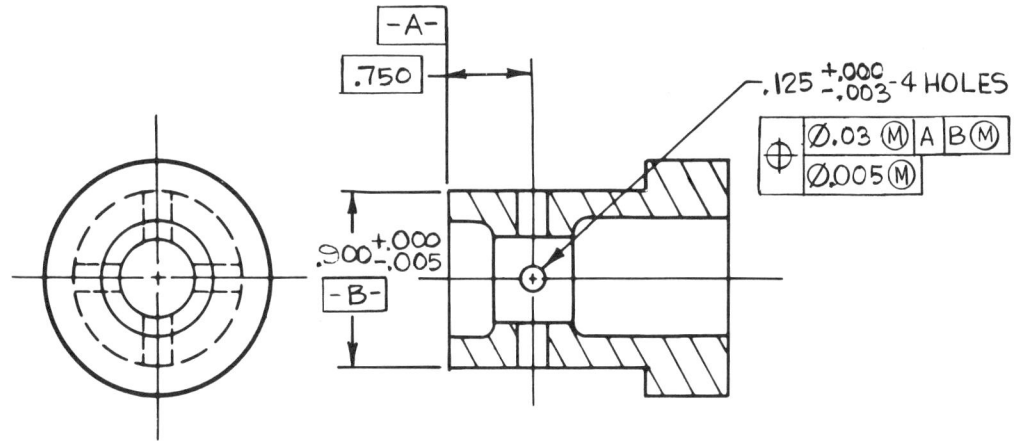

-A-

.750

.125 $^{+.000}_{-.003}$ - 4 HOLES

⊕ | Ø.03 Ⓜ | A | B Ⓜ |
    | Ø.005 Ⓜ |

.900 $^{+.000}_{-.005}$

-B-

MEANING

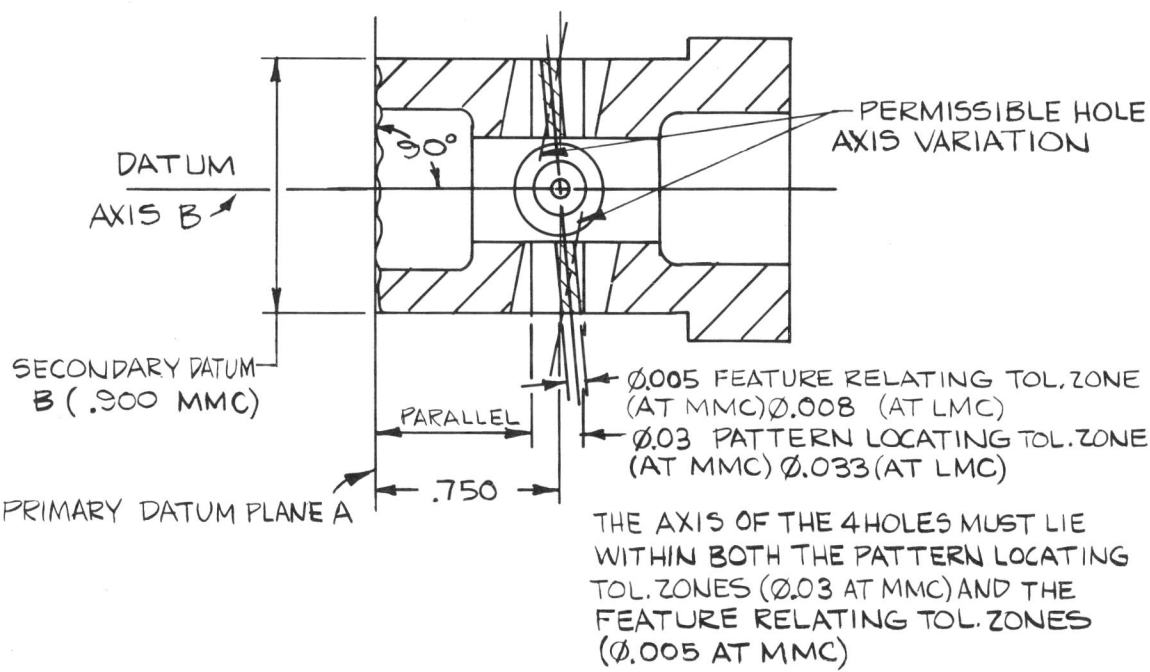

DATUM
AXIS B

90°

PERMISSIBLE HOLE
AXIS VARIATION

SECONDARY DATUM
B (.900 MMC)

PARALLEL

.750

Ø.005 FEATURE RELATING TOL. ZONE
(AT MMC) Ø.008 (AT LMC)
Ø.03 PATTERN LOCATING TOL. ZONE
(AT MMC) Ø.033 (AT LMC)

PRIMARY DATUM PLANE A

THE AXIS OF THE 4 HOLES MUST LIE
WITHIN BOTH THE PATTERN LOCATING
TOL. ZONES (Ø.03 AT MMC) AND THE
FEATURE RELATING TOL. ZONES
(Ø.005 AT MMC)

POSITION
EXTENDED
PRINCIPLES

POSITION TOLERANCING EXTENDED PRINCIPLES

This section discusses additional applications of position tolerancing and extension of the principles to more unusual situations. Since the position tolerance method may be applied to a wide variety of situations, our examples were chosen to illustrate the versatility of this technique.

Caution should be exercised in the more unusual applications to ensure uniform interpretation through consistent use of the position tolerance principles discussed earlier.

Coaxial Features—Zero Tolerance 183–184

Combination and Unique Applications 192–200

Mating Parts—Fixed Fasteners 186–191

Perpendicularity Control Within Position Tolerancing 174

Projected Tolerance Zone 175

Separate Patterns of Holes 173

Zero Tolerance—Round Holes 176–182

POSITION ⊕
SEPARATE PATTERNS OF HOLES

Patterns of holes (or features) on a part which are either shown by common basic or untoleranced dimensions or are related to common datums are normally assumed to be related and therefore considered a single composite pattern. Occasionally, convenience in the placement of dimensions or the need to show common dimension origination from related datum features on the drawing can unintentionally impose restrictions on production or inspection.

If certain holes (or features) in a BASIC dimensioned pattern are related to one another functionally but not to others which are also shown in the same pattern, or if we wish to produce or gage certain portions of the pattern separately, we may indicate this by the notation SEP REQT or SEPARATE REQUIREMENT beneath or adjacent to the concerned feature control symbol or symbols. Coding the appropriate holes ✦ on the drawing is helpful in clarifying the drawing intent.

Hole (or feature) patterns on a part separated by rectangular coordinates, related to *different* datums, or having no direct tie by dimensional arrangement, are considered to be separate patterns; no further notation is required in this case.

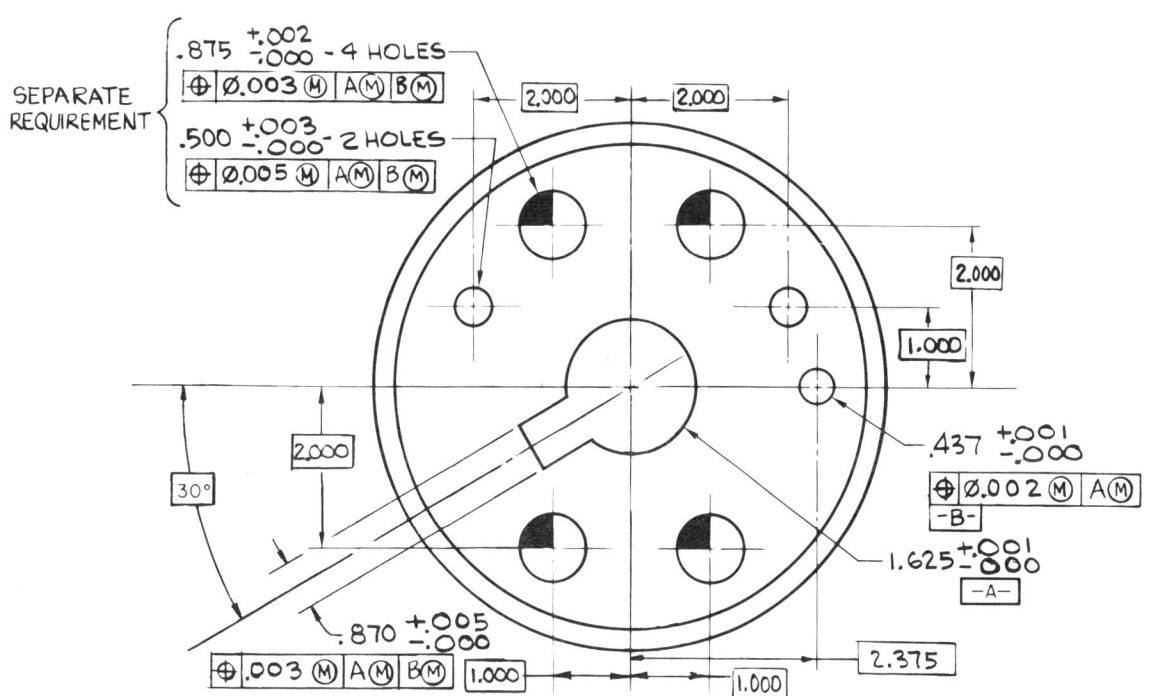

POSITION ⊕

PERPENDICULARITY (SQUARENESS) CONTROL
WITHIN POSITION TOLERANCING

Where out-of-perpendicularity of threaded or plain holes could cause inserted screws, bolts, studs, pins, or dowels to interfere with mating parts or result in cocked seating of screw or bolt heads, closer control of perpendicularity *within* the position tolerance may be required.

Perpendicularity control may be specified on the drawing, using one of the methods illustrated below.

PERPENDICULARITY SPECIFICATION ON DRAWING

AS DRAWN   **MEANING**

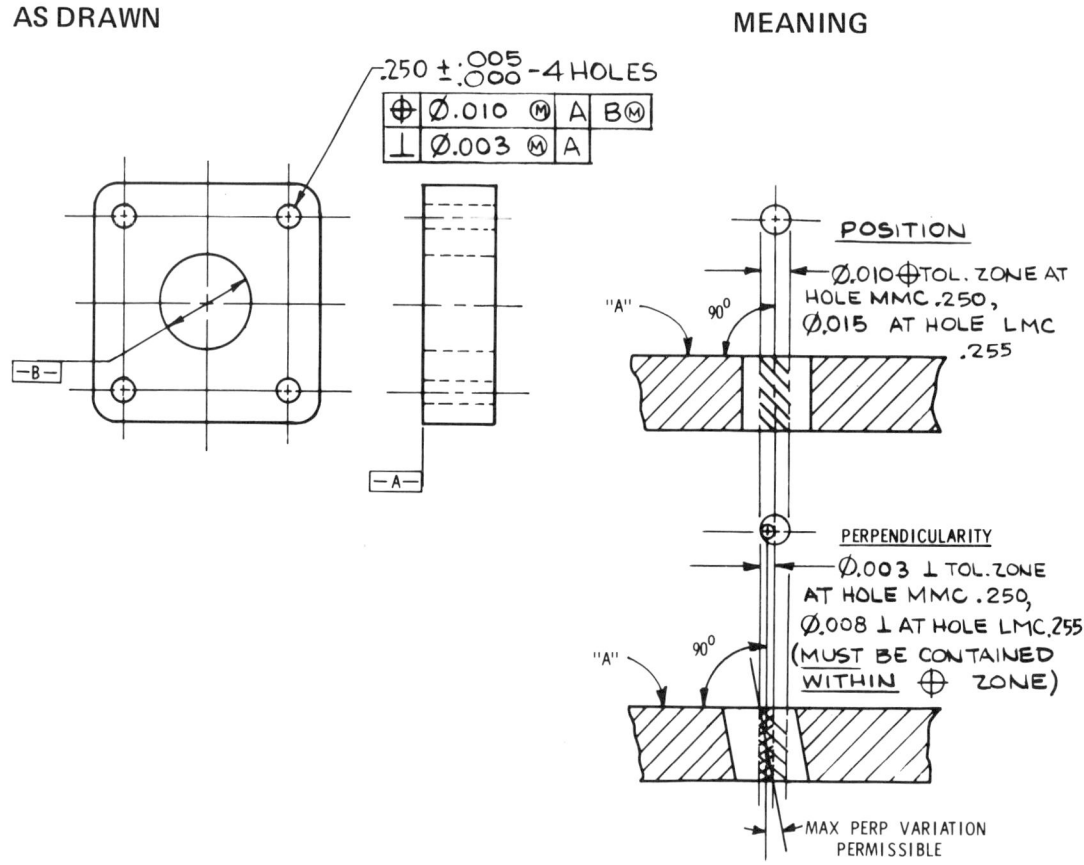

Note: Where the projected error of inserted screws or pins in such holes could cause mating part interference, the holes in the mating part may have to be enlarged to account for this tolerance variation unless the "projected tolerance zone" method is applied to the tapped holes.

PROJECTED TOLERANCE ZONE

The projected tolerance zone method prevents the condition shown in Fig. 1, where interference could possibly exist with conventional positional tolerancing. The variation from perpendicularity of the portion of the bolt passing through the mating part is of concern. Therefore the location and perpendicularity of the tapped hole is of importance insofar as it affects this extended portion of the bolt. The projected tolerance zone method (Fig. 2) eliminates this interference.

With this method, we can use conventional "fixed fastener" calculations to determine the position tolerance. Further, specifying by this method means that gaging techniques will simulate the mating part relationship, and the projected perpendicularity error will be accounted for in the tolerance and in the gaging.

FIGURE 1

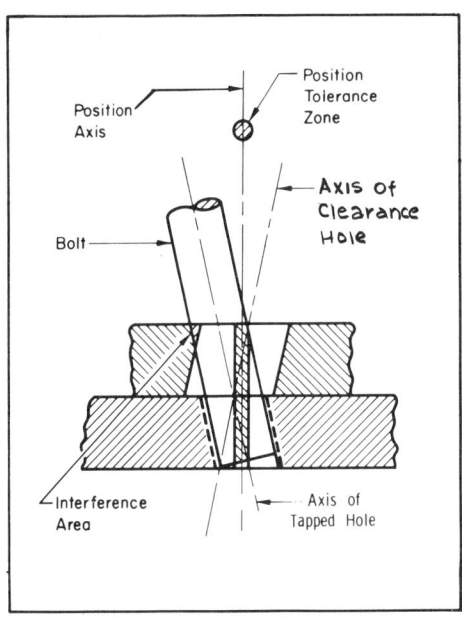

FIGURE 2

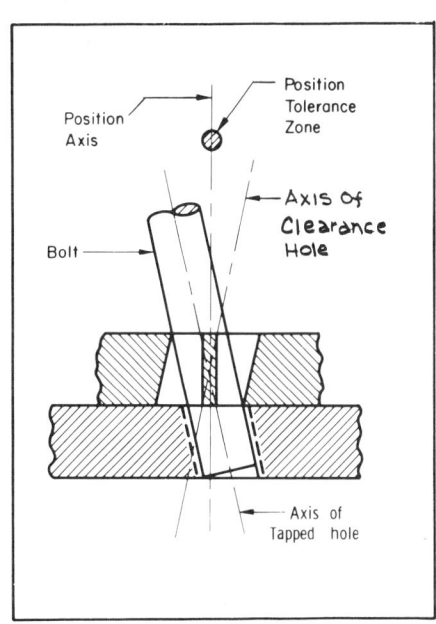

EXAMPLE   MEANING

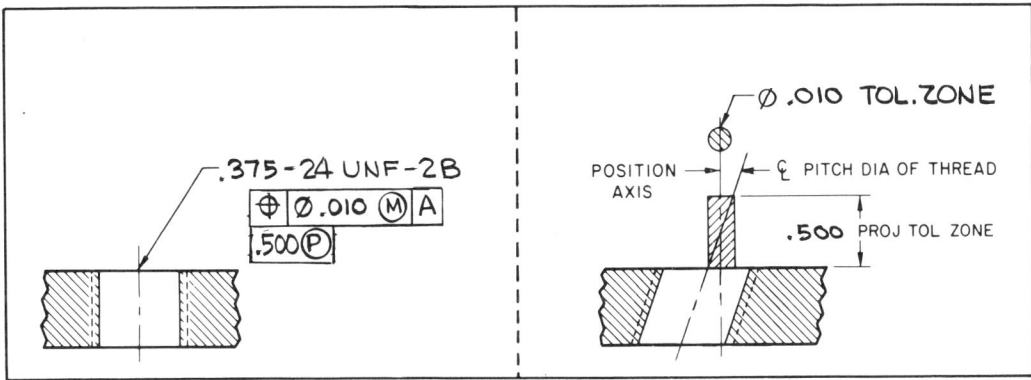

POSITION ⊕
ROUND HOLES—0 TOLERANCE (PERFECT POSITION AT MMC)

"Zero" (0) position tolerancing is a technique adaptable to situations requiring functional interchangeability and maximum tolerance advantage in the feature size, form, and position interrelationships. Where mating parts and features are simply to mate up or "GO" and tangent contact of the mating features could occur, zero tolerancing is technically acceptable.

However, under some conditions, zero position tolerancing is *not* appropriate. For example, where specific running clearances, fit, or similar special mating feature conditions are required, zero position tolerancing will not, in general, be technically applicable. There are other considerations, also, which require evaluation to determine whether or not zero position tolerancing is applicable.

For most position applications and examples shown elsewhere in this text, we could have used the zero position tolerancing method. It is an optional method for stating many common position mating part requirements.

As foregoing sections of this text have emphasized, position tolerances are usually established on the basis of MMC size relationships of mating part features. The feature sizes are the criteria with which the process of developing the position tolerances starts. The designed clearance between the mating components is the basis for the position tolerances which are stated on the drawing and applied in manufacture. When the features specified by position tolerances are actually produced, any size departure from the MMC size (e.g. enlarging size of a hole) adds to the permissible position tolerance.

In zero position tolerancing the same principles apply, except that the position tolerance stated is always a fixed "zero," with all the tolerance placed on the size dimension. This, of course, assumes that the actually produced feature will show some deviation from MMC, which is then added to the "zero" tolerance to give a working position/ form tolerance.

In either conventional or zero position tolerancing, size, form, and position variations are considered simultaneously as a composite value. This is really the fundamental principle (along with the MMC principle) on which functional position tolerancing is based. The reason for this is that related mating part features perform their function in the space limitations provided, regardless of whether that space is derived from size, form, or position variation.

There is some controversy about zero position tolerancing, with both proponents and opponents contributing valuable comments. However, a good understanding of the principles involved, their proper application, and an awareness of their limitations will go a long way toward ensuring proper use of this effective geometric tolerancing tool.

A forceful argument for the use of zero tolerancing arises in situations where a produced part with a positional hole pattern might be acceptable to a functional gage yet be rejectable on the basis of a low limit "GO" size violation, with the result that functionally good parts might be scrapped.

At this point, we wish to emphasize that in conventional position tolerancing, the stated size tolerance can be used for size, form, and position variations as the feature size departs from MMC, whereas a stated position tolerance may be used only for form and position variations. Size tolerance variations of the feature from MMC size can thus add to the position tolerance; but, according to standard practices, unused position variations cannot be added to size tolerances.

The above principle is best described by referring to the CONVENTIONAL POSITION TOLERANCE APPLICATION example which follows in this section of the text. The notation at the bottom of the illustration states that if the hole is produced in perfect location, its size will be permitted to exceed the low limit .255 (MMC) size down to the virtual condition size of .250. The virtual condition size is developed from the MMC size of the hole, .255, minus the stated position tolerance, .005. This is, of course, also the functional gage pin size, and represents the mating part feature at its extreme condition of assembly.

Obviously, not many parts will be produced in perfect position. However, the point is that when the position tolerance, or any portion of it, has *not* been used up, the *size* of the feature should reap the benefit (e.g. the hole should be allowed to get smaller an equal amount and yet be accepted as functional).

Note that the virtual condition size developed in the ZERO POSITION TOLERANCE example is also .250. In this method the virtual condition size and the MMC size are the *same*. The .005 position tolerance of the "conventional" example has been shifted to the size tolerance in the "0" method. The result for both parts, insofar as resulting position gaging is concerned, is the same; but the above-mentioned "rejectable" part (low limit size violation) will be acceptable under the zero method.

The .250 diameter, or virtual condition size, on both examples represents the most extreme condition of assembly at MMC, or the most extreme mating condition of the hole which would accept the mating part feature. Thus, the zero method recognizes, accounts for, and uses the full tolerance advantage. Size tolerances, in particular, are fully utilized with no loss of form or position tolerance.

Further analysis of zero tolerancing, however, provokes questions that tend to temper some of its advantages. First, for the less experienced or uninitiated user, zero tolerancing presents a psychological barrier: the zero may give a false impression of the "perfection" expected. Second, the designer may feel that he is relinquishing excessively broad discretion to the production departments, thus abdicating design responsibilities in favor of production conveniences such as larger size tolerances. Other comments heard from those hesitant to make full use of zero tolerancing are: it will be misunderstood, it costs more, it is only for gaging convenience, it may result in line-to-line fits.

These objections are primarily due to lack of understanding and experience. Improved understanding and repeated experience with practical applications will readily convince the user of the advantages of zero tolerancing.

Zero form tolerancing, occasionally used on perpendicularity, is not discussed here. However, examples of it are given in appropriate sections of the text. The reasoning underlying zero *form* tolerancing is the same as that on which zero position tolerancing is based except for the added consideration of position relationships. A position tolerance, of course, contains form tolerance. An understanding of the zero form control method, however, might contribute to better acceptance of the zero position principles.

POSITION ⊕
ZERO POSITION

To clarify the reasons underlying the use of the "zero" position method versus the "conventional" position method, further discussion is presented in Figs. 1 and 2.

In Fig. 1, one of the holes illustrated in the 0 method example is shown with reference to the gage pin (or simulated mating part component). It is seen that the zero (0) position specification requires a perfect part (perfect form and perfect position) when at MMC, or virtual condition size.

FIGURE 1

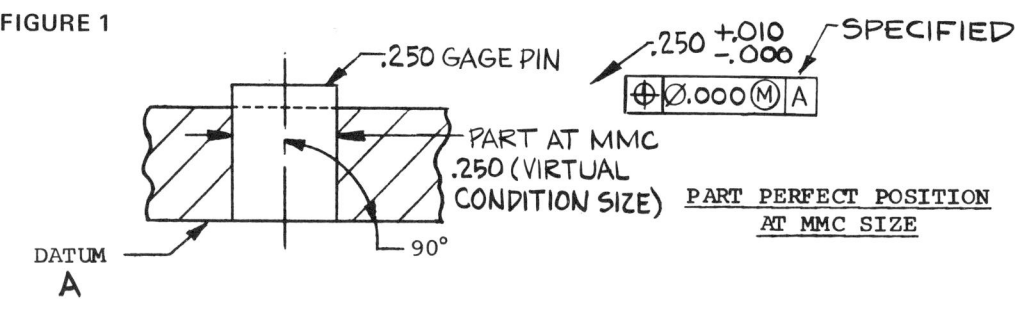

FIGURE 2

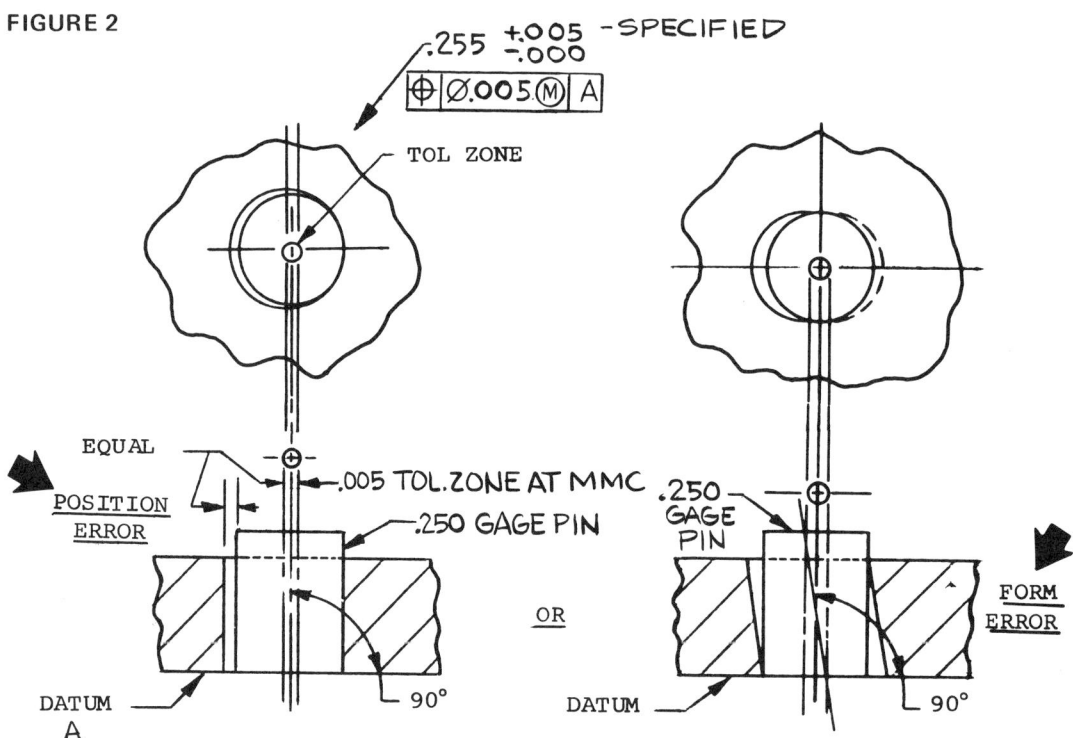

Since there must be *some* clearance between the hole and the inserted mating component or they will not assemble, there is an immediate deviation from the perfect "zero-clearance-zero interference" situation and some tolerance is acquired.

Further, since additional tolerance latitude is usually required to assure assembly, we may find it useful to establish a fixed position tolerance. Thus we calculate an acceptable tolerance on the basis of MMC sizes and use the "conventional" position method.

Figure 2 illustrates the "conventional" method and the established position tolerance. The tolerance of .005 will permit either position or form error (or a combination of both) to this extent, when the feature is at MMC. With the same size gage pin as in Fig. 1, we see that the position tolerance of .005, plus the size tolerance of .005, is equivalent to the .010 size tolerance obtained by the zero method in Fig. 1.

An application which will demonstrate the effectiveness of zero tolerancing is a very critical relationship of locating dowels relative to locating holes on a mating part.

As an example, imagine .2500 $^{+.0000}_{-.0015}$ on the locating dowels, and .2520 $^{+.0015}_{-.0000}$ on the locating holes. Using the conventional position "fixed fasteners" method, the calculations are:

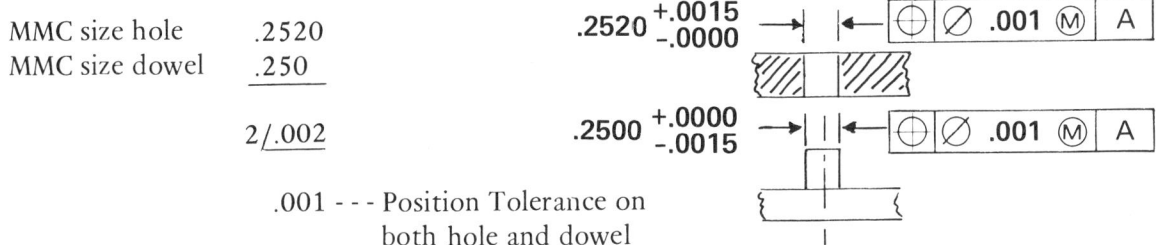

MMC size hole   .2520

MMC size dowel   .250

              2/.002

.001 - - - Position Tolerance on
both hole and dowel

The actual position tolerance in production on both parts would be somewhere between .001 and .0025 (increase due to MMC departure). A functional gage pin size to check the holes would be .251 (hole MMC .252, MINUS POSN TOL .001 = .251). Since the gage pin represents the worst condition (virtual condition size) of the mating dowel at .251, the hole size could be acceptable functionally at .251; yet this exceeds the stated hole size low limit.

The dowel size, too, could be functional at .251 (dowel MMC .250, PLUS .001 = .251) which represents the mating part hole at the extreme condition (virtual condition size). This exceeds the stated dowel size high limit.

However, the zero position method can provide more total tolerance and yet guarantee proper control if stated as:

.2510 $^{+.0000}_{-.0025}$ (dowel)   and   .2510 $^{+.0025}_{-.0000}$ (hole).

A comparison of the two methods in terms of the full tolerance range difference between the hole and dowel which determines usable size, form, and position tolerance is shown on the following page.

POSITION TOLERANCE ⊕

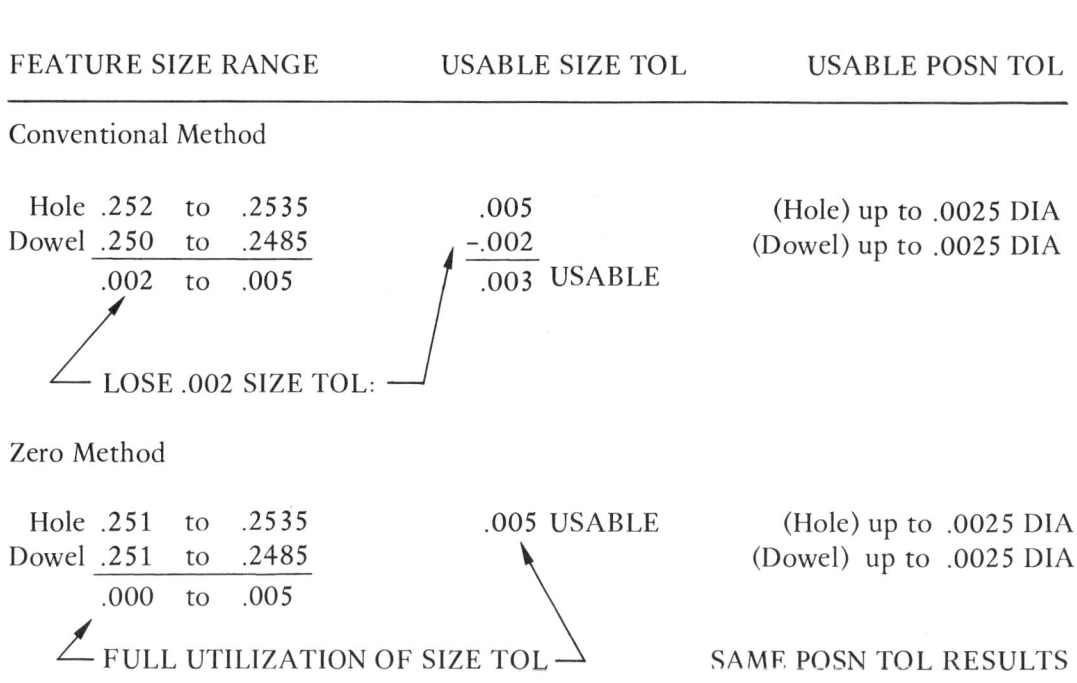

FEATURE SIZE RANGE	USABLE SIZE TOL	USABLE POSN TOL

Conventional Method

Hole .252 to .2535 — .005
Dowel .250 to .2485 — −.002
.002 to .005 — .003 USABLE

(Hole) up to .0025 DIA
(Dowel) up to .0025 DIA

LOSE .002 SIZE TOL:

Zero Method

Hole .251 to .2535 — .005 USABLE
Dowel .251 to .2485
.000 to .005

(Hole) up to .0025 DIA
(Dowel) up to .0025 DIA

FULL UTILIZATION OF SIZE TOL

SAME POSN TOL RESULTS

The foregoing example assumes that there is only a remote probability that both hole and dowel will actually be produced at exactly .251; hence some clearance will be present. However, if this should cause concern and some compensation is desired, a slight adjustment of specified size limits (and virtual condition size) can be made to eliminate the problem and ensure that no line-to-line condition will result.

As an alternative to the zero position tolerancing method, some companies have established in-house standards which permit the use of a functional gage as a hole "GO" size gage, thus allowing, for example, the low limit size of a hole to deviate simulating the zero method and the mating part relationship. However, transfer of work to outside sources would then require special documentation to assure that specifications are properly interpreted.

The illustrations on the following pages further illustrate the principles described on page 178. Note that the "conventional" method establishes a fixed position tolerance which presents the possibility of a part being *acceptable* to a functional gage but *rejectable* due to exceeding the low limit (MMC) size. The zero method on page 182, however, eliminates that problem. It does, however, retain the inherent considerations previously discussed. Use of zero position tolerancing should be based upon the best balance of design/production objectives.

CONVENTIONAL POSITION TOLERANCE APPLICATION FOR COMPARISON WITH ZERO POSITION TOLERANCE

POSITION TOLERANCE (CONVENTIONAL)

AS DRAWN

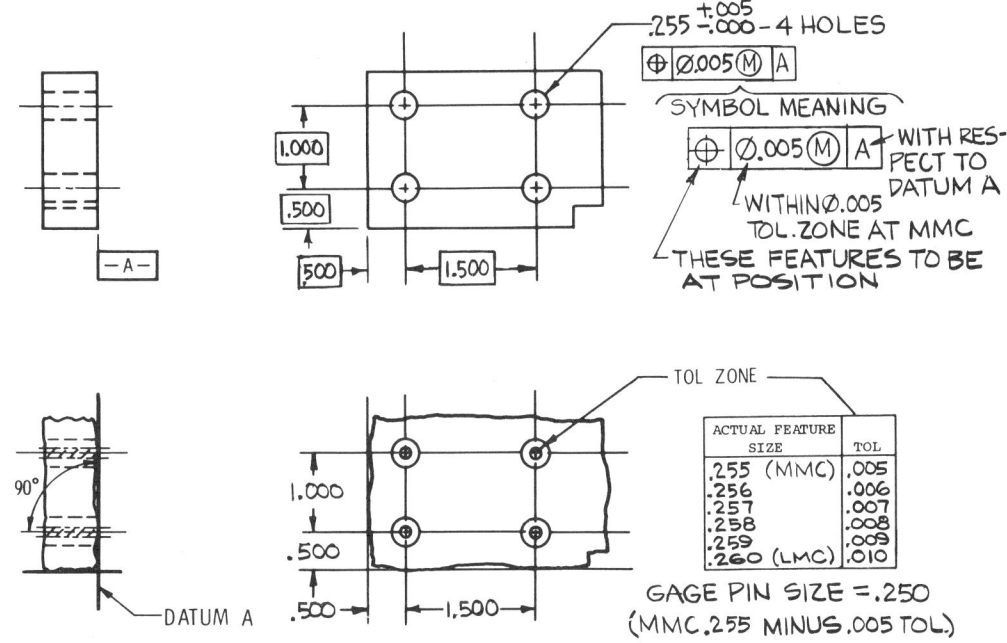

MEANING

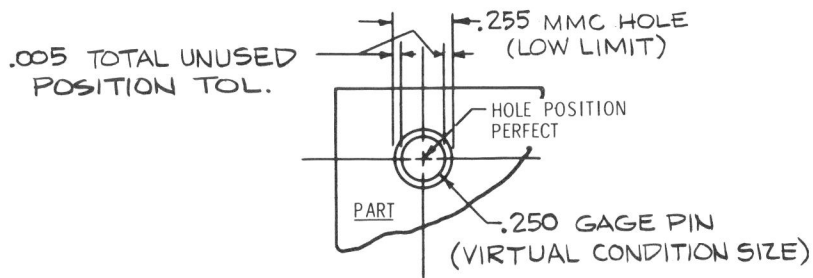

ASSUMING THAT THE GAGE PIN REPRESENTS THE MOST EXTREME MATING CONDITION, AS *POSITION* APPROACHES PERFECT, IT IS EVIDENT THAT THE HOLE SIZE COULD GO DOWN TO .250 (.005 BELOW .255 LOW LIMIT OF HOLE) AND STILL PASS THE GAGE PINS, HOWEVER, PARTS BELOW THE LOW (MMC) LIMIT HOLE SIZE OF .255 WOULD BE REJECTED ON *SIZE* YET ARE GOOD PARTS.

POSITION TOLERANCE ⊕
ZERO POSITION TOLERANCE

AS DRAWN

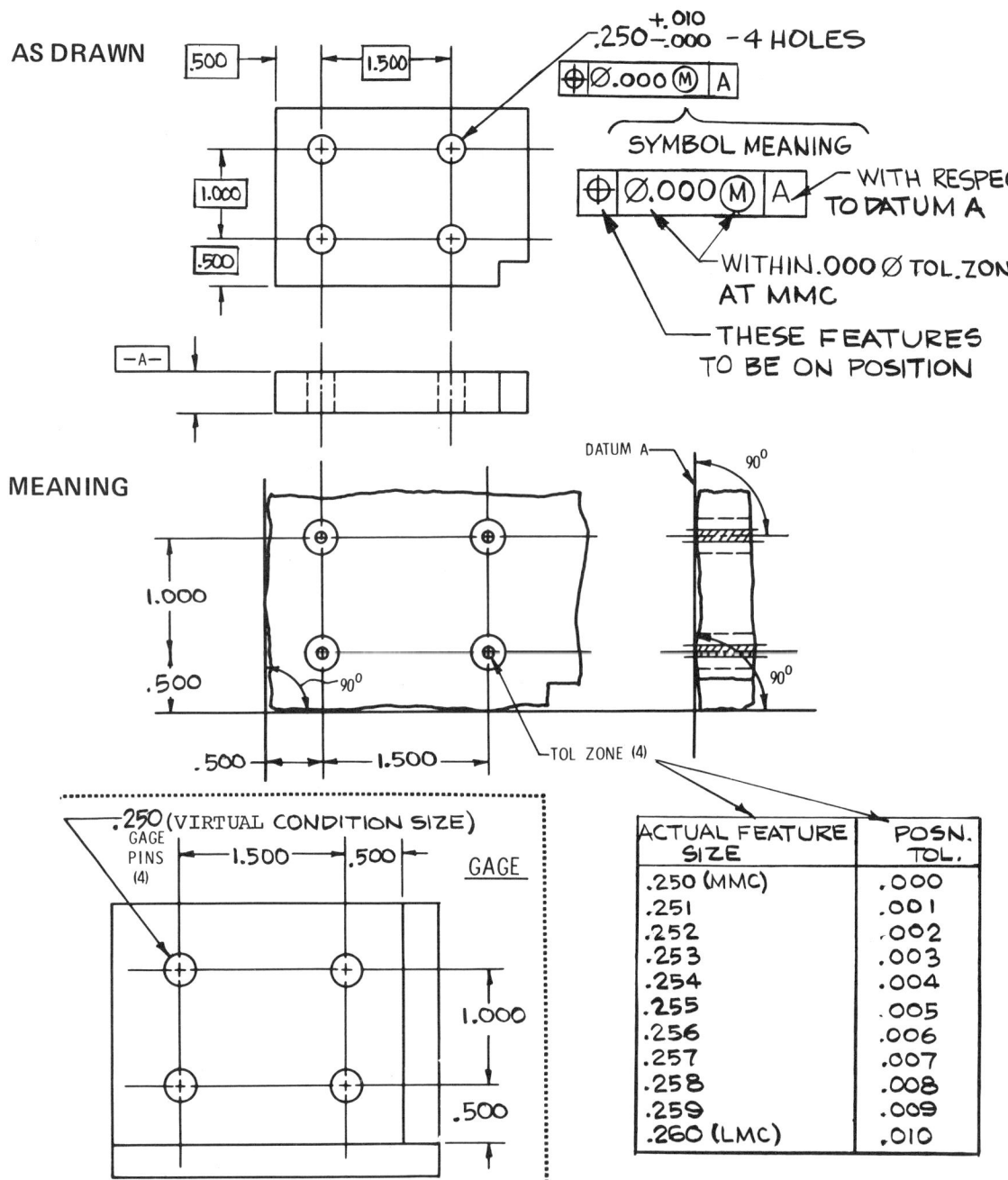

$.250^{+.010}_{-.000}$ – 4 HOLES

⊕ | Ø.000 Ⓜ | A

SYMBOL MEANING

⊕ | Ø.000 Ⓜ | A — WITH RESPECT TO DATUM A

— WITHIN .000 Ø TOL. ZONE AT MMC

— THESE FEATURES TO BE ON POSITION

MEANING

DATUM A — 90°

TOL ZONE (4)

ACTUAL FEATURE SIZE	POSN. TOL.
.250 (MMC)	.000
.251	.001
.252	.002
.253	.003
.254	.004
.255	.005
.256	.006
.257	.007
.258	.008
.259	.009
.260 (LMC)	.010

.250 (VIRTUAL CONDITION SIZE)
GAGE PINS (4)   1.500   .500   GAGE

ADVANTAGES OF ZERO POSITIONAL TOLERANCE

1. *Hole "GO" plug gage not needed* if using "GO" functional gage which checks positional location of pattern and hole low limit size simultaneously.

2. *No unused positional tolerance* when using zero positional tolerance method. As locations approach perfection under a conventional (specified pos tol e.g., .005) position application, the unused positional tolerance *cannot* be added to the *size* tolerance. Therefore, under some conditions good parts are rejected as they exceed the low limit size.

POSITION TOLERANCE
COAXIAL FEATURES-ZERO TOLERANCE

Zero position tolerancing as illustrated in the preceding examples may also be applied to coaxial features.

Where position relationships between coaxial features must be very exact, zero tolerancing may provide the necessary control. As in other position applications, it is appropriate for mating parts. It has the advantage of fully utilizing *size* tolerances as these interrelate with form and position relationships.

The example on the next page shows zero tolerancing applied to a simple part. Position tolerance is acquired as the feature sizes (feature and datum feature) depart from MMC. When the part is everywhere at its MMC size, it must be perfectly located. However, in normal applications, the sizes produced are somewhere within the size tolerance range and thus develop allowable positional tolerance. The actually produced sizes establish the position tolerance to the amount of size departure from MMC.

Note that the sizes of both the feature *and* the datum have an effect on the total relationship and the developed tolerance. Just as in all applications of position tolerancing, the part must be produced and its size established before the position tolerance is fixed.

A functional or receiver gage is also illustrated. The principles involved are identical to those in the previous examples. However, with the zero method, the high limit "GO" size of the diameter is also checked, along with the position requirement.

For this application, the use of maximum size tolerances and functional principles has been facilitated by the zero method.

POSITION TOLERANCE ⊕
COAXIAL FEATURES—ZERO TOLERANCE

AS DRAWN

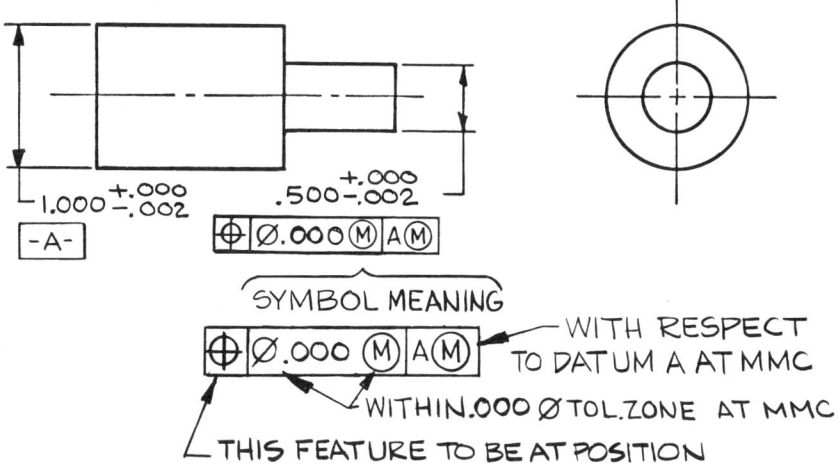

MEANING

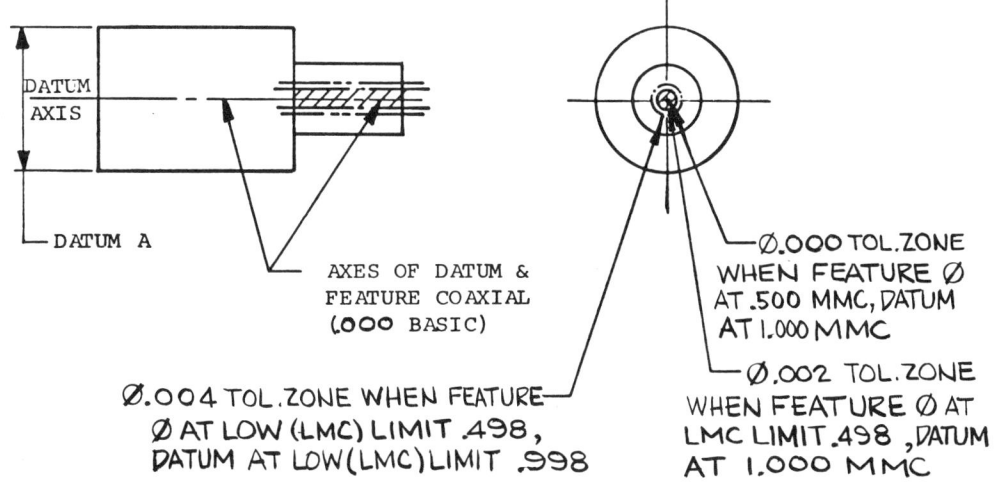

GAGE

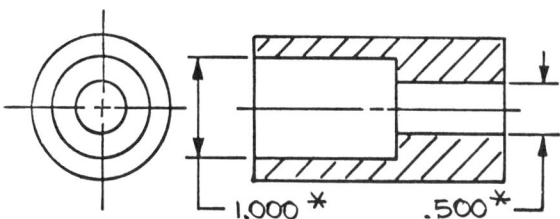

* GAGE MAKERS TOL APPLY

POSITION $\oplus$
MATING PARTS, FIXED FASTENERS

The following series of figures further illustrates mating part situations and the use of the fixed fastener method of calculation. It covers the typical applications listed below.

Application	Page
Base (tapped hole), gasket, cover (clearance hole)	186
Base (tapped hole), cover (clearance hole)	187
Base (tapped hole), cover (clearance hole, Ctsks.)	188
Base (tapped hole, vent hole), cover (clearance holes)	189
Housing (tapped hole), cover (clearance hole)	190
Housing (tapped hole, locating dowels), cover (locating holes)	191

POSITION ⊕
MATING PARTS, FIXED FASTENERS

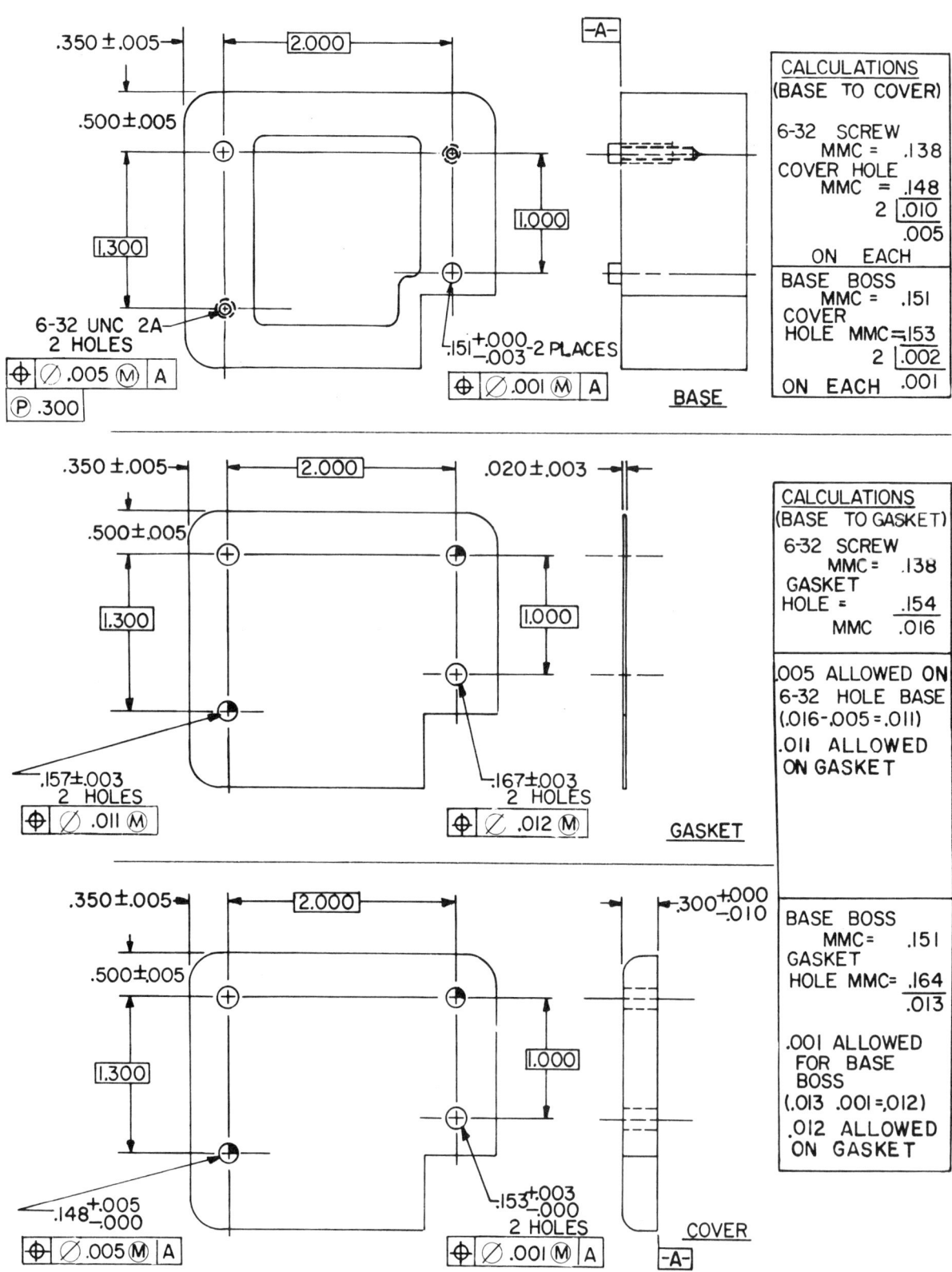

BASE

.350 ± .005 2.000 .500 ± .005 1.300 1.000

6-32 UNC 2A
2 HOLES

⊕ | ⌀ .005 Ⓜ | A
Ⓟ .300

.151 +.000 −.003 -2 PLACES

⊕ | ⌀ .001 Ⓜ | A

−A−

CALCULATIONS
(BASE TO COVER)

6-32 SCREW
  MMC = .138
COVER HOLE
  MMC = .148
        2 .010
          .005
  ON EACH

BASE BOSS
  MMC = .151
COVER
HOLE MMC = .153
        2 .002
ON EACH .001

GASKET

.350 ± .005 2.000 .020 ± .003 .500 ± .005 1.300 1.000

.157 ± .003
2 HOLES

⊕ | ⌀ .011 Ⓜ

.167 ± .003
2 HOLES

⊕ | ⌀ .012 Ⓜ

CALCULATIONS
(BASE TO GASKET)
6-32 SCREW
  MMC = .138
GASKET
HOLE =   .154
  MMC   .016

.005 ALLOWED ON
6-32 HOLE BASE
(.016 − .005 = .011)
.011 ALLOWED
ON GASKET

COVER

.350 ± .005 2.000 .300 +.000 −.010 .500 ± .005 1.300 1.000

.148 +.005 −.000

⊕ | ⌀ .005 Ⓜ | A

.153 +.003 −.000
2 HOLES

⊕ | ⌀ .001 Ⓜ | A

−A−

BASE BOSS
  MMC = .151
GASKET
HOLE MMC = .164
          .013

.001 ALLOWED
  FOR BASE
  BOSS
(.013 .001 = .012)
.012 ALLOWED
ON GASKET

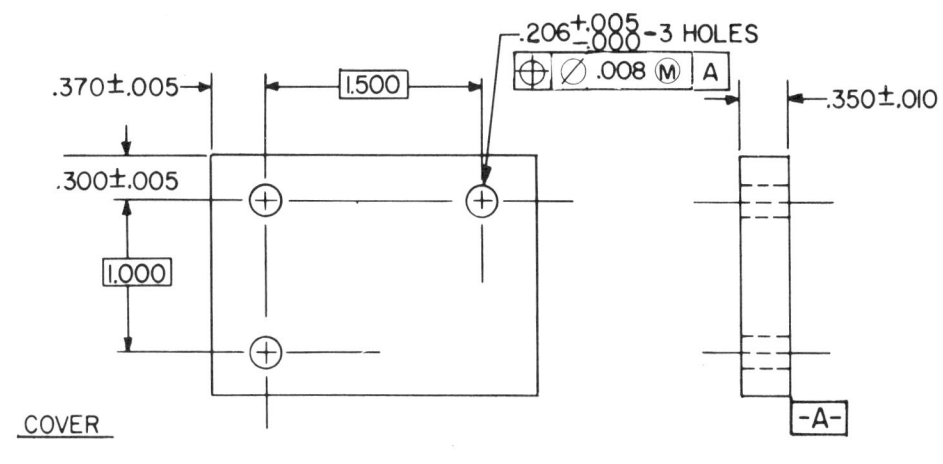

COVER

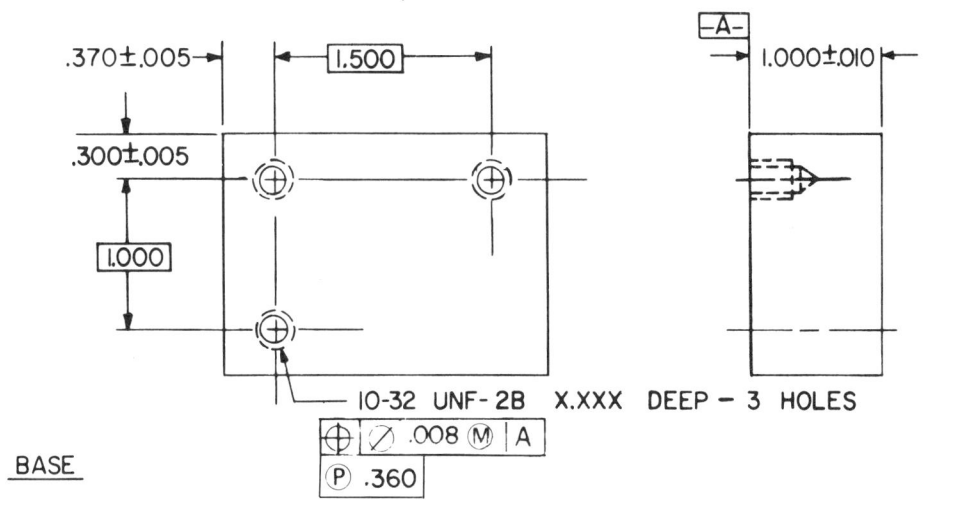

BASE

ASSEMBLY

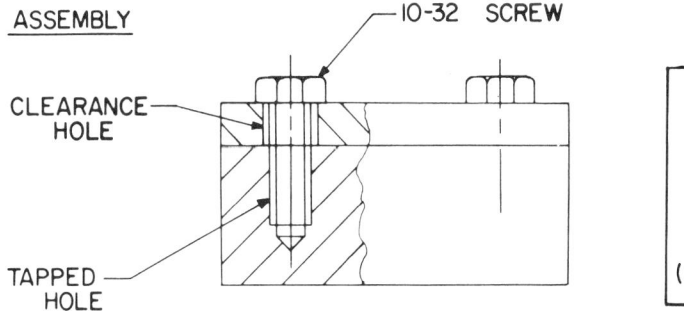

10-32 SCREW

CLEARANCE
HOLE

TAPPED
HOLE

CALCULATIONS	
MMC HOLE (COVER)	.206
MMC 10-32 (SCREW)	(-1.190
	2 .016
BOTH PARTS POS TOL .008	
(OR .010 & .006, .009 & .007 ETC.)	

CLEARANCE BETWEEN SCREW & TAPPED HOLE IS DISREGARDED
NORMALLY BECAUSE OF CENTERING EFFECT OF SCREW IN TAPPED HOLE.
CENTER (AXIS) OF SCREW & TAPPED HOLE CONSIDERED
COINCIDENTAL AT ASSEMBLY.

POSITION ⊕

MATING PARTS, FIXED FASTENERS

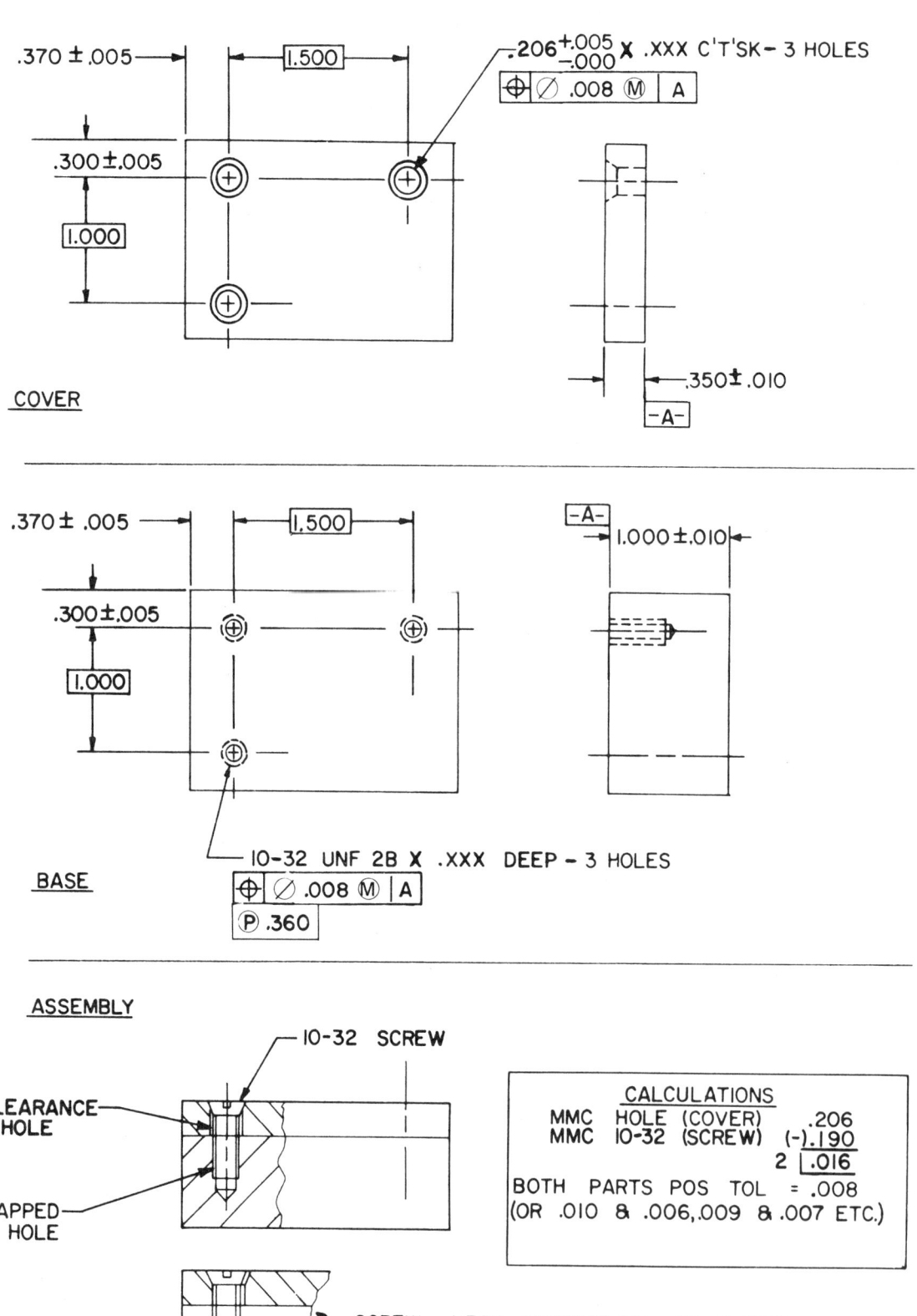

.370 ± .005

1.500

.206 +.005 -.000 X .XXX C'T'SK – 3 HOLES

⊕ | ∅ .008 Ⓜ | A

.300 ± .005

1.000

.350 ± .010

-A-

COVER

.370 ± .005

1.500

-A-

1.000 ± .010

.300 ± .005

1.000

10-32 UNF 2B X .XXX DEEP – 3 HOLES

⊕ | ∅ .008 Ⓜ | A

Ⓟ .360

BASE

ASSEMBLY

10-32 SCREW

CLEARANCE HOLE

TAPPED HOLE

CALCULATIONS

MMC HOLE (COVER)		.206
MMC 10-32 (SCREW)	(-)	.190
2		.016

BOTH PARTS POS TOL = .008
(OR .010 & .006,.009 & .007 ETC.)

SCREW HEAD PROBABILITY OF BOTTOMING AS SHOWN RECOGNIZED.

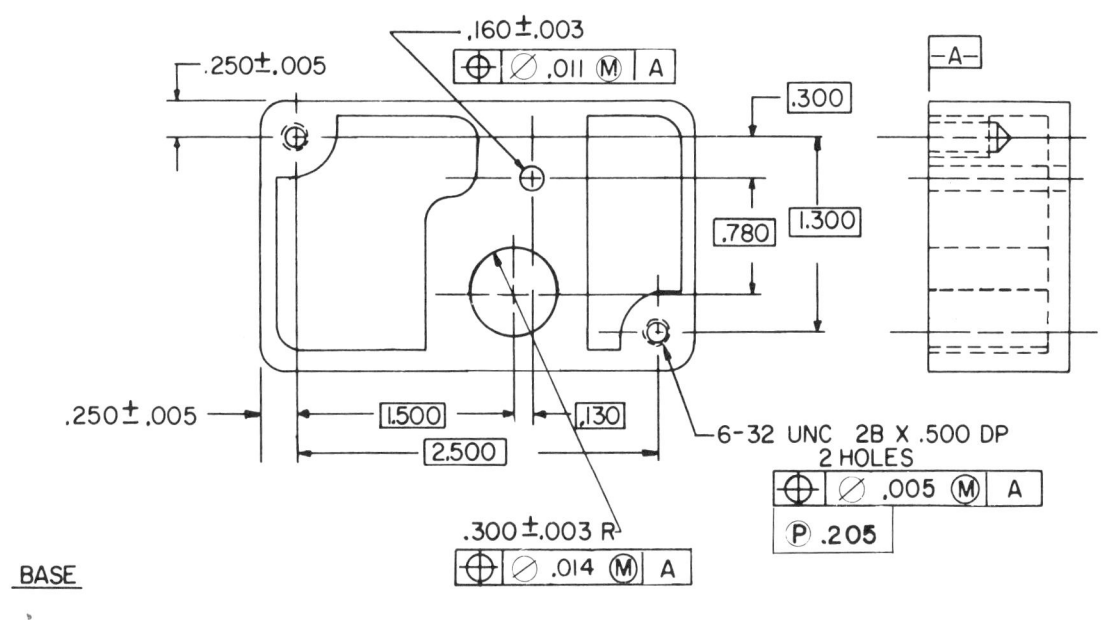

BASE

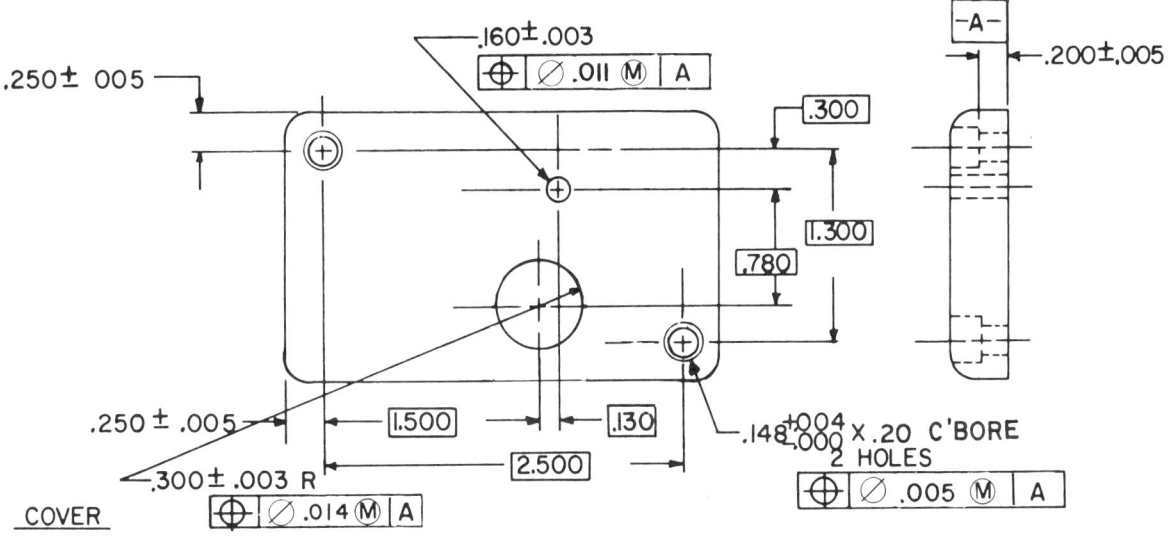

COVER

CALCULATIONS		

MMC 6-32 .138
MMC HOLE .148
2 |.010
.005 DIA
TP .005 ON EACH PART
(COULD BE .006 &.004,
.007 & .003, ETC.)
	AIR VENT HOLES TO
	LINE UP TO MINIMUM
	DIA. OF .146
	  MMC HOLES .157
	  MIN DIA. .146
	           .011 DIA
	ON EACH PART
	(COULD BE .009 &.013,
	ETC.)
		RADII TO CLEAR .580 DIA.
		MMC R .297= .594 DIA.
		           .580 MIN. DIA.
		           .014 DIA
		ON EACH PART
		(COULD BE .018 &.010,
		ETC.)

Imagine .146 & .580 DIAS as columns of air (floating fasteners).
Floating fastener method may be used since overlap of edges is
permissible; one dia need not clear the other so long as the
.146 & .580 DIAS will pass.

POSITION ⊕

MATING PARTS, FIXED FASTENERS AND FLOATING FASTENERS

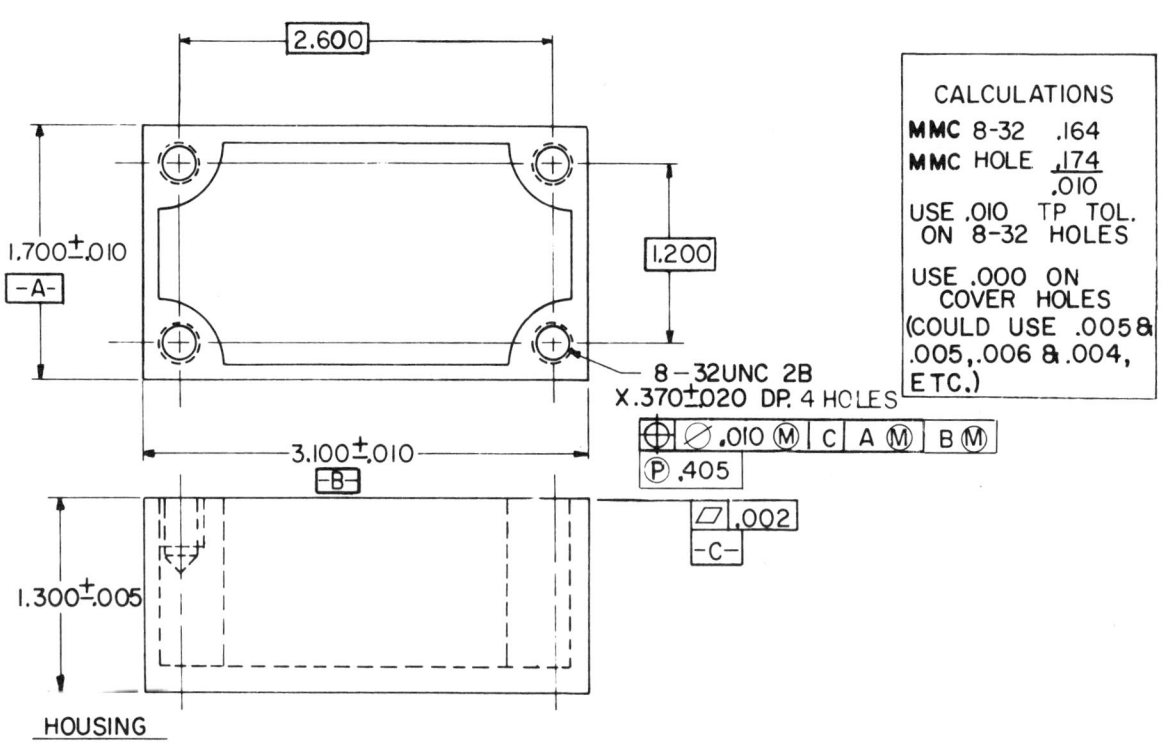

CALCULATIONS
MMC 8-32 .164
MMC HOLE .174
.010
USE .010 TP. TOL.
ON 8-32 HOLES

USE .000 ON
COVER HOLES
(COULD USE .005 &
.005,.006 & .004,
ETC.)

HOUSING

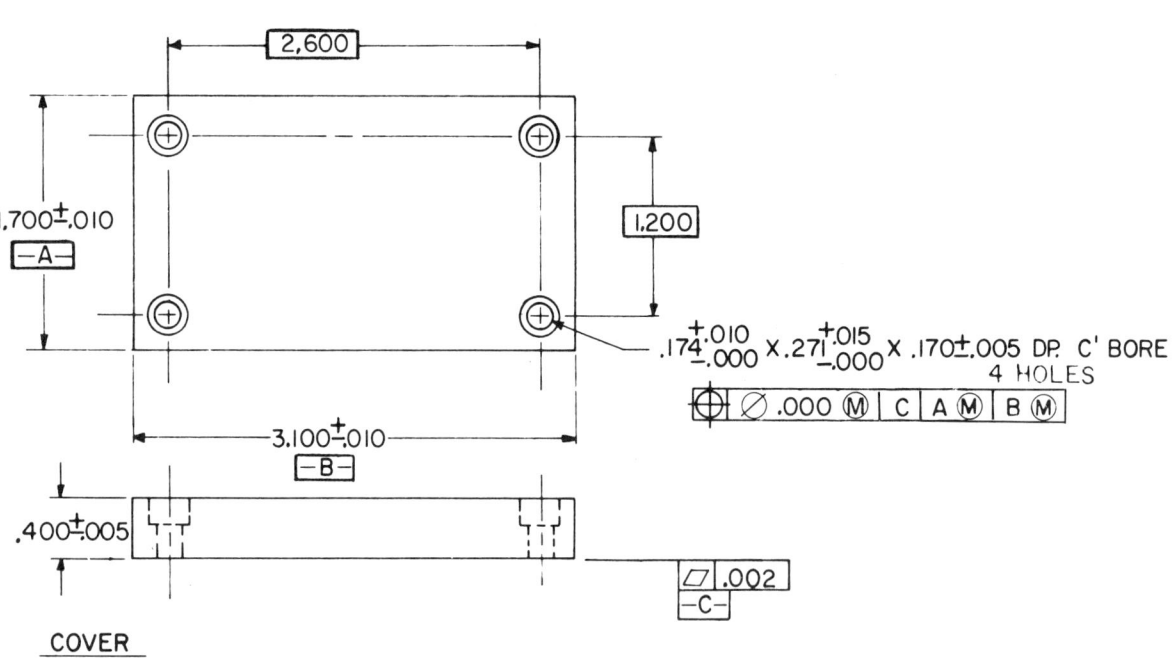

COVER

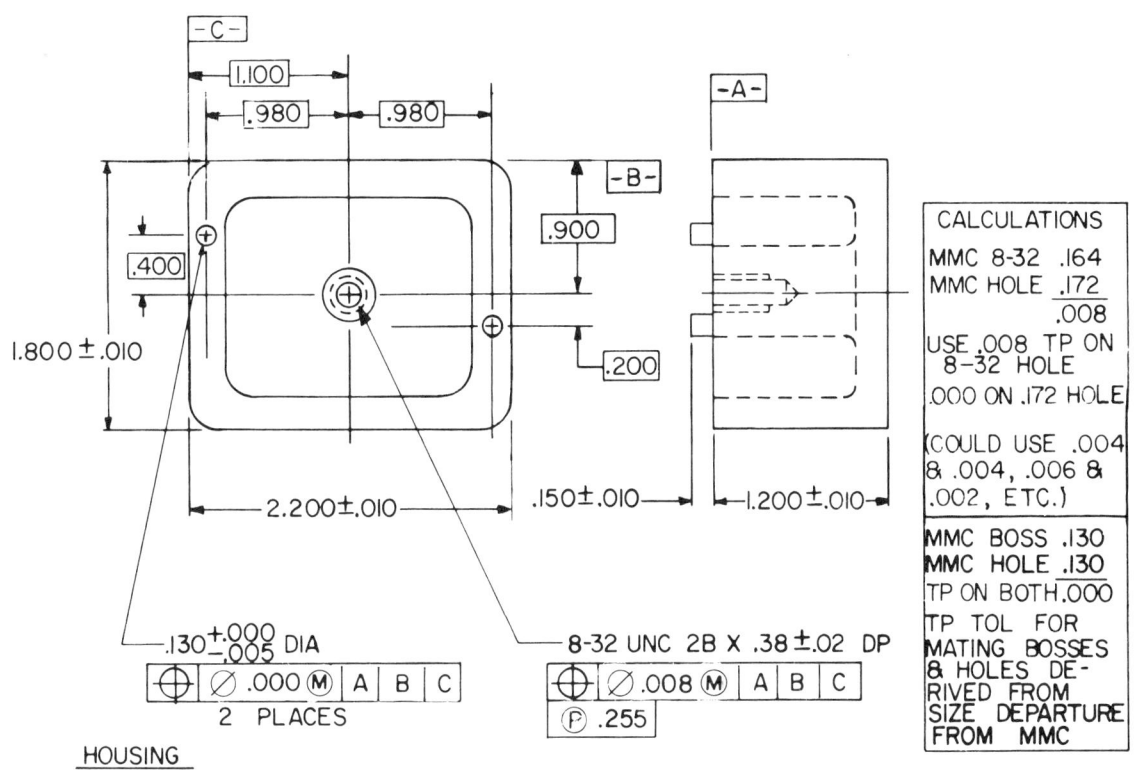

CALCULATIONS

MMC 8-32 .164
MMC HOLE .172
.008

USE .008 TP ON 8-32 HOLE
.000 ON .172 HOLE

(COULD USE .004 & .004, .006 & .002, ETC.)

MMC BOSS .130
MMC HOLE .130
TP ON BOTH.000
TP TOL FOR MATING BOSSES & HOLES DERIVED FROM SIZE DEPARTURE FROM MMC

.130 +.000 -.005 DIA

⊕ | ⌀ .000 Ⓜ | A | B | C
2 PLACES

8-32 UNC 2B X .38 ±.02 DP

⊕ | ⌀ .008 Ⓜ | A | B | C
Ⓟ .255

HOUSING

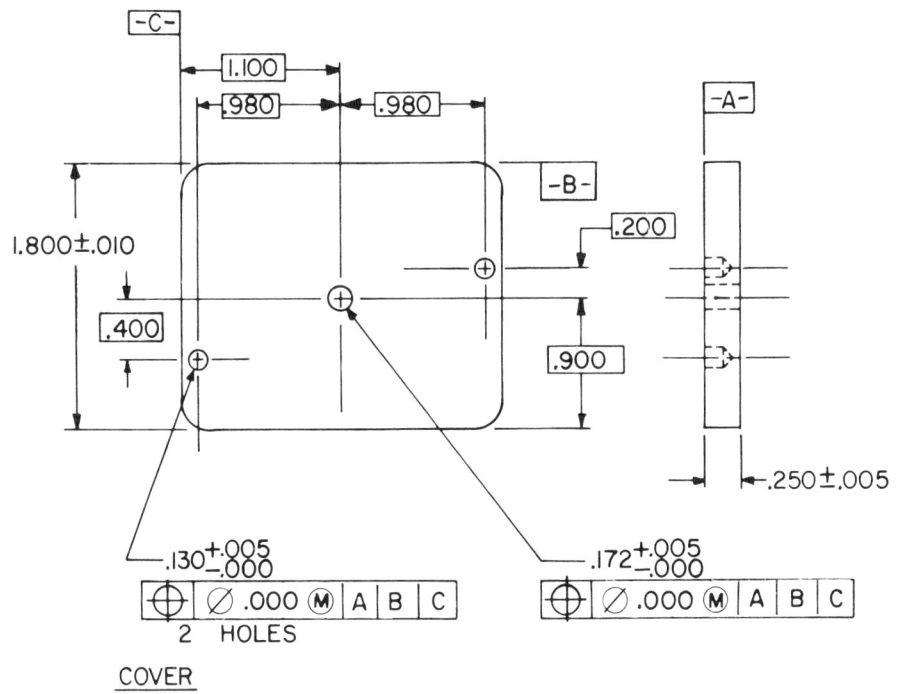

.130 +.005 -.000

⊕ | ⌀ .000 Ⓜ | A | B | C
2 HOLES

.172 +.005 -.000

⊕ | ⌀ .000 Ⓜ | A | B | C

COVER

191

POSITION ⊕
COMBINATION AND UNIQUE APPLICATIONS

Position tolerancing may be applied in numerous ways. The principles are extended in the following illustrations to unique combinations of DIA and TOTAL tolerances, square holes, etc. Note that the established principles can be adapted to many different situations.

Elongated holes 195
Features symmetrically located 197
Pattern and feature orientation from datum surfaces 196
Round holes and elongated holes 194
Round holes: greater positional tolerance in one direction
    than in the other 193
Square holes 198
Square holes: greater tolerance in one direction
    than in the other 199

ROUND HOLES, GREATER POSITIONAL TOLERANCE IN ONE WAY THAN IN THE OTHER

AS DRAWN

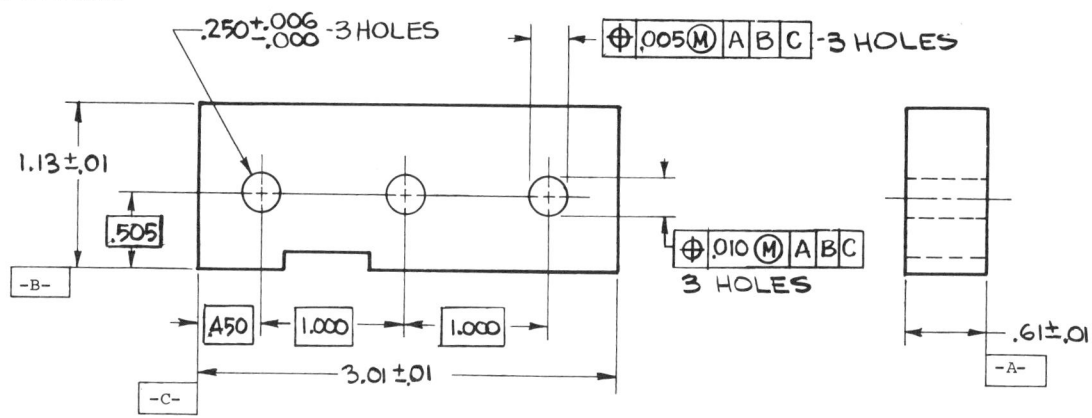

MEANING

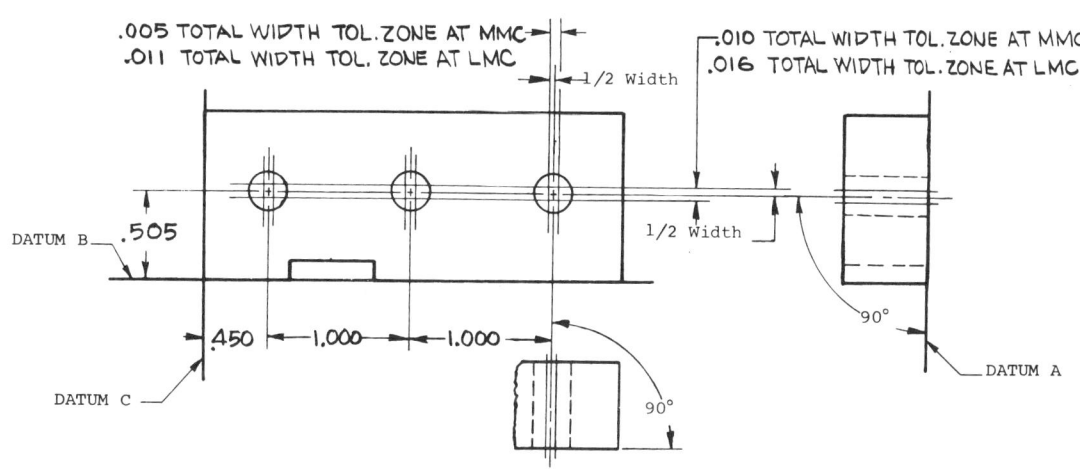

GAGE

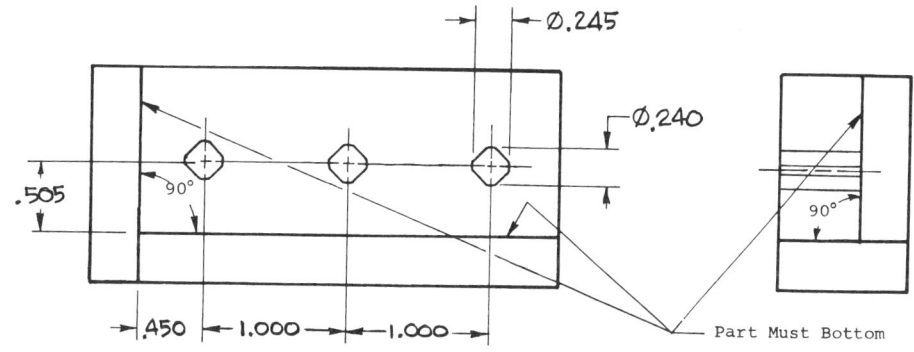

NOTE: GAGE MAKERS TOL APPLY

POSITION ⊕
DIAMETER AND TOTAL WIDE TOLERANCE ZONES, ROUND HOLES AND ELONGATED HOLES

AS DRAWN

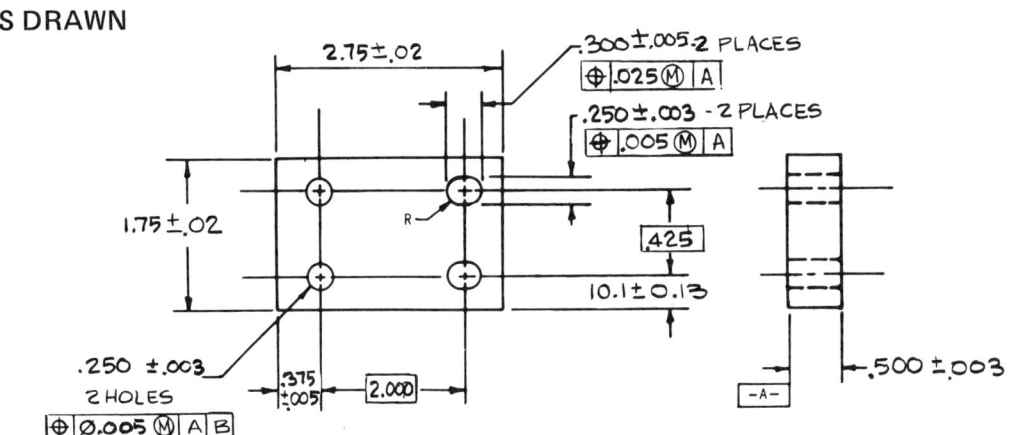

MEANING

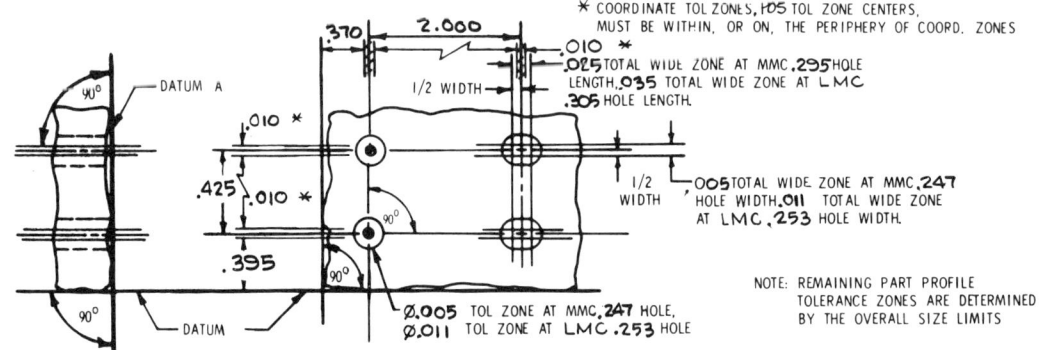

GAGE (FOR POSITION PATTERN)

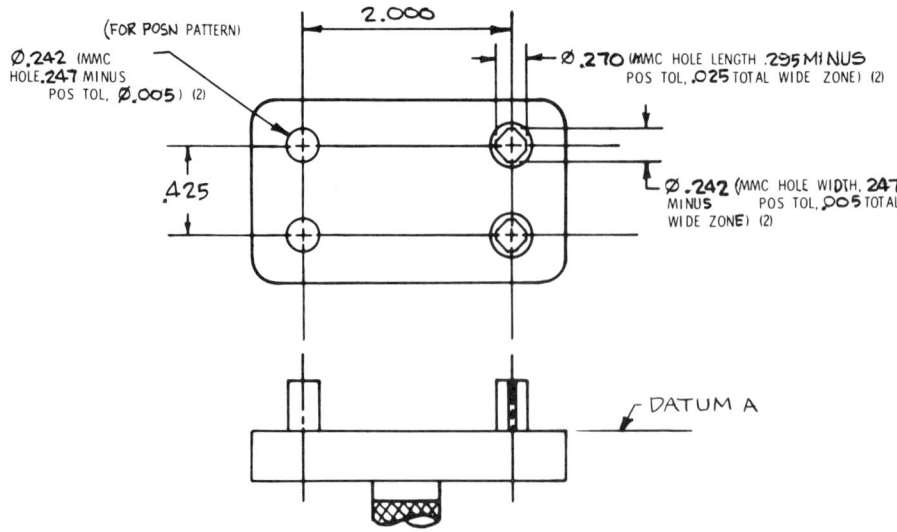

NOTE: GAGE MAKERS TOL. APPLY

194

ELONGATED HOLES, TOTAL WIDE TOLERANCE ZONES

AS DRAWN

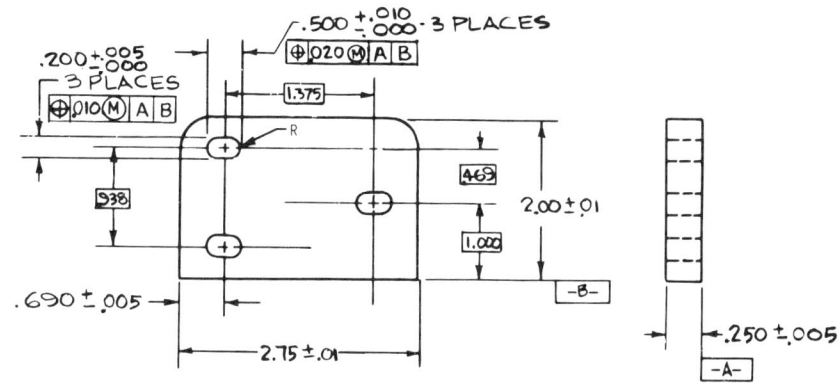

MEANING

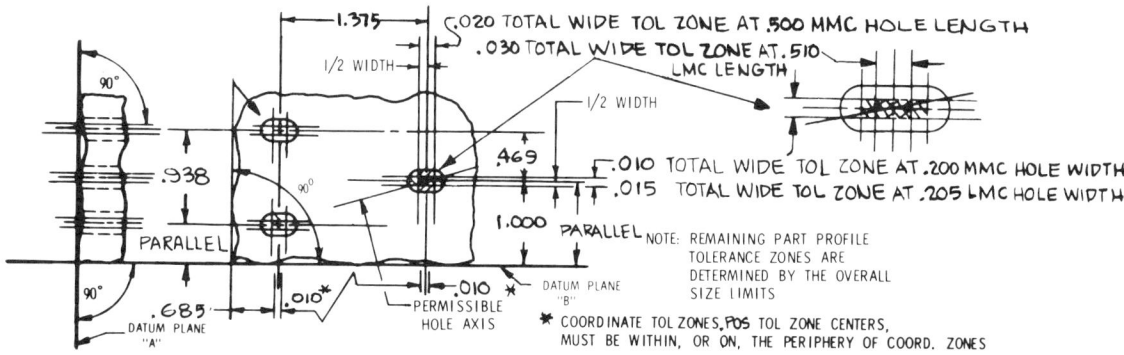

GAGE (FOR POSN. PATTERN)

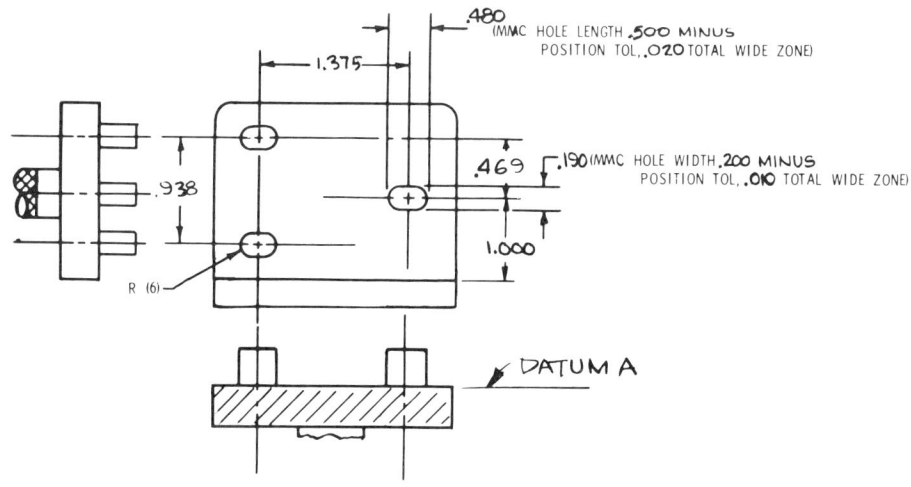

NOTE: GAGE MAKERS TOL. APPLY

POSITION ⊕

PATTERN AND FEATURE ORIENTATION FROM DATUM SURFACES

USING PART PROFILE AS DATUM

AS DRAWN

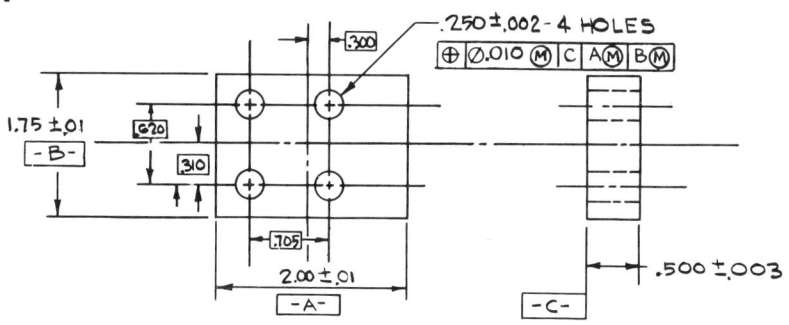

MEANING

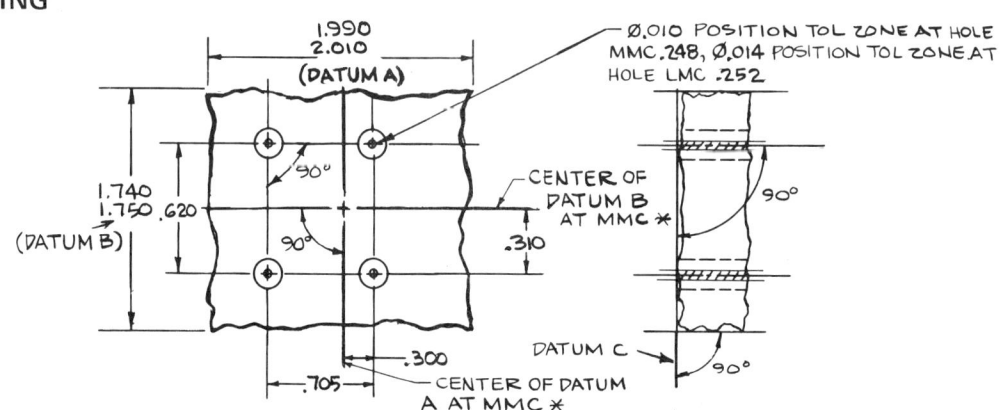

✳ CENTERS MAY SHIFT WITHIN A ZONE EQUAL TO THE DEPARTURE OF DATUM FEATURES A & B SIZES FROM MMC.

GAGE

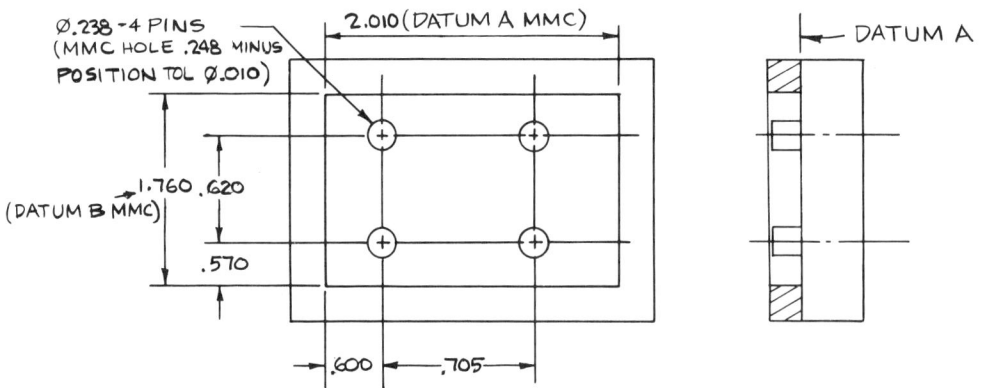

NOTE: GAGE MAKERS TOL APPLY

TOTAL WIDE TOLERANCE ZONES, FEATURES TO BE SYMMETRICALLY LOCATED

AS DRAWN

WITH GREATER TOL ONE WAY THAN THE OTHER

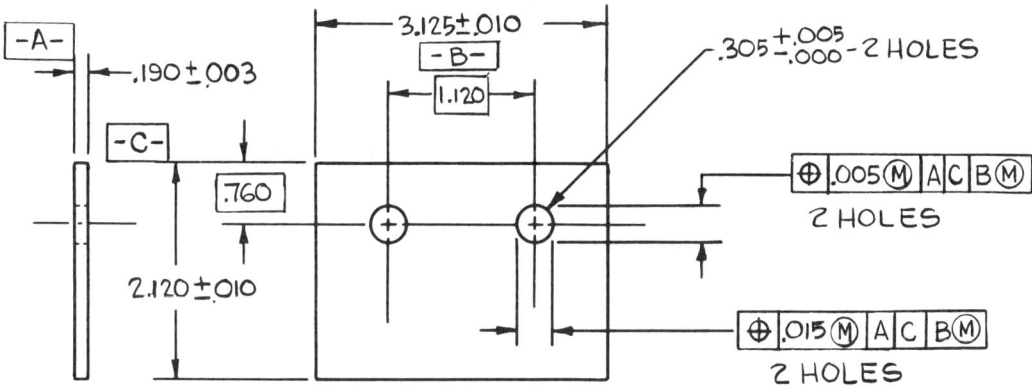

MEANING

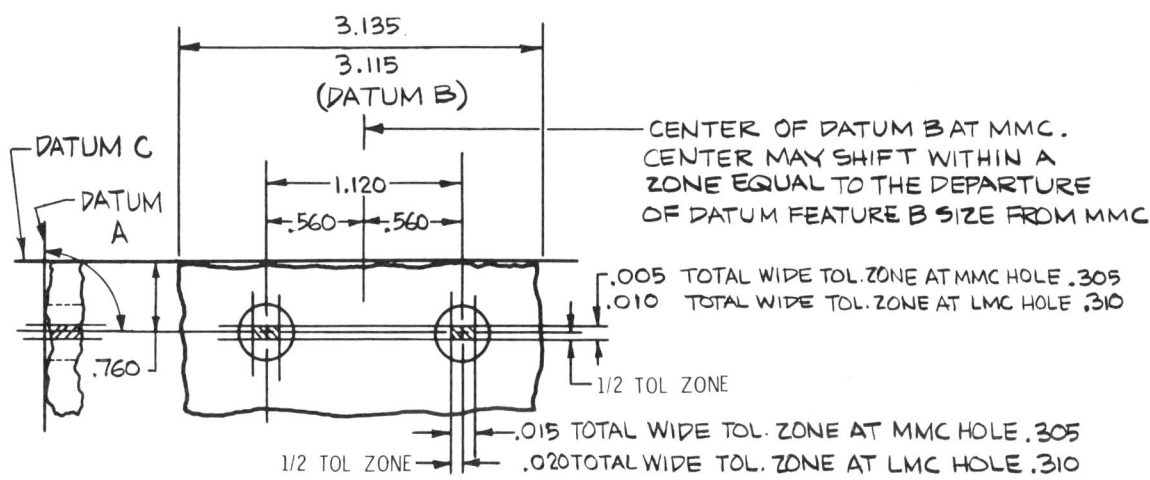

GAGE

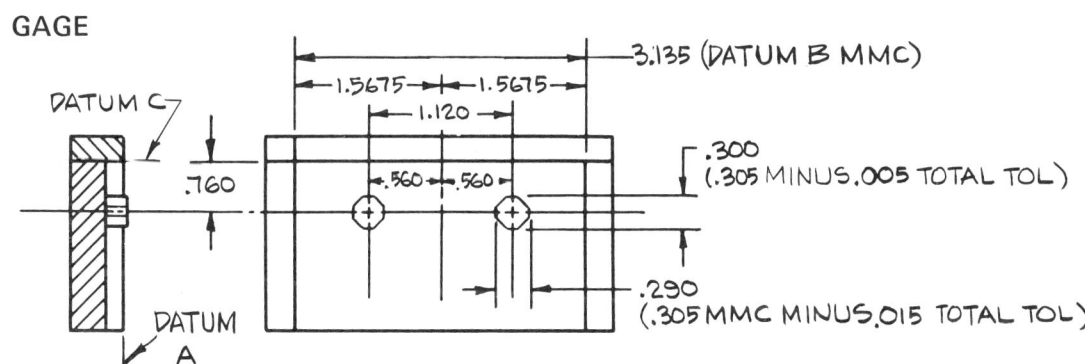

NOTE: GAGE MAKERS TOL. APPLY

POSITION ⊕

SQUARES HOLES, TOTAL WIDE ZONE POSITIONAL TOLERANCE

AS DRAWN

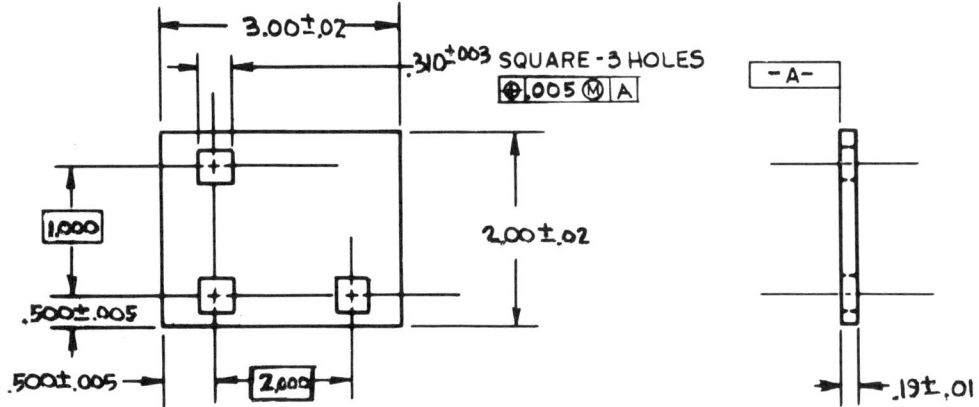

MEANING

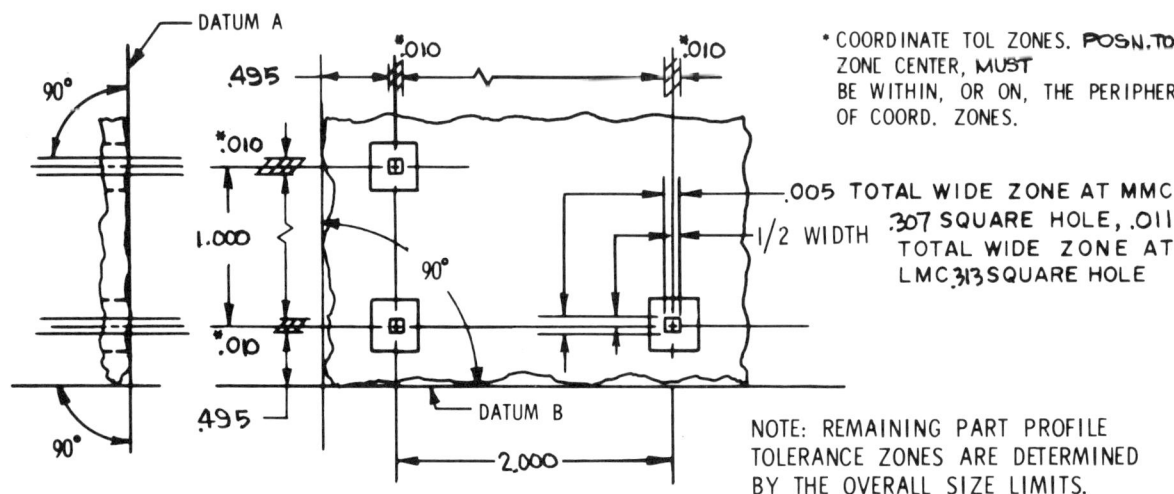

* COORDINATE TOL ZONES. POSN.TOL ZONE CENTER, MUST
BE WITHIN, OR ON, THE PERIPHERY
OF COORD. ZONES.

.005 TOTAL WIDE ZONE AT MMC
.307 SQUARE HOLE, .011
TOTAL WIDE ZONE AT
LMC .313 SQUARE HOLE

NOTE: REMAINING PART PROFILE
TOLERANCE ZONES ARE DETERMINED
BY THE OVERALL SIZE LIMITS.

GAGE (FOR POSN. PATTERN)

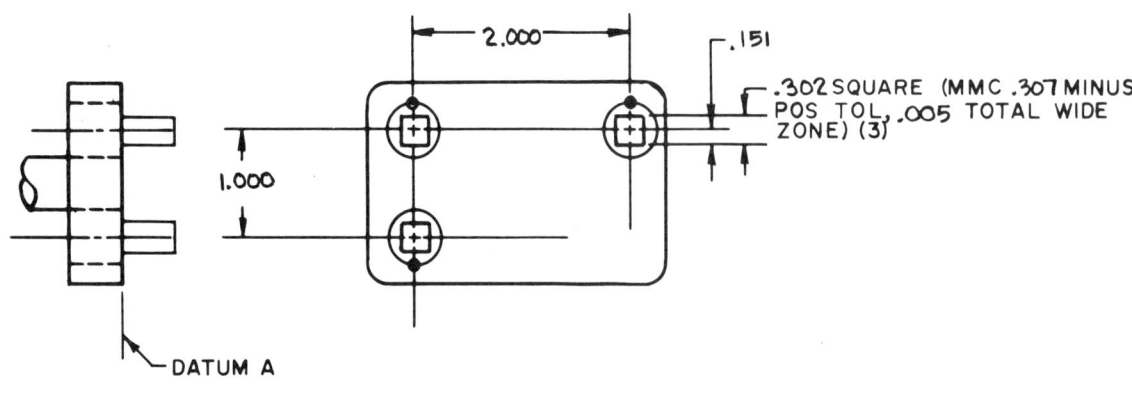

NOTE: GAGE MAKERS TOL APPLY

198

SQUARE HOLES, GREATER POSITIONAL TOLERANCE ONE WAY THAN ANOTHER

AS DRAWN

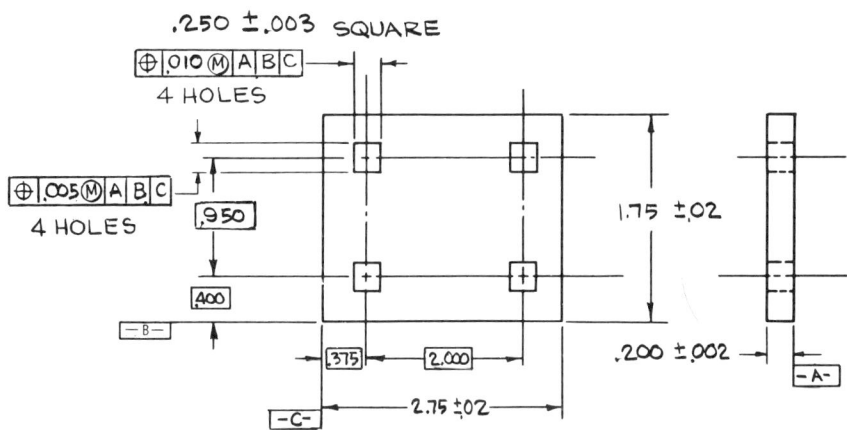

MEANING

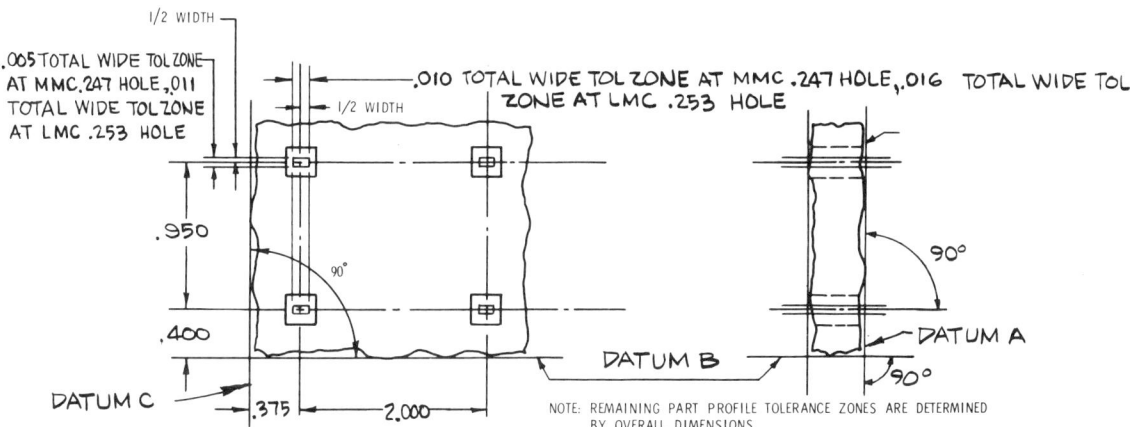

NOTE: REMAINING PART PROFILE TOLERANCE ZONES ARE DETERMINED BY OVERALL DIMENSIONS.

GAGE

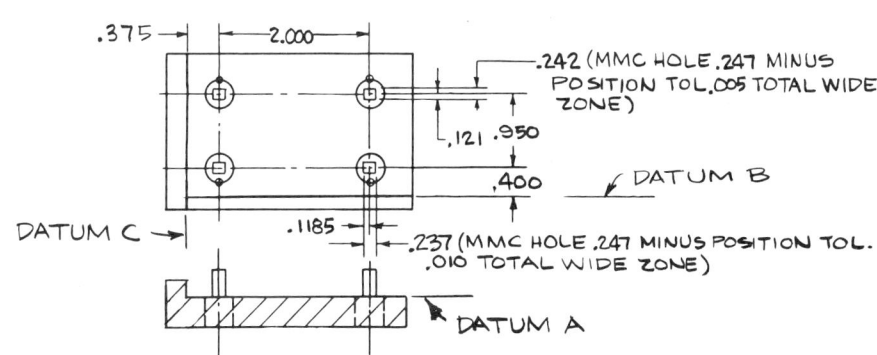

NOTE: GAGE MAKERS TOL. APPLY

DATUMS

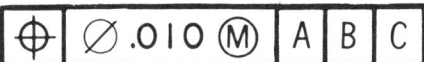

DATUMS

In the previous Sections we have made continual reference to datums and have implied the importance of their use. Intelligent and effective application of the principles of geometric dimensioning and tolerancing depends to a great extent on a good understanding of the kinds of datums used, their definitions, proper selections, and the interpretations implied by the various uses.

The first portion of this section on datums covers the various kinds of datums, their definitions, establishment, relationship to each other and to measuring planes, and their meaning.

The last portion of this section will cover applications, selection, and further sample part examples.

Accuracy of 221, 222

Areas – Establishing Datum Planes from 239-241

Axis – Definition 205, 208, 209

Cylinder – Definition 208, 209

Cylinders – Establishing Datum Surfaces Three Plane Concept 215-217

Equalizing Datums 207, 243, 245

Extended Principles 246-255

Feature – Definition 209, 291, 292

Features – Specified, Using Position Dimensioning and Tolerancing 226-235

Features (Surface Holes) – Specified, Using Perpendicularity, Flatness,
      Parallelism, and Position Dimensioning and Tolerancing 230, 231

Functional Tolerance to Datums 254, 255

Interrelated Datum Systems 250-253

Line – Definition 205, 206

Lines – Establishing Datum Planes from 212

Multiple Datum Systems 246, 247

Plane – Definitions 206, 207

Planes – Establishing Datum Surfaces from Three Plane Concept 210, 211

Planes Part Orientation to 206, 207, 209

Point – Definition 204, 205

Points – Establishing Datum Planes from 205, 212, 213, 232-238

Points on Surfaces, Specified 212, 232-238

Points on Surfaces – Specified, Using Position Dimensioning
      and Tolerancing 232-236

Selection of Datum 220, 221

Step Datums 243, 244

Surface – Definition 209

Surfaces – Implied, Using Position Dimensioning and Tolerancing 222-224

Surfaces, Multiple, Specified 242

Surfaces, Partial, Specified 240, 241

Surfaces – Specified, Using Perpendicularity and Position
    Dimensioning and Tolerancing 226-228

Surfaces – Specified, Using Perpendicularity, Flatness, and
    Position Dimensioning and Tolerancing 228, 229

DATUMS

A DATUM is a point, line, plane, cylinder, axis, etc., which, for purposes of computation or reference, is assumed to be theoretically exact, and from which the location of part features may be established (or related). Datums are established by, or relative to, actual part features or surfaces. Where form or position relationships are specified from a datum, the features involved are located with respect to this datum and not with respect to one another.

Every feature on a part can be considered a possible datum. That is, every feature shown on a drawing depicts a theoretically exact geometric shape as specified by the design requirements. However, a feature normally has no practical meaning as a datum unless it is actually used for some functional relationship between features. Thus a datum appearing on an engineering drawing can be considered to have a dual nature: it is (1) a "construction" datum, which is the geometrically exact representation of any part feature, and (2) a "relationship" datum, which is any feature used as a basis for a functional relationship with other features on the part. Since the datum concept is used to establish relationships, the "relationship" datum is the type always used on engineering drawings.

By the above definition, a datum on an engineering drawing is always assumed to be "perfect." However, since perfect parts cannot be produced, a datum on a physically produced part is assumed to exist in the contact of the actual feature surface with precise manufacturing or inspection equipment such as machine tables, surface plates, gage pins, etc. These are not perfectly true planes, cylinders, etc., but they are usually of such high quality that they adequately simulate true references. This contact of the actual feature with precise equipment is also assumed to simulate functional contact with a mating part surface.

DATUM POINT

A datum point is that which has position but no extent, such as the apex of a pyramid or cone, the center point of a sphere, or a reference point on a surface for functional, tooling, or gaging purposes.

The "apex of a pyramid or cone" and "the center point of a sphere," are considered *construction* datums, which are used to construct the geometry of the feature. "A reference point on a surface for functional, tooling, or gaging purposes," is considered a functional *relationship* datum and has meaning in the drawing specification. It may be implied or specified, dependent upon the extent of control desired.

An example of implied datum point and a specified datum point are shown on p.205. *Implied* datum points are usually the extremities or high points of an actual surface which establish primary, secondary, and tertiary planes of orientation for the part. They occur at random locations as a part of the actually produced surfaces. Using the principles of geometry, we establish a "primary" datum plane by three (minimum) datum points, a "secondary" datum plane by two (minimum) datum points, and a "tertiary," or third, datum plane by one (minimum) datum point.

Where datum planes are to be established for specific locations on the surface, be it for design, or for manufacturing and inspection repeatability, they may be *specified* as datum target points, as shown. Location dimensions are required with *specified* datum target points.

DATUM POINTS

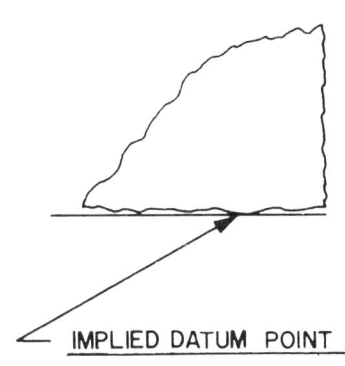

IMPLIED DATUM POINT

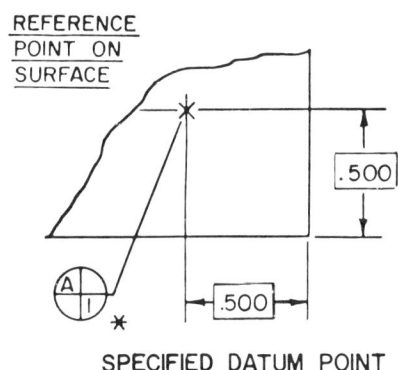

REFERENCE POINT ON SURFACE

.500

.500

SPECIFIED DATUM POINT

* SPECIFIED DATUM TARGET. SEE FOLLOWING PAGES FOR EXPLANATION.

DATUM LINE

A datum line is that which has length but not breadth or depth, such as the intersection line of two planes, the center line or axis of holes or cylinders, or a reference line for functional, tooling, or gaging purposes.

Examples showing an *implied* datum line and *specified* datum lines are shown on p. 206. Note that a hole (part a) contains a datum center line (as every round hole does) which is derived from the intersection of two imaginary planes at an angle of 90°. The datum center line is also an axis as derived from the geometric center of a cylinder. These are construction datums inherent in the geometry.

However, unless the hole is *specified* as a datum (part c) with a relationship to other features, it has no meaning as a datum. When stated as datums, the same construction datums that are shown in part (a) become part of the requirement as established from the actual datum hole surface. (See also DATUM CYLINDER, DATUM AXIS in following text.)

Where a specific datum line is to be established on a surface for design function, or manufacturing and inspection repeatability, it may be *specified* as a datum target line, as shown in part (b).

DATUMS
DATUM LINE

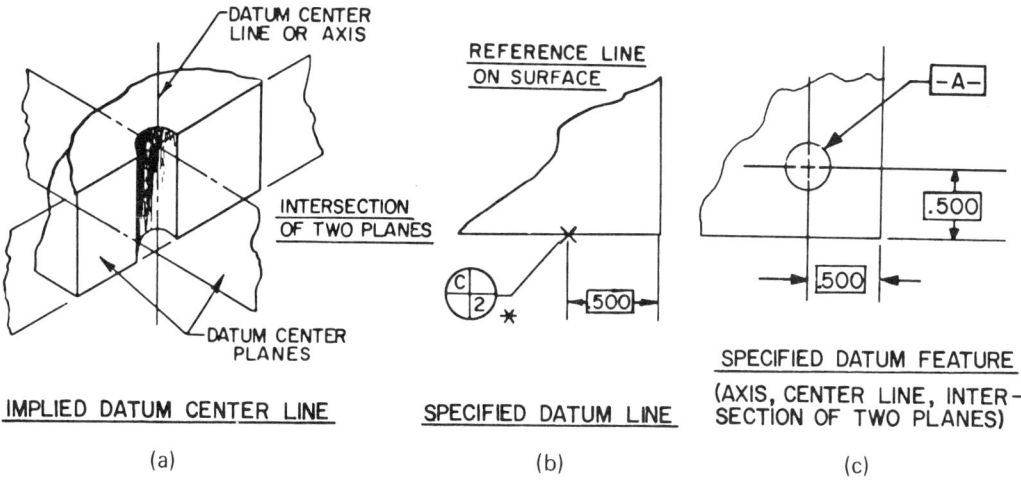

IMPLIED DATUM CENTER LINE

(a)

SPECIFIED DATUM LINE

(b)

SPECIFIED DATUM FEATURE
(AXIS, CENTER LINE, INTER-
SECTION OF TWO PLANES)

(c)

* SPECIFIED DATUM TARGET. SEE FOLLOWING PAGES FOR EXPLANATION

DATUM PLANE

A datum plane when developed from a primary datum feature which is nominally flat is a theoretically exact plane established by the extremities or contacting points of the *actual* feature surface with a reference plane (surface plate or other checking device).

Since measurements or references cannot be made from *theoretical* planes, the planes are therefore assumed to exist, not in the part itself, but in the contact of the part with more precise manufacturing or inspection equipment. Machine tables, surface plates, fixture surfaces, etc., are *not* true planes, but they are usually of such high quality that they adequately simulate true planes and therefore are considered true references. Occasionally reference planes are developed by coordinate movements of tooling or gaging equipment.

Datum planes are also established as theoretically exact references in relating form or position tolerances with actual features such as the center planes in a position toleranced hole pattern.

In example (a) below, the desired surface is shown as an exact drawing representation and as a datum plane. At right (b) is shown the establishment of the datum plane from the actual part surface through contact of the surface extremities with the tool or gage reference. A relationship of this datum plane to other features on the part would be found in actual application.

Since in the lower right example a primary datum is shown, three (minimum) points of contact extremities would establish the plane from the actual surface.

DATUM PLANE

EXAMPLE   MEANING

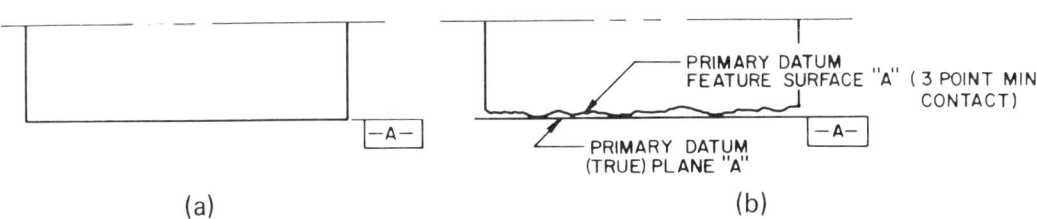

(a)   (b)

When a primary datum plane is established from a surface of irregular or unusual configuration (rough, curved, convex, etc.) and the part has a tendency to rock or "teeter-totter," it may be necessary to orient the part about the surface extremities until the inaccuracies of the reference surface are equalized and minimized as much as possible with respect to the datum plane.

AS DRAWN

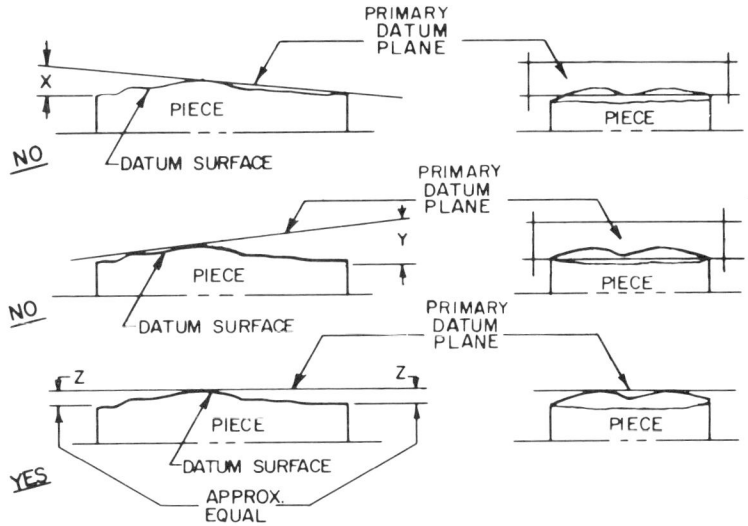

"X" IS GREATER THAN "Z"   "Y" IS GREATER THAN "Z"
THEREFORE "Z" GIVES MINIMUM VARIATION FROM DATUM SURFACE TO DATUM PLANE AND IS THE CORRECT METHOD TO BE USED IN ESTABLISHING A PRIMARY DATUM PLANE WITH A PART OF IRREGULAR OR UNUSUAL CONFIGURATION.

DATUMS

DATUM CYLINDER—DATUM AXIS

A datum cylinder (or other geometric form) is a theoretically exact or true form profile established by the extremities or contacting points of the actual datum feature surface (such as a cylindrical surface, etc.). The datum axis is the theoretically exact center line of the datum cylinder as established by the extremities or contacting points of the actual datum feature cylindrical surface or the axis formed by the intersection of two datum planes.

Since measurements or reference cannot be made from *theoretical* cylinders, etc., they are therefore assumed to exist, not in the part itself, but in the contact of the part with more precise manufacturing or inspection equipment. Fixture or gage pins, or gage cylinders, are *not* true cylinders, but they are usually of such high quality that they adequately simulate true cylinders, etc., and therefore are considered "true."

EXAMPLES

AS DRAWN   MEANING

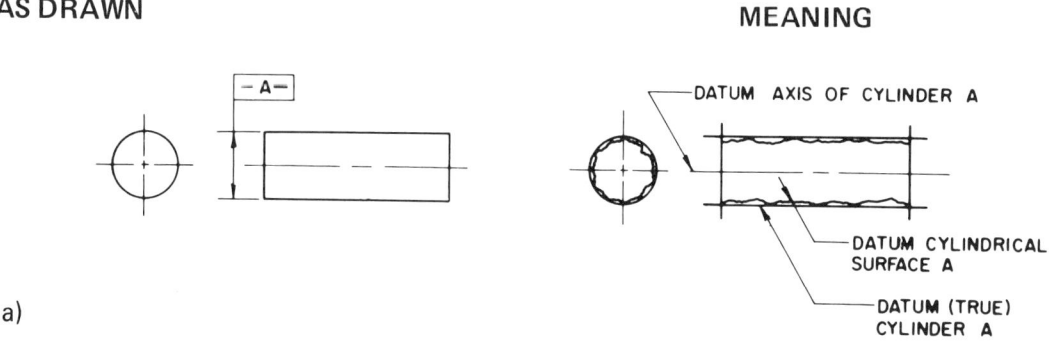

(a)

AS DRAWN   MEANING

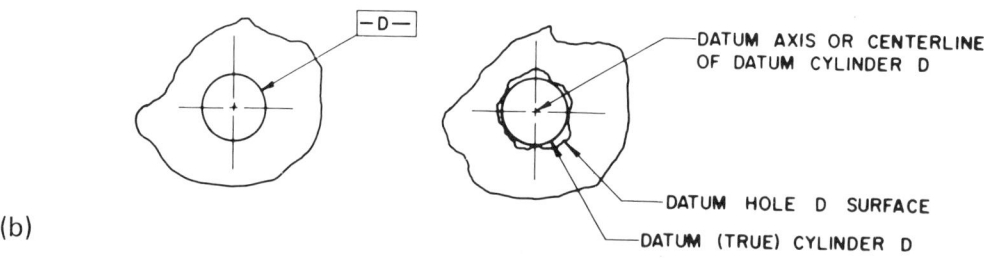

(b)

In both examples (a) and (b) the perfect datum feature is delineated on the drawing (shown at left); the geometric construction of these datums is established. In actually establishing the datum cylinder (RFS) or axis in example (a), we can imagine that the *minimum circumscribed* cylinder is in contact with the actual cylinder surface high points or extremities. A precision collet or chuck may be assumed to represent this principle. Simultaneously, by virtue of the geometry involved, we may imagine the intersection of two construction datum planes that are perpendicular to each other and pass through the derived datum cylinder axis.

Example (b) can be explained in the same manner except that the datum cylinder (RFS) or axis is established by the *maximum inscribed* cylinder in contact with the actual hole surface high points or extremities. A precision expanding pin or mandrel may be assumed to represent this principle.

Where the MMC principle is used on datum cylinders, the size of the datum is based on the MMC size of the feature concerned.

Although not shown, the datums of the examples are assumed to be in some functional relationship with other features.

DATUM SURFACE OR DATUM FEATURE (hole, slot, diameter, etc.)

By the definition of a DATUM, which is assumed to be theoretically exact, a "datum surface" or a "datum feature" is technically *not* a datum. These terms are, however, commonly used in referring to the actual part feature or surface *from which* a datum is established when the feature or surface is used as a reference or contact with a tool, surface plate or other checking device. A "datum surface" or "datum feature" refers to the actual part feature or surface coincidental with, relative to, and/or used to *establish* a datum. Since "datum surfaces" and "datum features" are actual things, they cannot be perfect, and thus they include all the real irregularities and inaccuracies of the surface or feature. (See example under DATUM PLANE.)

Generally, when a surface or feature is delineated (implied or specified) as a datum it is considered a datum (i.e., exact point, line, plane, cylinder, axis, etc.) when reference or measurement is taken *from* it. If the accuracy of the surface or feature delineated as datum is itself being checked, or if it is referred to in establishing a datum, the terms, "datum surface" and "datum feature" are commonly used.

ESTABLISHING DATUMS

ESTABLISHING DATUM PLANES FROM DATUM SURFACES—THREE PLANE CONCEPT RELATIONSHIP OF DATUM PLANES AND RELATED DIMENSION OR CENTER LINES

Datum planes are theoretically perfect reference planes. Datum planes are always established by, or are relative to, actual or physical features. The most common datum plane is the type established from a datum surface.

The example under "DATUM PLANE" illustrated the conventional establishment of a primary datum plane from a primary datum surface. In establishing datum planes in relation to defining or measuring a part, at least two and usually three datum and measuring planes are considered in locating features. In part configuration, other than cylindrical, there are usually three planes of orientation. These three planes conform to the relationship of the conventional geometric X, Y, and Z axes and resulting planes of orientation.

The three planes are referred to as primary, secondary, and tertiary (third) datum planes and are established from the appropriate actual datum surfaces. Unless otherwise specified or controlled, the largest or most important surface is usually selected as the primary datum, the next largest or most important as the secondary datum, and the remaining surface as the tertiary plane. Design functional requirements should be the first criterion for the establishment of datum priorities. Where the datums are to be specified, they are identified with appropriate letters as previously discussed.

All datum planes contained within a datum reference framework are 90° BASIC, or perpendicular, to each other by interpretation. The dimension lines or center lines related to and/or shown perpendicular to these datum planes are implied as 90° BASIC, or perpendicular, to the datum planes. The resulting measuring planes are therefore considered to be mutually perpendicular to one another.

The illustration shows the establishment, orientation, and relationship of three datum planes from the part surfaces to determine the datum reference system framework. The relationship of the measuring planes to the datum planes is also illustrated.

The PRIMARY datum surface A establishes the first relationship of the part for proper orientation. The three (minimum) extremities or contact points of the surface establish the primary datum plane A.

The SECONDARY datum surface B establishes a further relationship of the part for proper orientation. The two (minimum) extremities or contact points of the surface establish the secondary datum *plane* B at 90° BASIC to datum plane A.

The TERTIARY (third) datum surface C completes the part orientation. One (at least) extremity or contact point of the surface establishes the third datum *plane* C at 90° BASIC to both datum planes A and B.

Auxiliary datums, specified with respect to feature relationships not related to or functional with the part main datum reference system framework, would not be implied as 90° BASIC to the main datum system, but rather as 90° BASIC to the auxiliary datum reference system.

ESTABLISHING DATUM PLANES FROM DATUM SURFACES
THREE PLANE CONCEPT

AS DRAWN

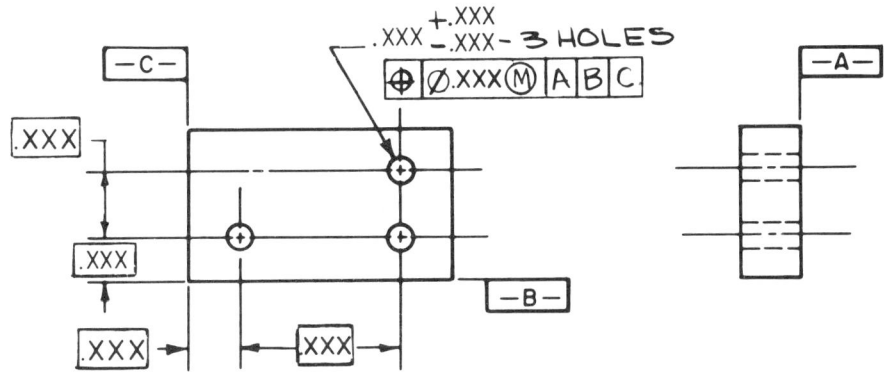

ESTABLISHING THE DATUM PLANES

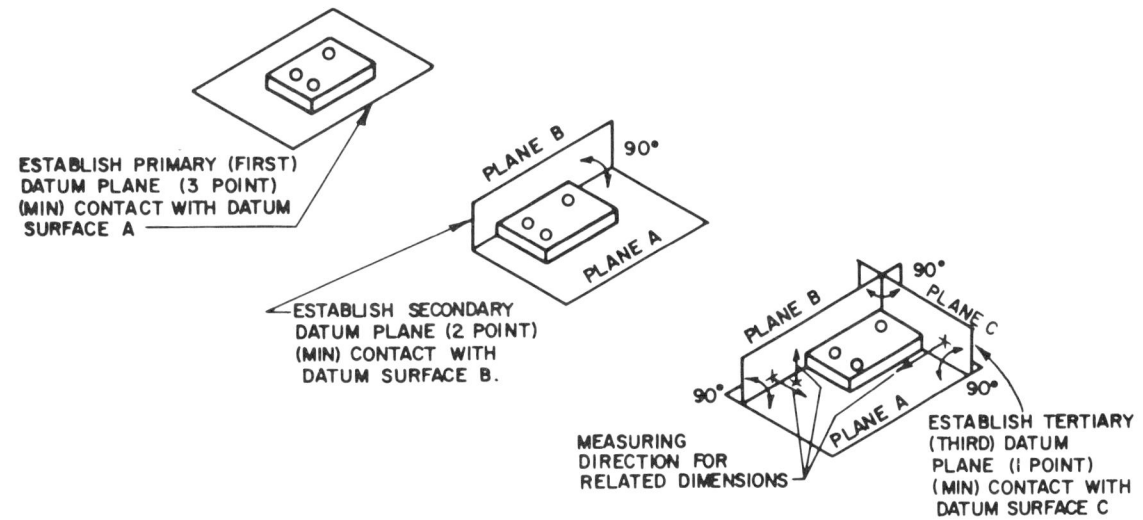

ESTABLISH PRIMARY (FIRST) DATUM PLANE (3 POINT) (MIN) CONTACT WITH DATUM SURFACE A

ESTABLISH SECONDARY DATUM PLANE (2 POINT) (MIN) CONTACT WITH DATUM SURFACE B.

MEASURING DIRECTION FOR RELATED DIMENSIONS

ESTABLISH TERTIARY (THIRD) DATUM PLANE (I POINT) (MIN) CONTACT WITH DATUM SURFACE C

ESTABLISHING DATUMS

ESTABLISHING DATUM PLANES FROM DATUM POINTS, LINES, OR AREAS (PARTIAL DATUM SURFACES) USING DATUM TARGETS

Where datum orientation is required on parts of irregular contour, such as castings, forgings, sheet metal, etc., datum targets provide a valuable tool. Specified datum targets which serve as means of constructing special datum planes of orientation can be of three types: points, lines, or areas. Datum targets establish the necessary datum system framework and, in addition, ensure repeatable part location for manufacturing and inspection operations.

Datum targets are also used to indicate special, or more critical, design requirements where functional part feature relationships are to be indicated from specific points, lines, or areas on the part surface.

Datum points and lines have been previously defined. A datum area is a datum established from a partial datum surface. On a drawing, a datum area is outlined with phantom lines and identified by diagonal slash lines. It may be of any shape.

The previously mentioned geometric rules of establishing the datum planes of orientation (primary, secondary, tertiary) must be observed. However, where datum targets are used, particularly on irregular castings or forgings, the number of targets may vary occasionally from the conventional geometrical 3-2-1 point orientation as determined by the part configuration.

The locations and/or sizes of datum points, lines, or areas are controlled by BASIC or untoleranced dimensions and imply exactness within standard tooling, gaging, or shop tolerances. Where necessary, toleranced locations or sizes may be used with datum target symbols.

The drawings below illustrate the use of datum target symbols* to establish datum planes and part orientation.

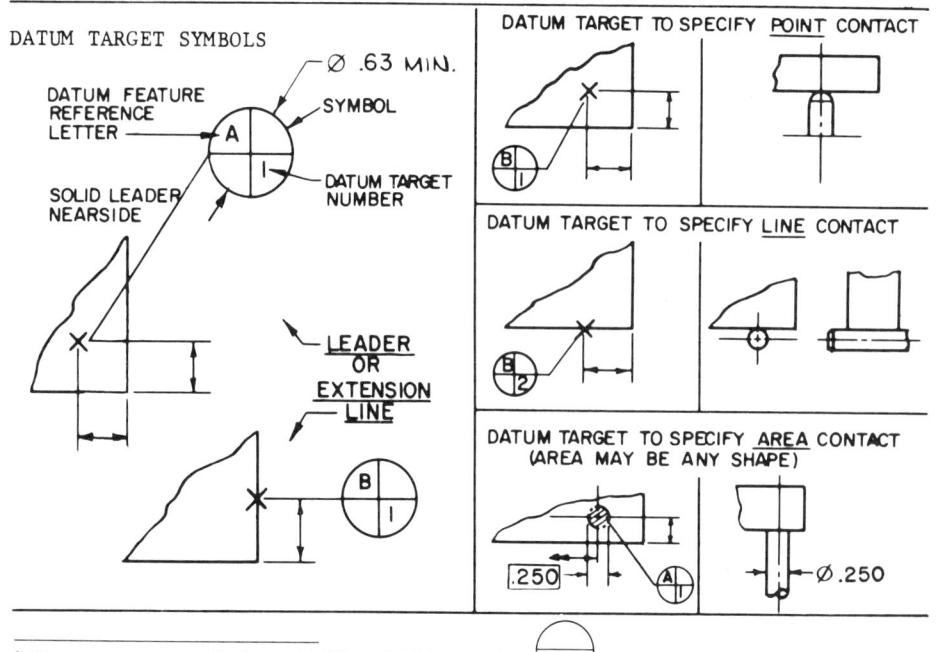

* Future target symbol per ANSI and ISO may be (A1).

212

PART ORIENTATION TO DATUM PLANES AS ESTABLISHED BY POINTS, LINES, OR AREAS

AS DRAWN

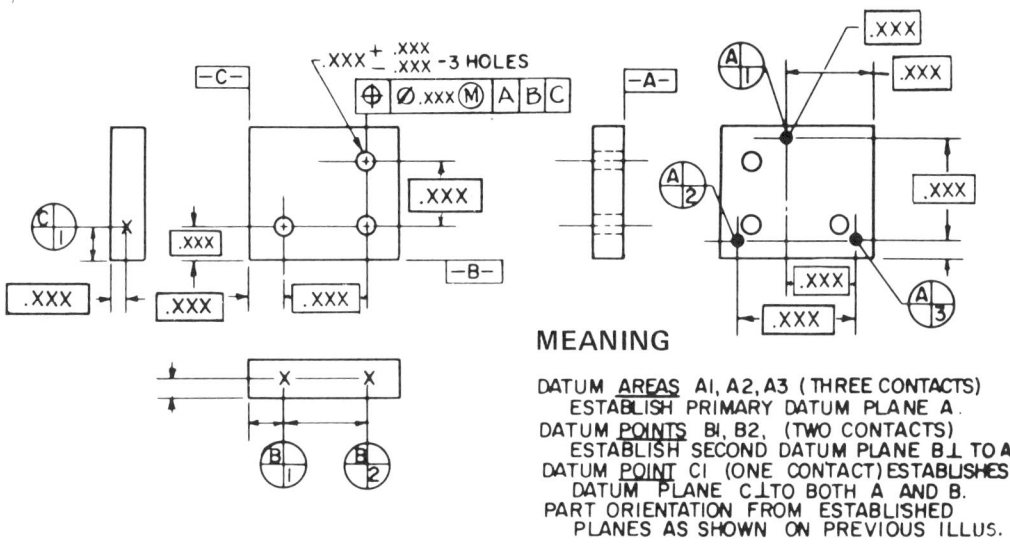

DATUM PLANES ESTABLISHED FROM DATUM POINTS AND AREAS ON NOMINALLY FLAT DATUM SURFACES:

MEANING

DATUM AREAS A1, A2, A3 (THREE CONTACTS) ESTABLISH PRIMARY DATUM PLANE A.
DATUM POINTS B1, B2, (TWO CONTACTS) ESTABLISH SECOND DATUM PLANE B⊥ TO A
DATUM POINT C1 (ONE CONTACT) ESTABLISHES DATUM PLANE C⊥TO BOTH A AND B.
PART ORIENTATION FROM ESTABLISHED PLANES AS SHOWN ON PREVIOUS ILLUS.

AS DRAWN

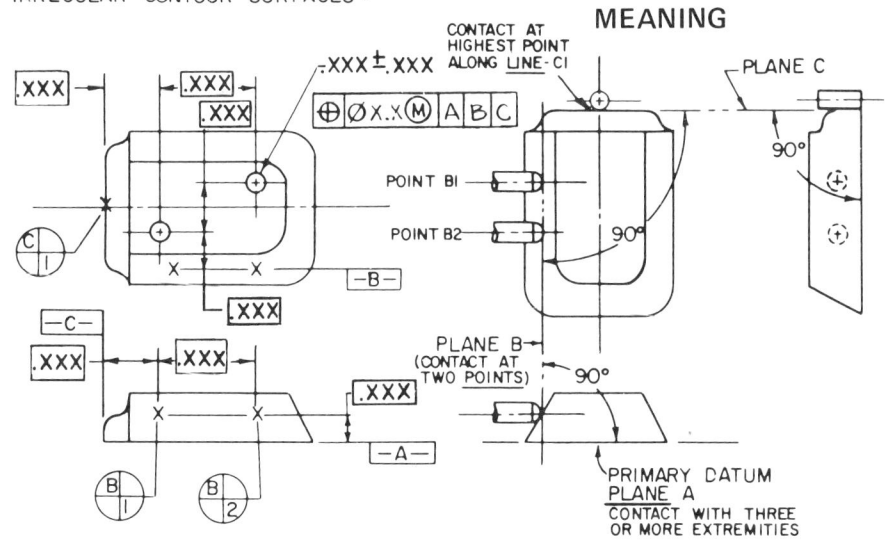

DATUM PLANES ESTABLISHED FROM DATUM POINTS AND LINES ON IRREGULAR CONTOUR SURFACES:

MEANING

ESTABLISHING DATUMS

For a rectangular part, the three reference planes are readily visualized, since the flat surfaces of the part resemble datum planes, and orientation to the three geometric planes X, Y, and Z is obvious. However, this is not true for cylindrical part surfaces since they bear no resemblance to datum planes and their relationship to the X, Y, and Z planes of orientation is not so obvious. Due to the geometry of a cylinder, a cylindrical part may be assumed to be simultaneously located in two planes of orientation. These two planes may be visualized as center planes intersecting 90° BASIC to each other at the datum axis. As discussed in the section *Datum Cylinders*, on outside cylinders the datum axis is established from the minimum circumscribed datum cylinder contacting the actual cylindrical surface (RFS). On inside cylinders, it is established from the maximum inscribed datum cylinder contacting the actual cylindrical surface (RFS).

Where the MMC principle is to be applied, the datum axis of a cylindrical part is established by the center line, or axis, of the datum surface cylinder at MMC size. Under this condition, the actually produced feature axis may vary within a tolerance zone equal to the departure from MMC size of the feature. This situation is illustrated in the example, PRIMARY DATUM FEATURE MMC.

The illustrations following expand upon the application of datums to cylindrical parts. The examples and the illustrated interpretation of each should be self-explanatory. Note that a cylindrical feature, due to its geometry, establishes orientation in two directions simultaneously: the two imaginary datum construction planes intersect at the datum axis. However, only one datum letter identification is used to establish the datum cylinder.

The examples also illustrate the use of a secondary datum plane with the primary datum cylinder and the use of the cylinder as *secondary* datum. In the latter instance, note that the secondary datum cylinder (perpendicular, 90° BASIC to the primary datum plane) may have its axis at variance with the *actual* feature axis. This possibility, however, is recognized and is a part of the design criterion whenever datums are used in this sequence on a part of cylindrical shape.

Where rotational orientation of the cylindrical part relative to features is required, a tertiary datum is necessary. The example, TERTIARY (THIRD) DATUM FEATURE RFS illustrates this type of requirement.

ESTABLISHING DATUM CYLINDERS FROM DATUM SURFACES—THREE PLANE CONCEPT RELATIONSHIP OF DATUM PLANES AND RELATED DIMENSIONS, DATUM PRECEDENCE

PRIMARY DATUM FEATURE RFS

PART

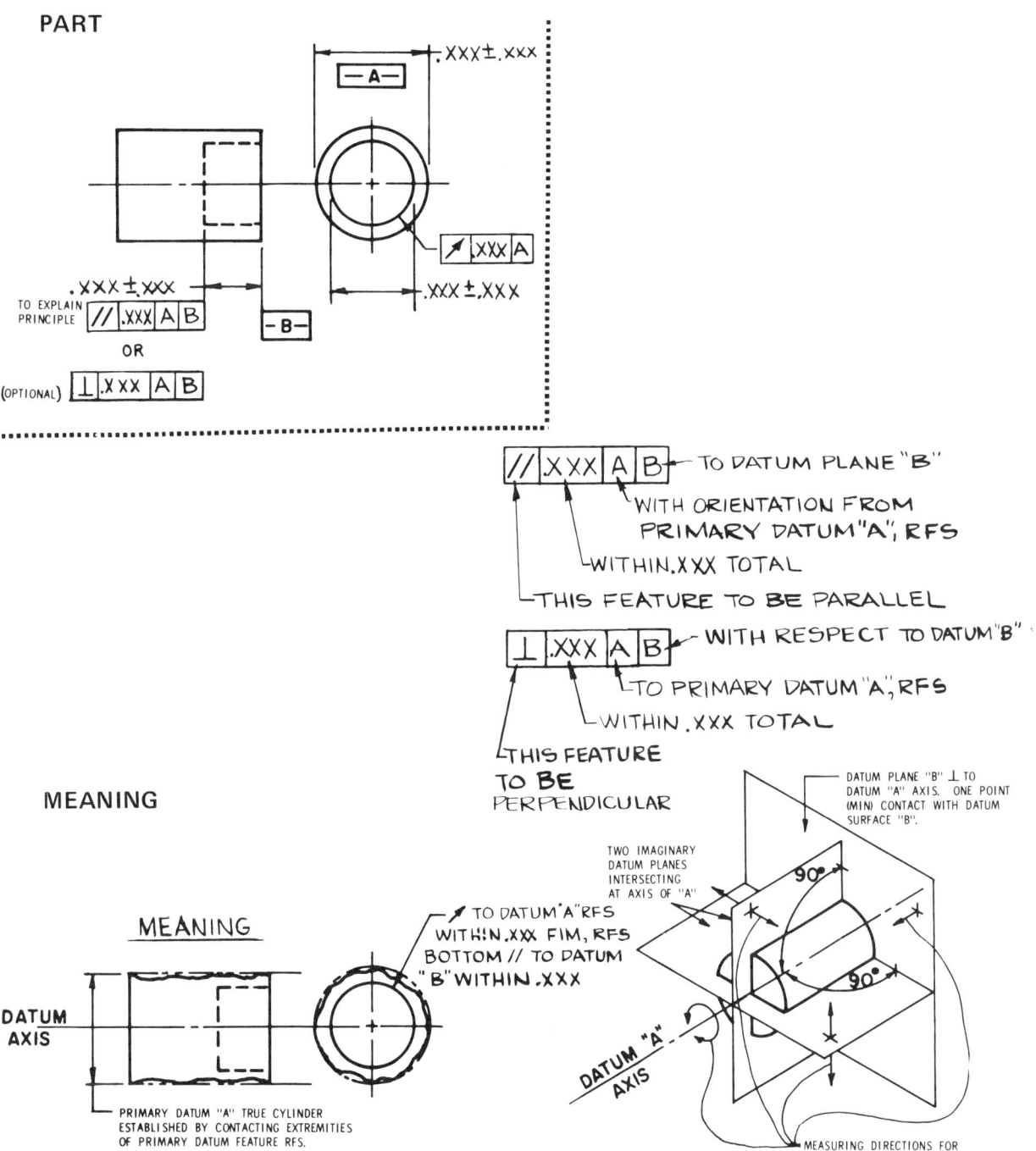

TO DATUM PLANE "B"

WITH ORIENTATION FROM PRIMARY DATUM "A", RFS

WITHIN .XXX TOTAL

THIS FEATURE TO BE PARALLEL

WITH RESPECT TO DATUM "B"

TO PRIMARY DATUM "A", RFS

WITHIN .XXX TOTAL

THIS FEATURE TO BE PERPENDICULAR

MEANING

DATUM PLANE "B" ⊥ TO DATUM "A" AXIS. ONE POINT (MIN) CONTACT WITH DATUM SURFACE "B".

TWO IMAGINARY DATUM PLANES INTERSECTING AT AXIS OF "A"

MEANING

TO DATUM "A" RFS WITHIN .XXX FIM, RFS BOTTOM // TO DATUM "B" WITHIN .XXX

DATUM AXIS

PRIMARY DATUM "A" TRUE CYLINDER ESTABLISHED BY CONTACTING EXTREMITIES OF PRIMARY DATUM FEATURE RFS.

DATUM "A" AXIS

MEASURING DIRECTIONS FOR RELATED FEATURES

215

ESTABLISHING DATUMS

PRIMARY DATUM FEATURE MMC
PART

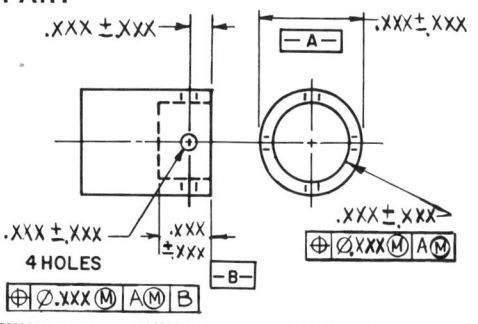

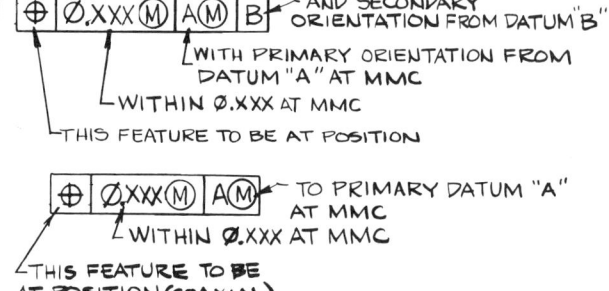

MEANING

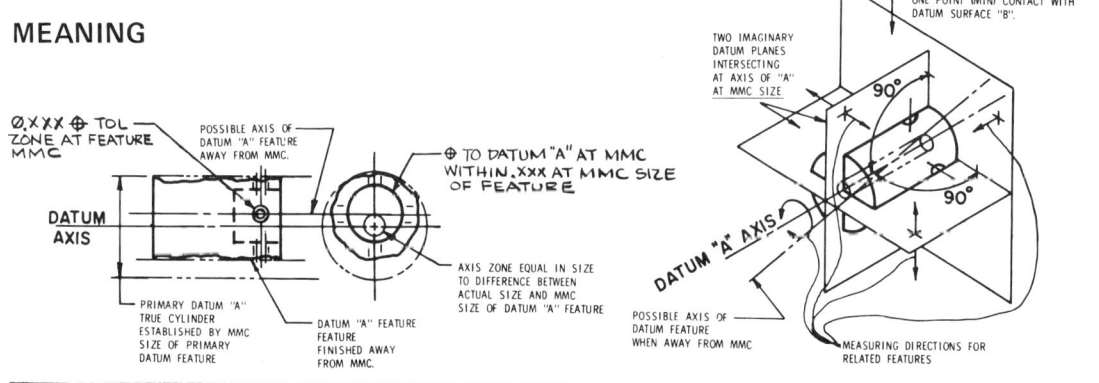

SECONDARY DATUM FEATURE RFS
PART

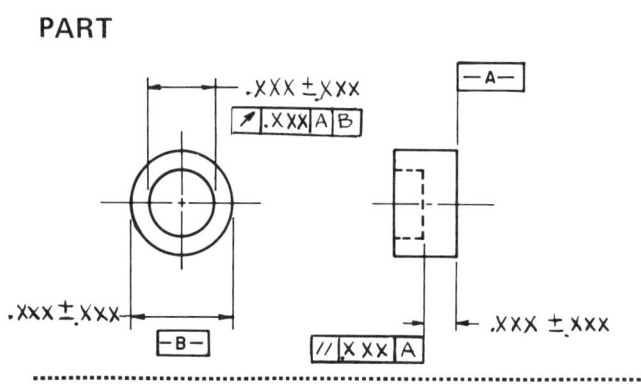

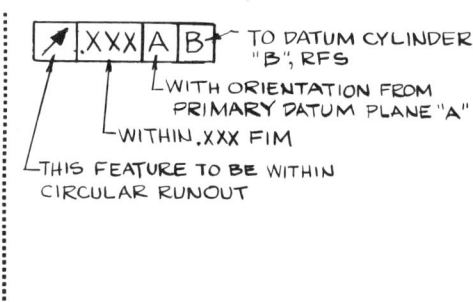

MEANING

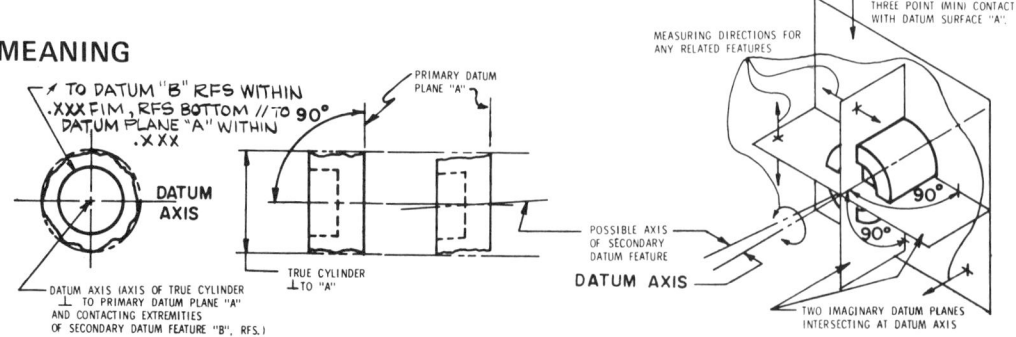

SECONDARY DATUM FEATURE MMC

PART

MEANING

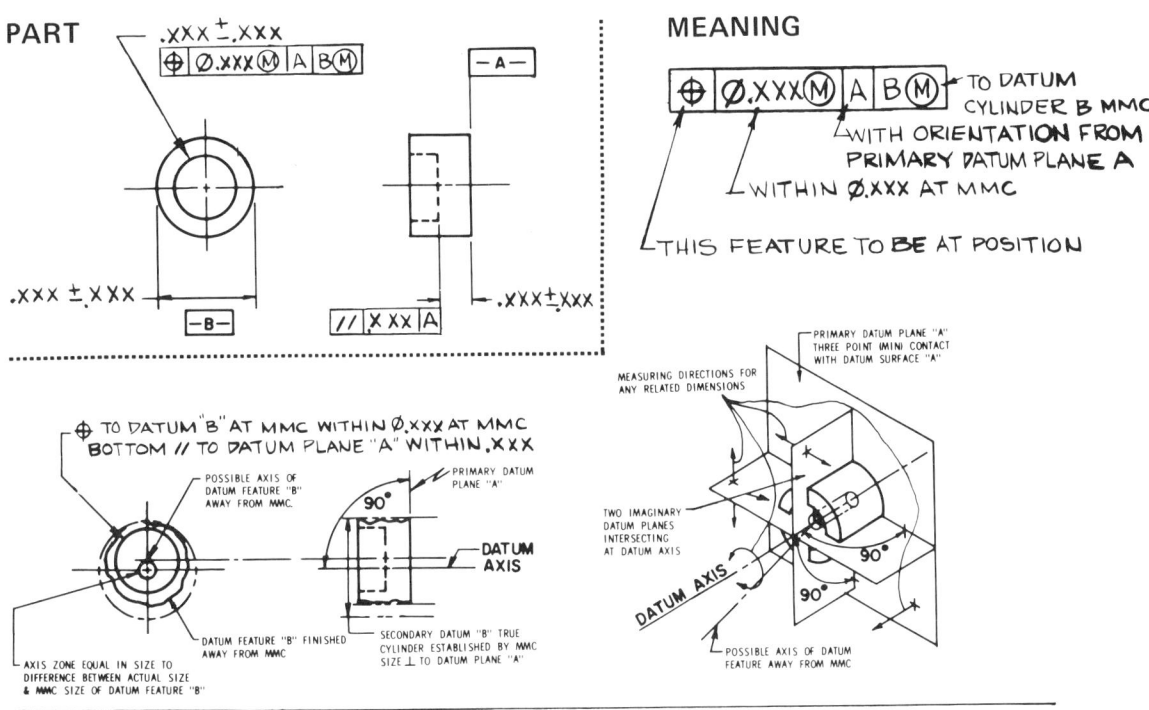

TO DATUM "B" AT MMC WITHIN Ø.XXX AT MMC
BOTTOM // TO DATUM PLANE "A" WITHIN .XXX

TERTIARY (THIRD) DATUM FEATURE RFS

PART

MEANING

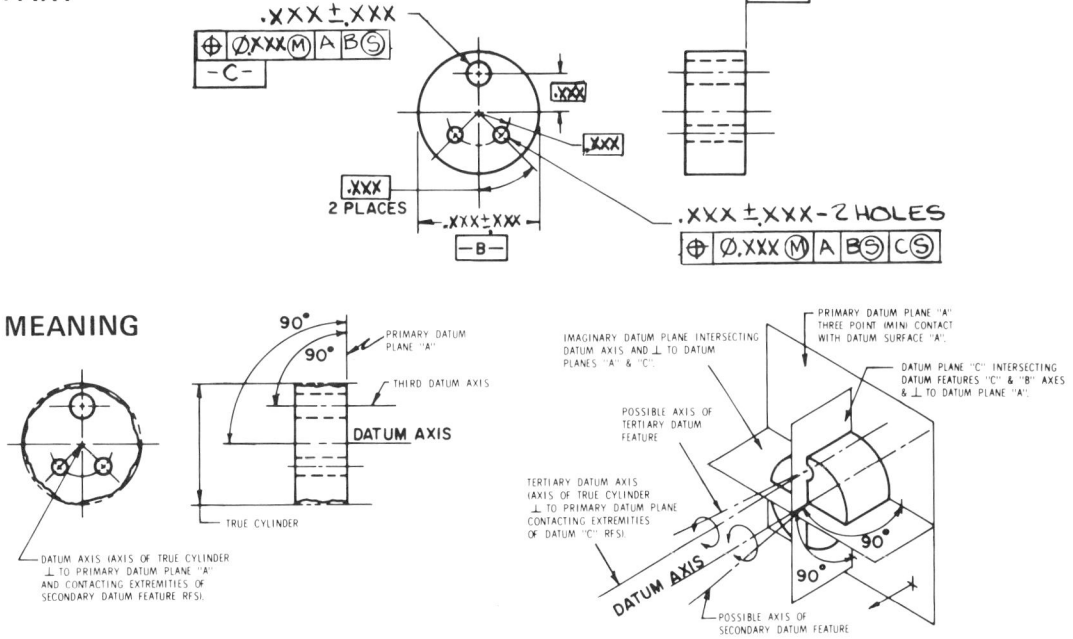

WHEN THE TERTIARY DATUM FEATURE IS SPECIFIED MMC, THE ACTUAL AXIS OF THE
FEATURE MUST BE WITHIN A ZONE CENTERED AROUND THE THIRD DATUM AXIS WHICH
IS EQUAL IN SIZE TO THE DIFFERENCE BETWEEN THE ACTUAL SIZE OF THE DATUM
FEATURE AND ITS SPECIFIED MMC SIZE.

ESTABLISHING DATUMS

ESTABLISHING DATUM CENTER PLANES FROM DATUM FEATURES— THREE PLANE CONCEPT DATUM PRECEDENCE

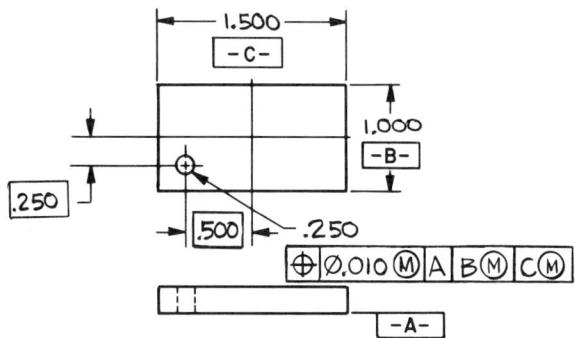

MEANING NO. 1

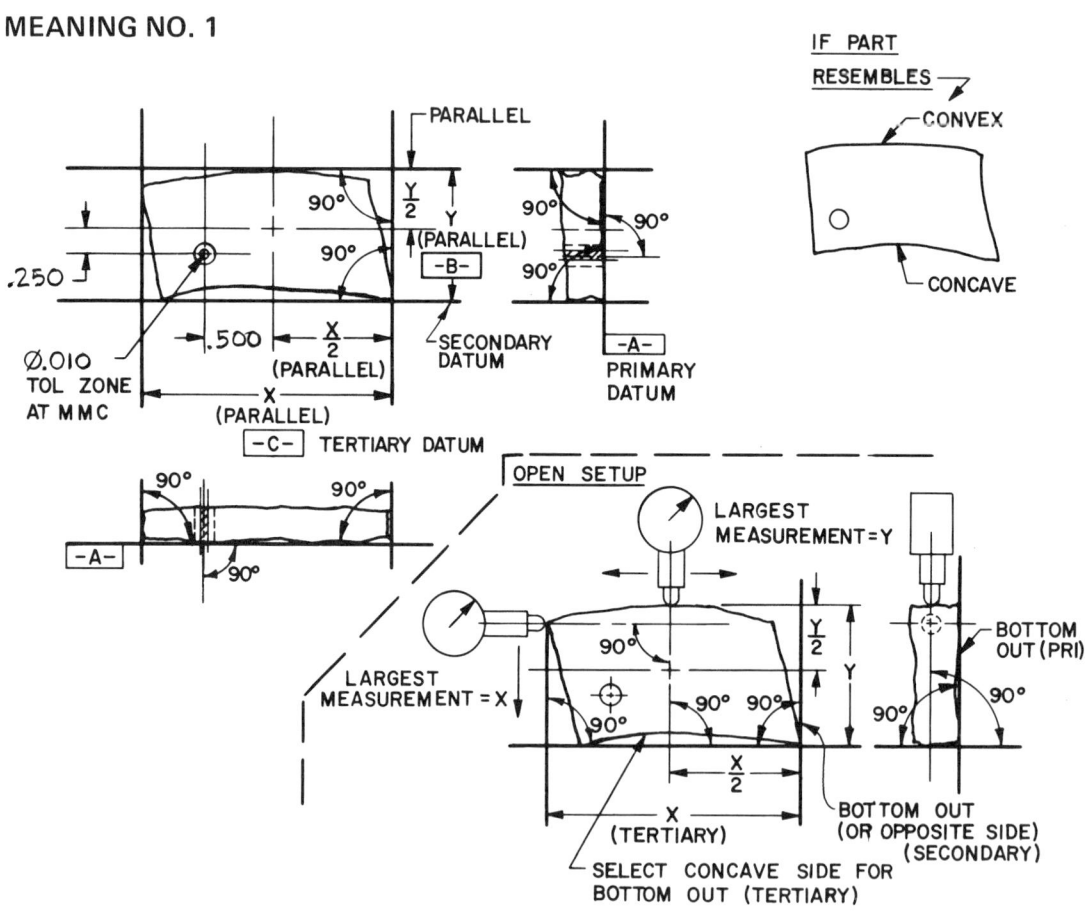

MEANING NO. 2

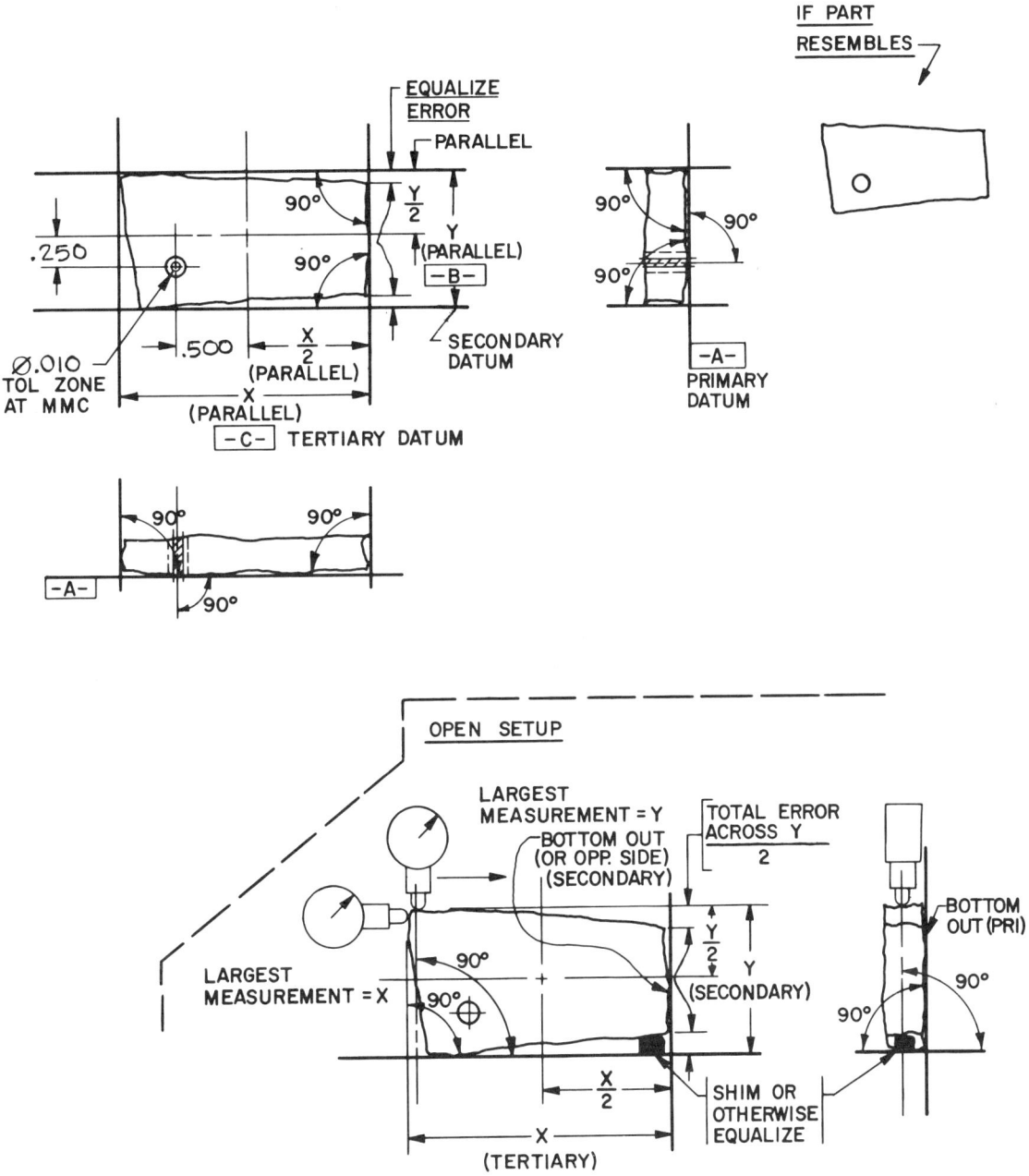

NOTE: The "open set-up" interpretations above and on the preceding page demonstrate the principles involved. Production gaging methods, or special gages patterned after these techniques, can facilitate more effective inspection, e.g., equalizing mechanisms can be used in lieu of the measuring and shimming method (see MEANING NO. 2).

DATUM APPLICATIONS

The following section presents details of selection, application, and interpretation of datums under sample part conditions. Both implied and specified datums are illustrated. Not all conditions are covered, but the examples shown are representative of a variety of typical datum applications.

The proper selection of datums is a very important aspect of geometric tolerancing. As previously emphasized, datums may be either implied or specified. In determining datums, we must consider first of all the part and feature relationships within the framework of the design requirements.

Where the relationships of the features are not of critical importance to part function or design intent, specific datums may not be required in the drawing specifications. However, usually some form of datum orientation for part manufacture and inspection is necessary. Therefore, whenever implied datums are invoked in the absence of specified engineering drawing datum detail, the designer depends on the discretion of the production and inspection departments.

Where specific relationships of features, datum precedence, and part orientation are of critical concern to design and manufacture, datums should always be specified. Of prime concern is the clarity with which the drawing conveys the design intent. Ambiguity should always be avoided, since it causes confusion and leads to misinterpretations.

Datum specification may appear to be a costly approach. Not so—on the contrary, it provides considerable advantages that pay off in economic production. It provides for clear instructions, protects the design intent, and ensures production and inspection follow-through in keeping with the intent.

The examples shown highlight the importance of adequate and proper datum specification. The illustrations on the left are not provided with datums, and the feature relationships are in doubt. The addition of datums in the examples on the right clarifies the relationships and thus ensures uniform interpretation.

SELECTION OF DATUMS

1. – *Features* which are selected to establish *datums* must be clearly identified and/or easily recognizable. Datums may be either implied or specified, but must clearly represent necessary design intent.

2. – *Corresponding features* on mating parts should be used in establishing datums to facilitate calculations and ensure proper part assembly.

3. – To be useful for measuring, a datum on an actual piece should be accessible during manufacture, so that *measurements* from it *can be made* readily.

4. – *Avoid ambiguity* of datums by specifying datums where necessary for clarity.

SELECTION OF DATUMS

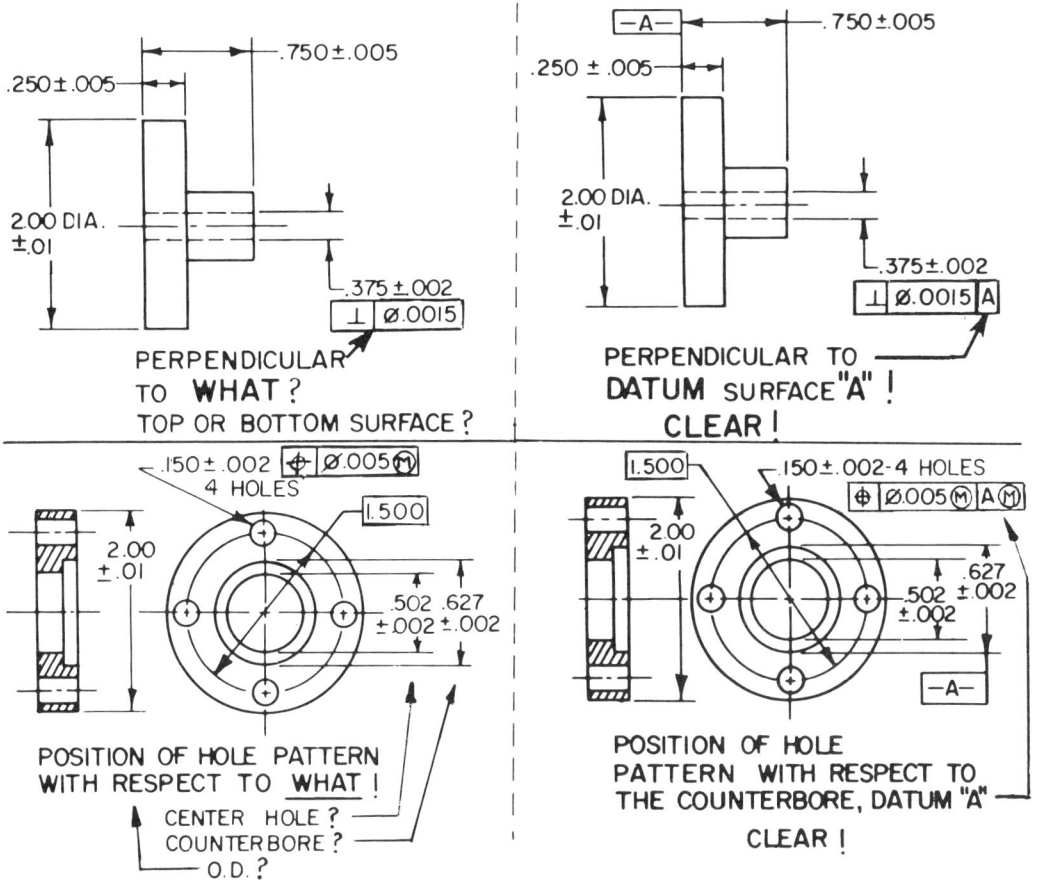

PERPENDICULAR TO **WHAT** ?
TOP OR BOTTOM SURFACE ?

PERPENDICULAR TO
DATUM SURFACE "A" !
CLEAR !

POSITION OF HOLE PATTERN
WITH RESPECT TO **WHAT** !
CENTER HOLE ?
COUNTERBORE ?
O.D. ?

POSITION OF HOLE
PATTERN WITH RESPECT TO
THE COUNTERBORE, DATUM "A"
CLEAR !

ACCURACY OF A DATUM SURFACE ON AN ACTUAL PIECE

The accuracy or quality of surfaces used for datums depends on the design requirement. The desired control, such as the form tolerance of flatness, may be specified where necessary.

Earlier U. S. standards have indicated the need to control datum surfaces to a finer degree of accuracy than that provided by dimensions or relationships taken *from* the datum surfaces. Often design requirements may call for this refinement and the need for it should therefore be considered. However, as an overall rule, a request for specific accuracy of the datum surface is not always valid and is no longer considered a requirement of current U. S. standards.

By the definition of datum, a datum surface may have inaccuracies which will have no effect on the features located from the datum plane (as established from the extremities of that surface).

DATUM APPLICATIONS

The illustration shows the two conditions of datum accuracy for features located from the datum, one where datum surface accuracy is *less* than that for feature location related to the datum (a), and one where the datum surface is held *more* accurate than the feature location related to the datum (b). Either requirement could be valid as dependent upon the design requirement.

Unless otherwise specified, part surface quality is normally a result of the type of machining process used to achieve size tolerances. Often, workmanship or local manufacturing tolerances will be the determining factors. Thus, where surface accuracy is of no critical concern, it need not be specified, yet that surface can serve as a datum reference from which other relationships can be specified.

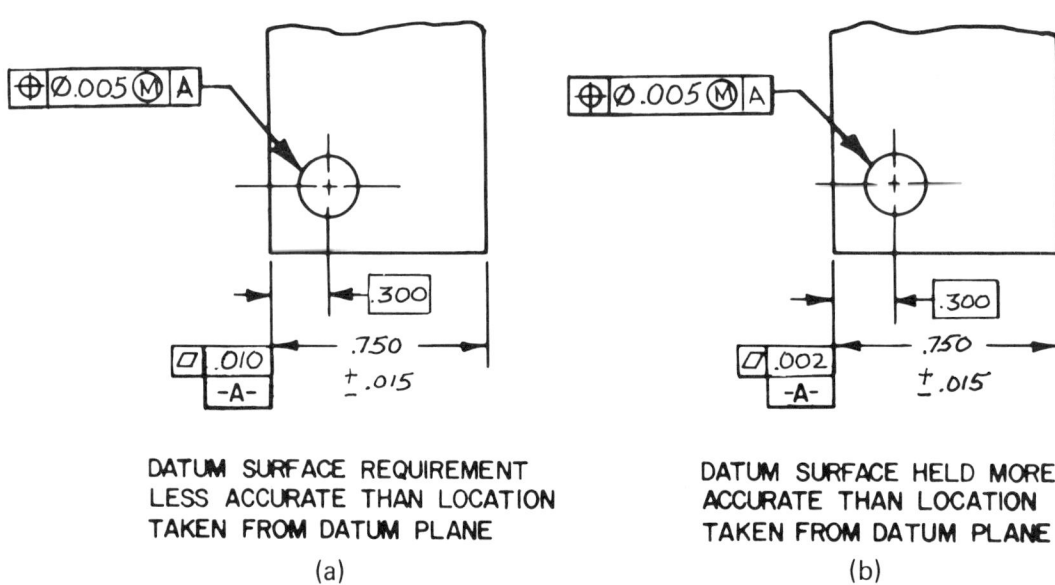

DATUM SURFACE REQUIREMENT
LESS ACCURATE THAN LOCATION
TAKEN FROM DATUM PLANE

(a)

DATUM SURFACE HELD MORE
ACCURATE THAN LOCATION
TAKEN FROM DATUM PLANE

(b)

IMPLIED DATUM SURFACES
USING POSITION ⊕ DIMENSIONING AND TOLERANCING

Position location of a pattern of features, such as is indicated on the three holes in the example on the next page, may be referenced from the edge surfaces of a part. In this case, the .500 ± .005 dimension from the left edge and the .630 ± .005 dimension from the lower edge establish the relationship of the hole pattern to the edges. These edges are *implied* datum surfaces. This is evident because the dimensions that interrelate with the position pattern originate from these surfaces. These datum *surfaces* establish datum *planes* and thus the position pattern is related to the datum planes.

Although the .500 ± .005 and the .630 ± .005 dimensions are drawn to the center lines of the lower holes of the position pattern, they locate the *complete* hole pattern. The directly toleranced dimensions are interpreted to apply to the pattern center lines and hence to the entire hole pattern as an entity.

AS DRAWN

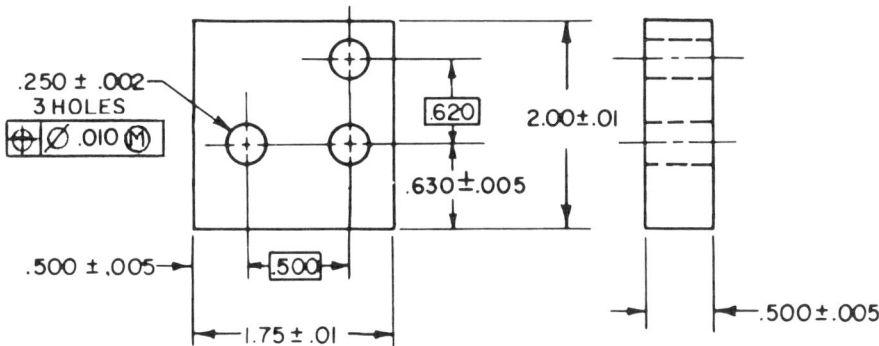

.250 ± .002
3 HOLES

⊕ | Ø .010 Ⓜ

.500 ± .005

.500

1.75 ± .01

.620

.630 ± .005

2.00 ± .01

.500 ± .005

MEANING

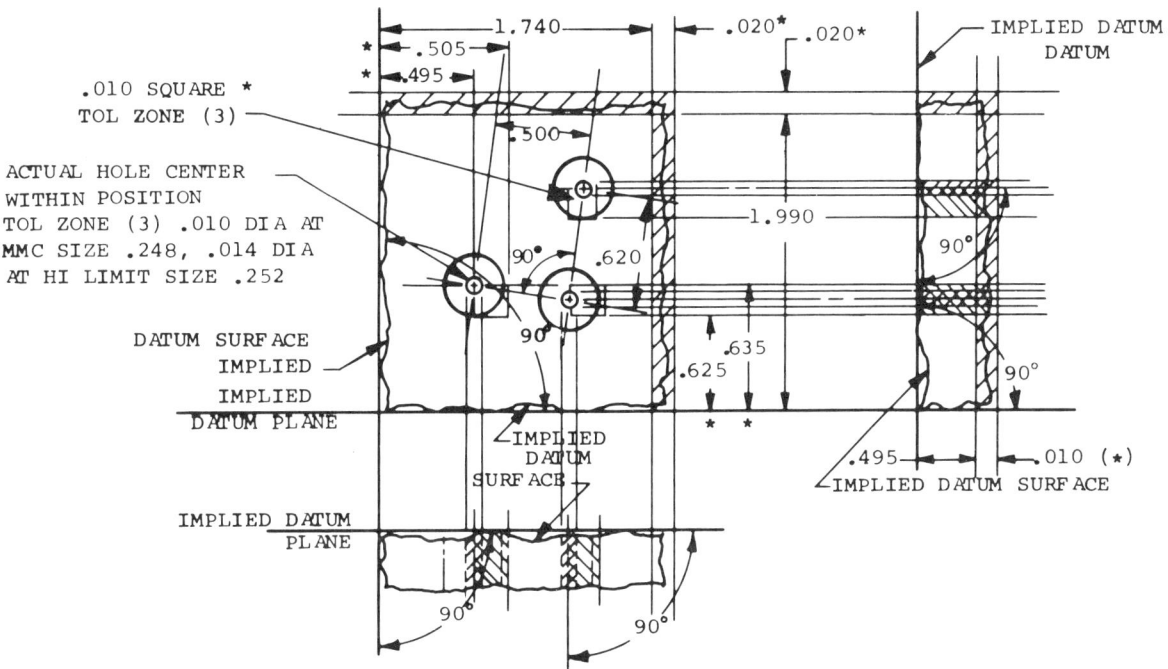

.010 SQUARE *
TOL ZONE (3)

ACTUAL HOLE CENTER
WITHIN POSITION
TOL ZONE (3) .010 DIA AT
MMC SIZE .248, .014 DIA
AT HI LIMIT SIZE .252

DATUM SURFACE
IMPLIED
IMPLIED
DATUM PLANE

IMPLIED DATUM
PLANE

1.740
.505
.495
.500
.620
90°
90°
.635
.625
IMPLIED
DATUM
SURFACE
90°
90°

.020*   .020*
1.990
90°
90°
.495
.010 (*)
IMPLIED DATUM SURFACE
IMPLIED DATUM
DATUM

NOTE: (*) TOLERANCE ZONES APPLY AS NORMALLY MEASURED FROM THE
DATUM PLANES. IF MEASURED BY SOME METHOD WHERE REFERENCE
ON THE DATUM SURFACE IS NOT COINCIDENTAL WITH THE DATUM
PLANE, TOL ZONES ARE MINUS DATUM SURFACE INACCURACY
(DEVIATION BETWEEN DATUM SURFACE AND DATUM PLANE) AT
THAT POINT. EDGE TOLERANCE ZONE INTERRELATIONSHIP TO
EACH OTHER ALSO SUBJECT TO CONSIDERATIONS OF GENERAL
RULE #1.

(**) ACTUAL HOLE LOCATION MAY VARY FROM PERPENDICULAR
ORIENTATION TO IMPLIED PRIMARY DATUM PLANE WITHIN TRUE
POSITION TOLERANCE ZONE ONLY.

DATUM APPLICATIONS

The directly toleranced coordinate dimensions from the implied datum surfaces to the hole pattern establish square or rectangular tolerance zones equal to the amount of the coordinate tolerances. Since the directly toleranced dimensions are interpreted to apply to the entire hole pattern as an entity, each hole in the pattern is considered to have an identical coordinate tolerance zone. The position pattern may shift as an entity within these coordinate zones. Simultaneously, each of the centers of the holes in the pattern may shift an additional amount from its true position as established by the basic dimensions within the positional tolerance zones.

The actual centers of the produced holes then must fall within the $\varnothing$.010 positional tolerance zone when at MMC (or smallest hole) size. As previously shown, the positional tolerance zone increases as the hole size increases or departs from MMC size to LMC size.

The positional tolerance zones are basically parallel to each other and the center planes of the pattern are perpendicular and parallel to each other.

Implied datums do not establish datum precedence. Thus, discretion must be used in implementing datum orientation of a part of this type. The largest surface, such as the top or bottom, would normally serve as the primary datum. The illustration shows the meaning adopted as an unofficial consensus, using the top view surface of the drawing as the primary datum.

The actual hole center (axes) are permitted to vary to the extent of the coordinate tolerance zones and the position tolerance zones relative to the implied secondary and tertiary datum references. However, the actual hole centers relative to the implied primary datum plane are permitted to vary from 90° (perpendicular) only to the extent of the position tolerance zone.

The part edge surface tolerance zones illustrated (at *) are intended to relate the edge extremities to the total requirements on this part. However, note that the interrelationship (perpendicularity) of these edge surfaces to each other is *not* controlled by the stated requirements, but is subjected to the considerations of Rule 1. The edge tolerance zones shown are representative of the size tolerance relationship of the individual 1.75 ± .01 and 2.00 ± .01 sizes only. Control of these surfaces in relationship to one another would require specified form tolerances (e.g., perpendicularity) or would have to be based on local workmanship or shop tolerances.

AS DRAWN

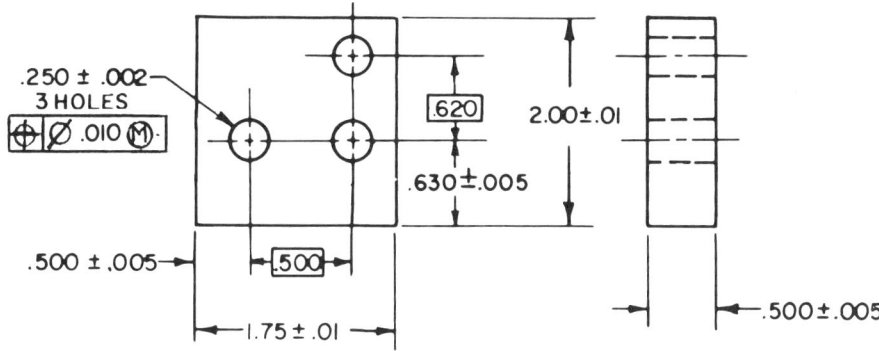

.250 ± .002
3 HOLES

⊕ | ⌀ .010 Ⓜ |

.620

2.00 ± .01

.630 ± .005

.500 ± .005

.500

1.75 ± .01

.500 ± .005

MEANING

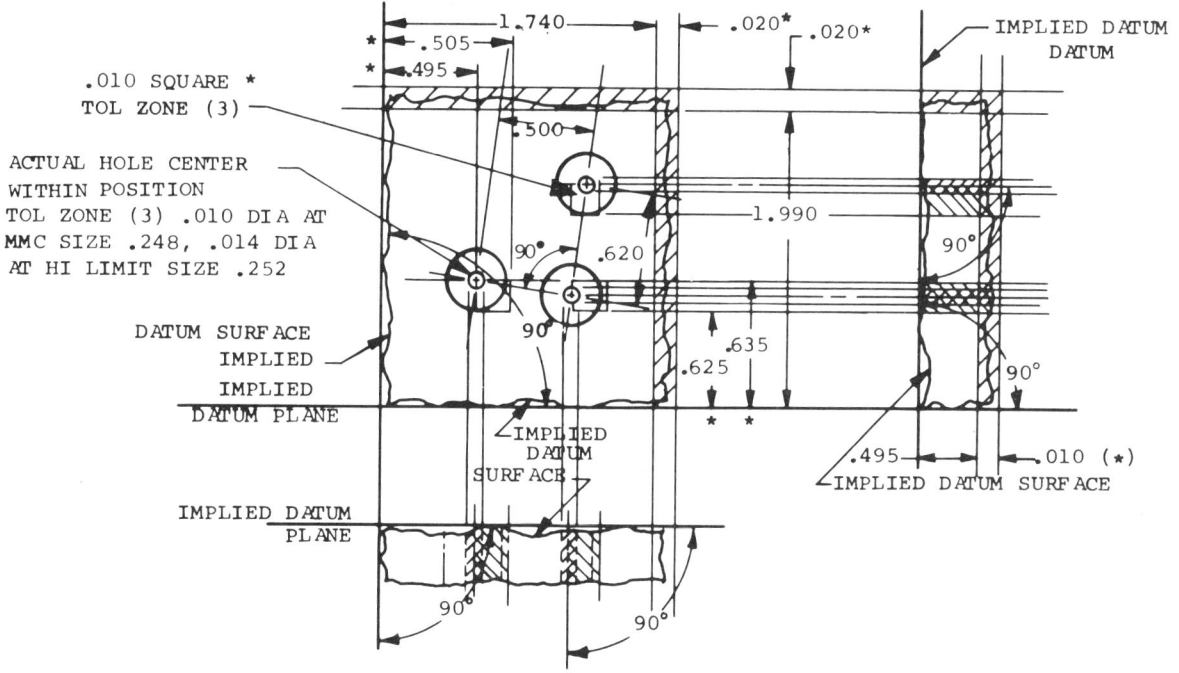

1.740

.020* .020*

IMPLIED DATUM
DATUM

* .505
* .495

.010 SQUARE *
TOL ZONE (3)

.500

ACTUAL HOLE CENTER
WITHIN POSITION
TOL ZONE (3) .010 DIA AT
MMC SIZE .248, .014 DIA
AT HI LIMIT SIZE .252

1.990

90°

.620

90°

.635

DATUM SURFACE
IMPLIED

.625

IMPLIED

IMPLIED
DATUM PLANE

* *

IMPLIED
DATUM
SURFACE

.495

.010 (*)

IMPLIED DATUM SURFACE

IMPLIED DATUM
PLANE

90°

90°

90°

NOTE: (*) TOLERANCE ZONES APPLY AS NORMALLY MEASURED FROM THE
DATUM PLANES. IF MEASURED BY SOME METHOD WHERE REFERENCE
ON THE DATUM SURFACE IS NOT COINCIDENTAL WITH THE DATUM
PLANE, TOL ZONES ARE MINUS DATUM SURFACE INACCURACY
(DEVIATION BETWEEN DATUM SURFACE AND DATUM PLANE) AT
THAT POINT. EDGE TOLERANCE ZONE INTERRELATIONSHIP TO
EACH OTHER ALSO SUBJECT TO CONSIDERATIONS OF GENERAL
RULE #1.

(**) ACTUAL HOLE LOCATION MAY VARY FROM PERPENDICULAR
ORIENTATION TO IMPLIED PRIMARY DATUM PLANE WITHIN TRUE
POSITION TOLERANCE ZONE ONLY.

DATUM APPLICATIONS

SPECIFIED DATUM SURFACES USING THE FORM TOLERANCE ⊥ AND POSITION ⊕ DIMENSIONING AND TOLERANCING

When feature interrelationships involving form and position tolerances are to be more accurately and directly controlled, it becomes necessary to specify datums on the drawing.

In this example, the top surface of the part is identified as datum A, the left edge as datum B, and the lower edge as datum C. Once these identifications are made, the specific relationships between surfaces and features can be stated.

As previously described, datum planes are established from the extremities of the actual datum surfaces. Each relationship is thus taken from the datum plane. The datum surface inaccuracies, not coincidental with the datum plane, are not involved in the relationship.

The interpretation illustrates the establishment of the datum planes from the specified datum surfaces. The specified datums were used in this example to control attitude or perpendicularity of certain surfaces to precise limits. They were also used to give datum precedence orientation from the part surfaces to the position toleranced pattern.

The left datum surface is specified as perpendicular to datum plane A within .002. The lower datum surface is specified as perpendicular to datum plane A within .003 and to datum plane B within .004.

In view (b) in the lower interpretation the .003 wide perpendicularity tolerance zone is established between two parallel planes exactly 90° BASIC (or perpendicular) to datum plane A. The actual surface must lie within this .003 wide tolerance zone when it is measured relative to datum plane A anywhere along the entire length of its surface.

In (a), the .004 wide perpendicularity tolerance zone is established between two parallel planes exactly 90° BASIC (perpendicular) to datum plane B. The actual surface must be within this .004 wide tolerance zone when it is measured relative to datum plane B anywhere along the entire length of its surface. In (c), datum surface B is to be held perpendicular to datum A within a .002 width tolerance zone 90° BASIC to datum plane A.

Note that the measurements are taken from the datum planes which are theoretically perfect as established by the extremities of the actual datum surfaces. The perpendicularity tolerance zones establish the limit of the perpendicularity or form inaccuracies of these surfaces.

As in the note to view (a) at left, the actual hole centers must fall within the position tolerance zone of ⌀.010 when the hole is at MMC size of .248. The tolerance zone can increase to ⌀.014 if the holes are produced to the high limit (LMC) size of .252. The position tolerance includes both form (attitude) and position error control.

The position tolerance zones are basically parallel to each other and the center planes of the pattern are basically perpendicular and parallel to each other. In addition, the position tolerance zones are basically perpendicular to the primary datum plane A, and parallel to the secondary and tertiary datum planes B and C.

SPECIFIED DATUM SURFACES USING THE FORM TOLERANCE ⊥ AND POSITION ⊕ DIMENSIONING AND TOLERANCING

AS DRAWN

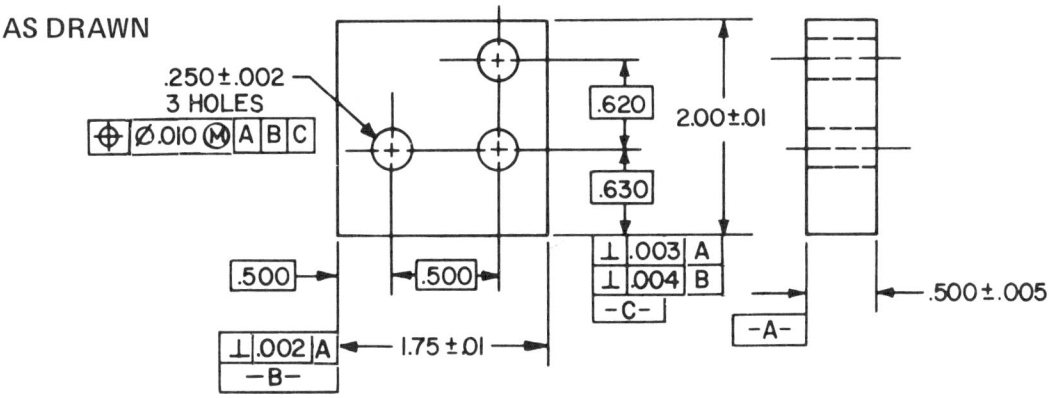

MEANING

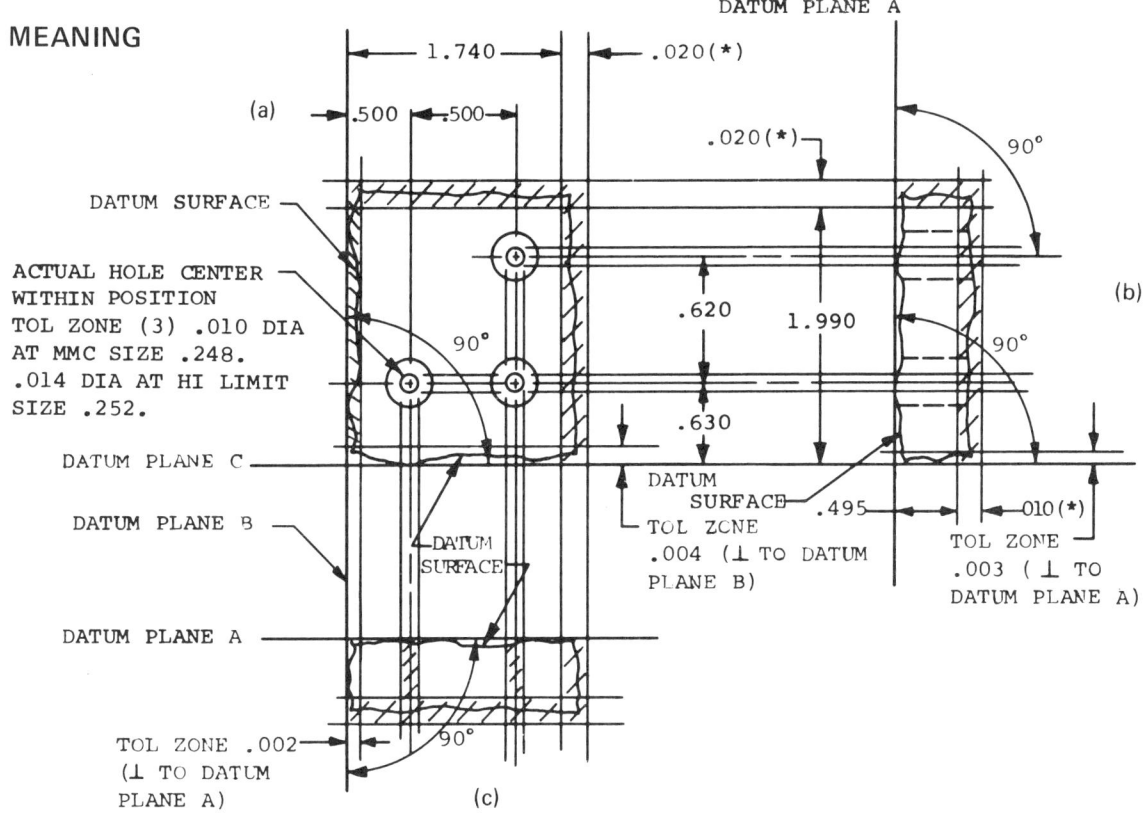

NOTE: (*) TOLERANCE ZONES APPLY AS NORMALLY MEA-
SURES FROM THE DATUM PLANES. IF MEASURED BY
SOME METHOD WHERE REFERENCE ON THE DATUM
SURFACE IS NOT COINCIDENTAL WITH THE DATUM
PLANE, TOL. ZONES ARE MINUS DATUM SURFACE AND
AND DATUM PLANE) AT THAT POINT. CONSIDERA-
TIONS OF GENERAL RULE 1 ALSO APPLIES.

DATUM APPLICATIONS

SPECIFIED DATUM SURFACES USING THE FORM TOLERANCE ⊥
AND POSITION ⊕ DIMENSIONING AND TOLERANCING

The edge surface tolerance zones illustrated (at * in the preceding figure) are intended to relate the edge extremities to the total part requirements. However, note that the interrelationship (perpendicularity) of these edge surfaces to each other is not controlled by the stated requirements; instead, it is subject to the considerations of Rule 1. The edge tolerance zones shown are representative of the size tolerance relationship of the individual 1.75 ± .01 and 2.00 ± .01 sizes only. Control of these surfaces in the relationship to one another would require specified form tolerances (e.g., perpendicularity), or it may depend on local workmanship or shop tolerances.

SPECIFIED DATUM SURFACES USING FORM TOLERANCES ⊥, ▱
AND POSITION ⊕ DIMENSIONING AND TOLERANCING

In the example on the facing page, the stated requirements and interpretation are identical to those in the preceding one except for the added flatness control of each of the datum surfaces. The accuracy or quality of the datum surfaces is specified.

Note that the flatness control is a refinement of other controls and is contained *within* the other form tolerance zones. Where combined form tolerances are applied, the predominant control (in this case perpendicularity) confines the tolerance latitude of additional controls (such as flatness).

Datum planes are established from the datum surfaces as previously explained; the orientation to these planes is as shown in the preceding example.

SPECIFIED DATUM SURFACES USING FORM TOLERANCES ⊥, ⊘
AND POSITION ⊕ DIMENSIONING AND TOLERANCING

AS DRAWN

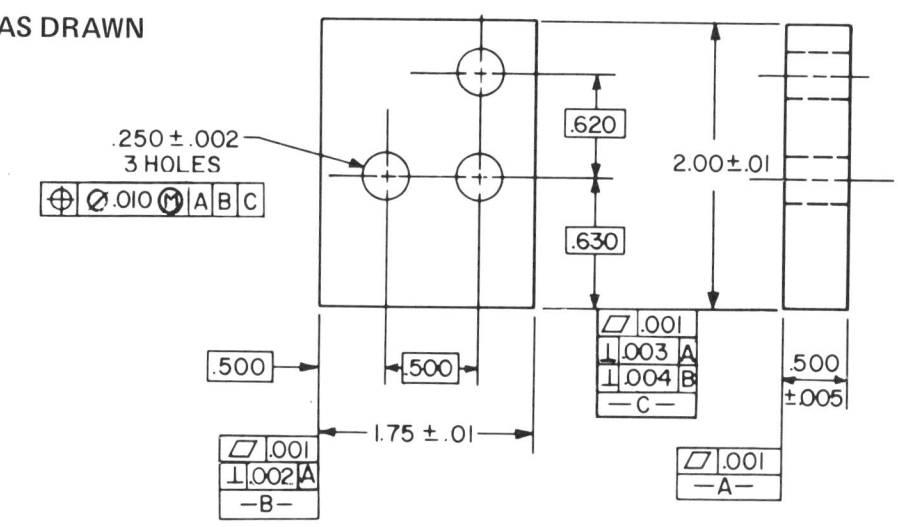

.250±.002
3 HOLES

| ⊕ | ⌀.010 Ⓜ | A | B | C |

.620
.630
2.00±.01

.500   .500
.500
±.005

⌀ .001
⊥.003 A
⊥.004 B
—C—

1.75±.01

⊘.001
⊥.002 A
—B—

⌀.001
—A—

MEANING

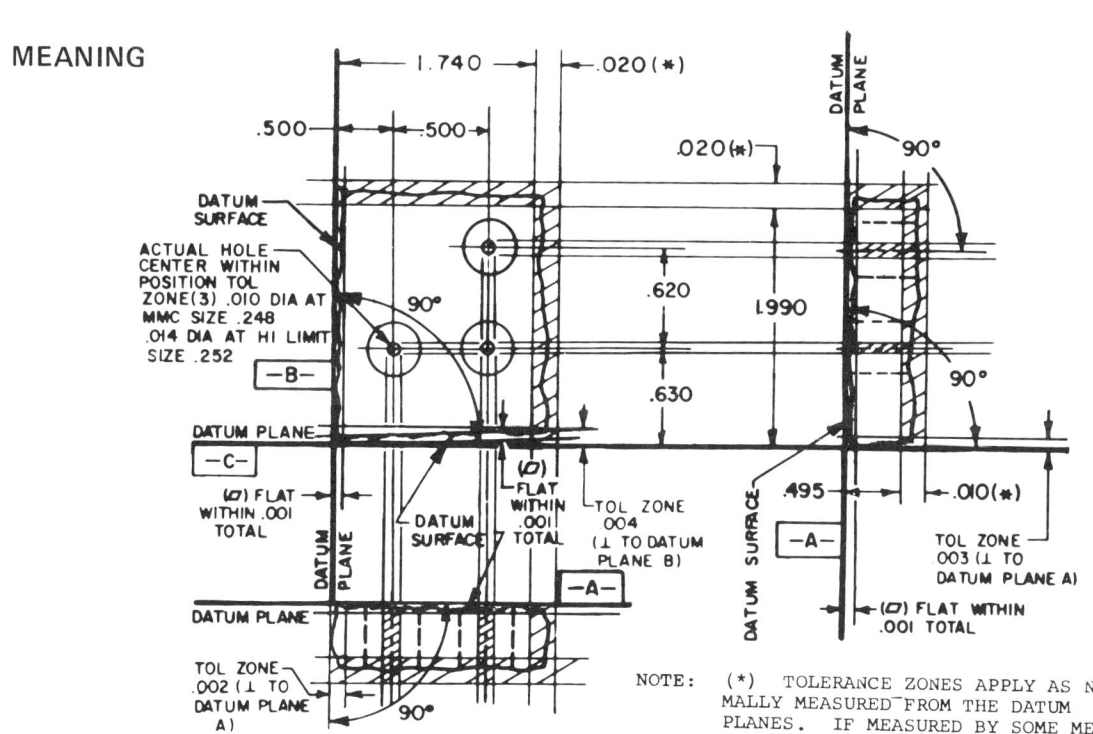

NOTE: (*) TOLERANCE ZONES APPLY AS NOR-
MALLY MEASURED FROM THE DATUM
PLANES. IF MEASURED BY SOME METHOD
WHERE REFERENCE ON THE DATUM SUR-
FACE IS NOT COINCIDENTAL WITH THE
DATUM PLANE, TOL ZONES ARE MINUS
DATUM SURFACE INACCURACY (DEVIA-
TION BETWEEN DATUM SURFACE AND
DATUM PLANE) AT THAT POINT. EDGE
TOLERANCE ZONE INTERRELATIONSHIP
TO EACH OTHER SUBJECT TO CONSI-
DERATIONS OF GENERAL RULE #1.

229

DATUM APPLICATIONS

SPECIFIED DATUM FEATURES (SURFACE, HOLES) USING FORM TOLERANCES ⊥, ▱, //, AND POSITION ⊕ DIMENSIONING AND TOLERANCING

In this example we use a variety of geometric controls. Note the manner in which datums are applied so that direct relationships may be controlled.

The large base surface in Fig. (a) is to be held flat within .001 and is also identified as datum plane A. Note that the .750 center hole diameter is to be perpendicular to datum plane A within ∅ .0015 and is also established as datum B.

Also in Fig. (a), the 1.250 diameter is to be held to a positional tolerance of ∅ .004 at MMC with respect to datum B at MMC; the 1.4375-16 UN–2A thread is to be held to a positional tolerance of ∅ .004 at MMC with respect to datum B at MMC; the face adjacent to the thread undercut is to be parallel within .003 with respect to datum A; and the 1.750 counterbored hole at left is to be held in total runout within .001 FIM with respect to datum B and is also established as datum C.

The six holes shown in Fig. (b) are to be located at position within ∅ .005 at MMC with respect to datum plane A and datum C at MMC. Datum plane A identifies the surface from which the positional pattern is related and orients the hole center tolerance zones as basically perpendicular in attitude to this plane. The theoretical center of datum hole C establishes the center of the 3.375 BASIC diameter and the 60° BASIC angles.

SPECIFIED DATUM FEATURES USING FORM TOLERANCES ⊥, ▱, ∥, AND POSITION ⊕ DIMENSIONING AND TOLERANCING

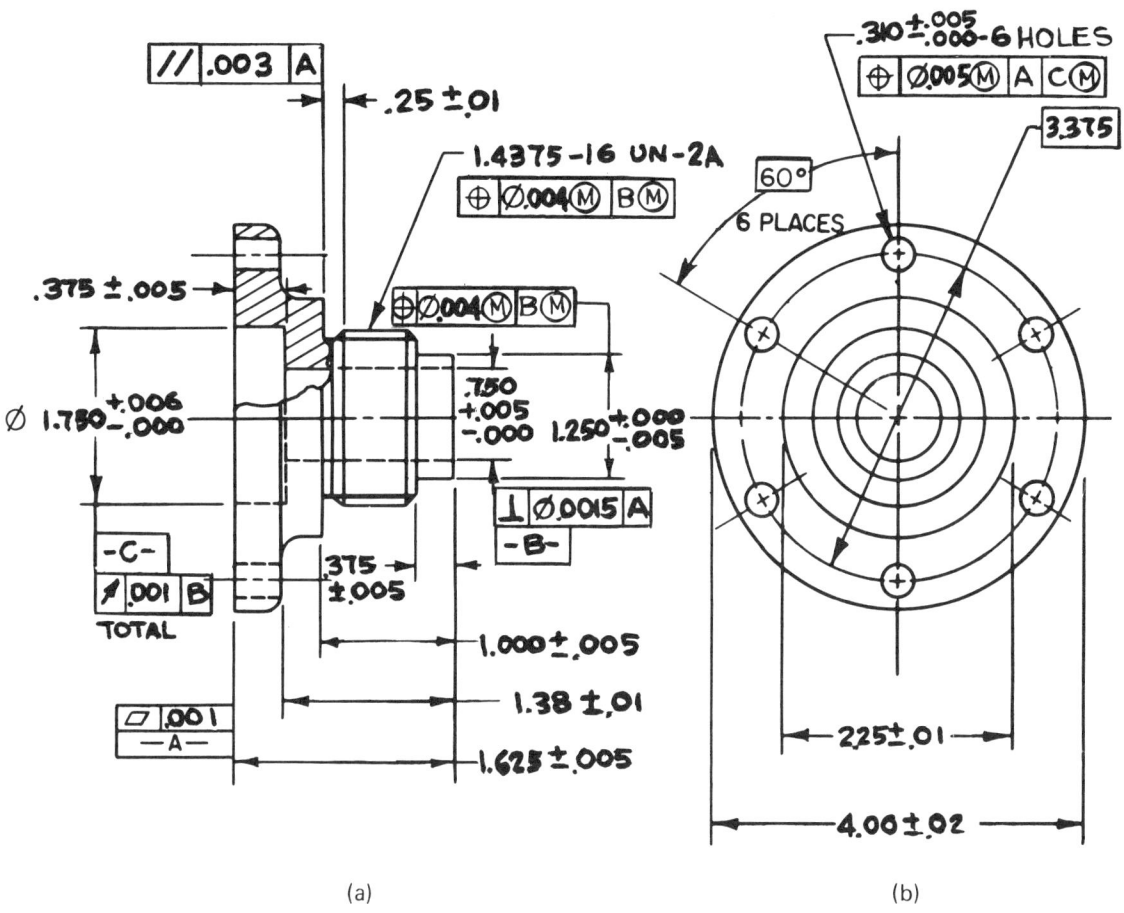

(a)   (b)

231

DATUM APPLICATIONS

SPECIFIED DATUM POINTS ON SURFACES USING POSITION ⊕ DIMENSIONING AND TOLERANCING

This example illustrates a part whose function requires that specific locating or reference points be established on the edge surfaces. These points are *datum* points and are identified by datum target symbols.

Usually, surface datum points are used either to simulate the desired location of contact with the mating parts and thus hold the related dimensions and tolerances with respect to these points, or to assure repeatable location and orientation of the part for successive fabrication and inspection operations on castings, forgings, or sheet metal parts that have irregular surfaces.

Note that the datum points are dimensioned with BASIC or untoleranced dimensions. By interpretation, BASIC dimensions are theoretically exact and, in this case, represent the ideal location for which to strive in order to relate the holes with the part edges. In these circumstances and in actual applications, the BASIC dimensions are assumed to represent exactness within tooling, gaging, or shop tolerances. Where necessary, toleranced locations or sizes may be used with datum target symbols.

Actually, as previously explained, the datum points establish datum planes. Thus, the position pattern is simultaneously related to these points and the planes as established from these points.

The two holes on the left, the $.300 \pm .002$ holes, are to be located at position within $\varnothing .006$ at MMC with respect to datum planes A, B, and C. The third hole at the right, the $.130 \, ^{+.002}_{-.000}$ hole, which also is a part of the hole pattern, is to be located at position within $\varnothing .004$ at MMC with respect to datum planes A, B, and C.

The meaning illustrates the relationship of the hole pattern to the datum points and established planes.

Datum points B1 and B2 in the left view establish a datum plane B at 90° BASIC to the top datum plane A. From this plane B, the vertical dimensions are taken. Thus, these dimensions are related *only* to points B1 and B2 and their common plane and *not* to the entire surface.

Datum point C1 establishes a third plane, C, 90° BASIC to both datum planes A and B. Only one point is, of course, necessary on the third plane. The horizontal dimensions are related *only* to point C1 and its common plane, C, and *not* to the entire surface.

Note that the positionally toleranced pattern of holes is oriented with respect to the established datum planes, since, by interpretation, all center planes are basically parallel and perpendicular to the edge datum planes.

The actually produced hole centers are permitted to vary within the tolerance zones stated on the drawing. However, as the positional relationships are specified at MMC, the positional tolerance of the two larger holes may possibly increase from $\varnothing .006$ (at MMC size of .298) to $\varnothing .010$ when the holes are at LMC size, or largest size, of .302. The positional tolerance of the third hole may possibly also increase from $\varnothing .004$ (at MMC size of .130) to $\varnothing .006$ at LMC size, or largest size, of .132.

AS DRAWN

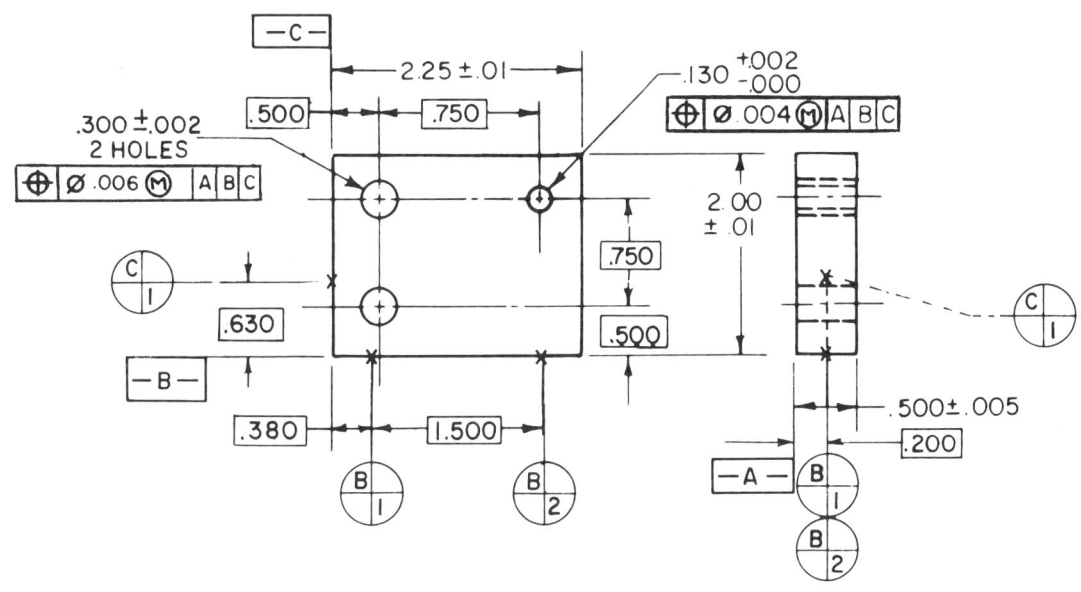

MEANING

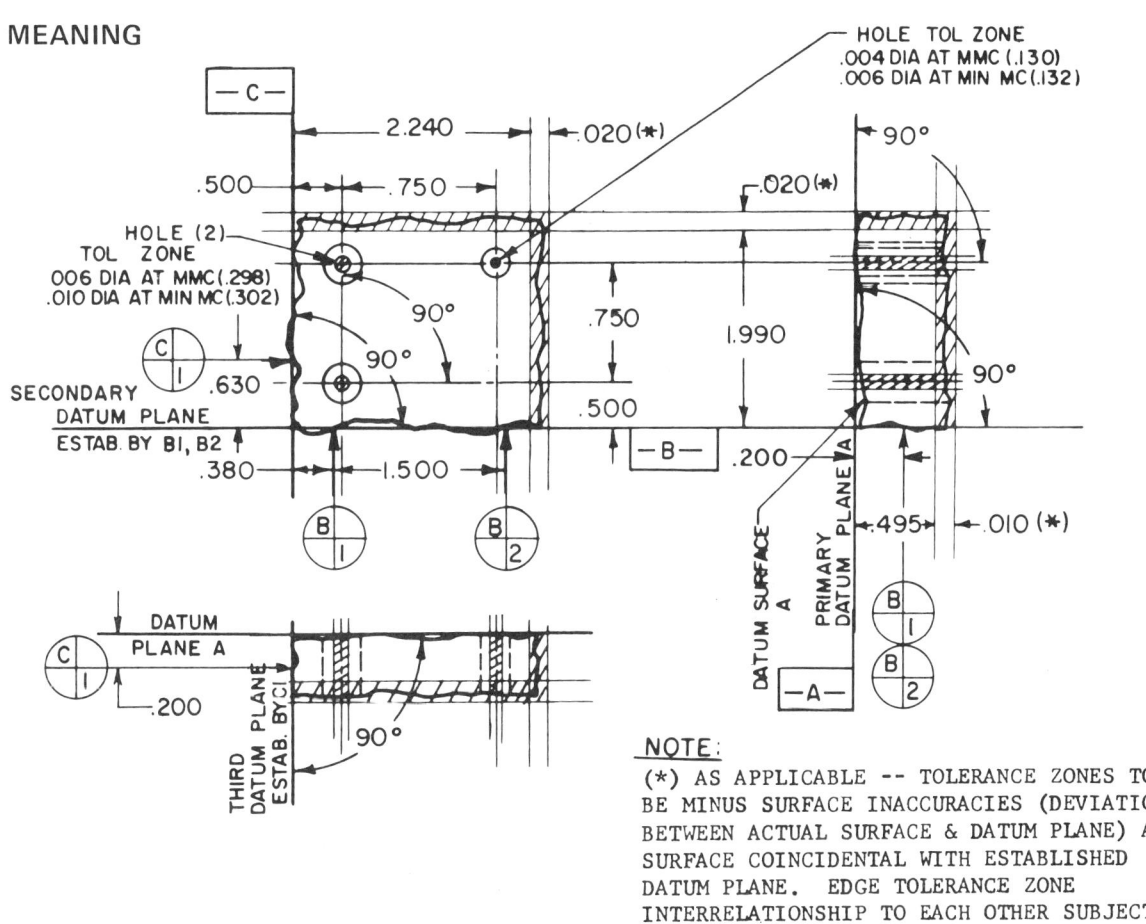

NOTE:
(*) AS APPLICABLE -- TOLERANCE ZONES TO
BE MINUS SURFACE INACCURACIES (DEVIATIONS
BETWEEN ACTUAL SURFACE & DATUM PLANE) AT
SURFACE COINCIDENTAL WITH ESTABLISHED
DATUM PLANE. EDGE TOLERANCE ZONE
INTERRELATIONSHIP TO EACH OTHER SUBJECT
TO CONSIDERATIONS OF GENERAL RULE #1.

DATUM APPLICATIONS

SPECIFIED DATUM POINTS ON SURFACES USING POSITION ⊕ DIMENSIONING AND TOLERANCING

Note again that when using this form of position dimensioning and tolerancing with datum points, the position pattern and thus the individual holes are related to the datum points and their established planes only—they are *not* related to the entire surface.

The deviations or imperfections of the remainder of the actual surfaces on which the specified datum points are established are controlled by the overall coordinate dimensions. The edge surface tolerance zones illustrated are intended to relate the edge extremities to the total part requirements. However, note that the relationship (perpendicularity) of the edge surfaces to each other is *not* controlled by the stated requirements. This interrelationship is subject to Rule 1. The edge tolerance zones shown are representative of the size tolerance relationship of the individual 2.00 ± .01 and the 2.25 ± .01 sizes only. Control of these surfaces in relationship to one another would require specified form or attitude tolerances (e.g. perpendicularity), or may depend on local workmanship or shop tolerances.

Typical applications of this form of dimensioning and tolerancing are an irregular surface, such as a casting, which requires specific tooling or gaging pickup for repeatability, or a part whose function requires specified location of points of contact or relationship with a mating part. When we locate these points and identify them, we ensure that specific design or tooling and gaging requirements are clearly stated and that the manufacturing and inspection processes will follow well-defined instructions in agreement with the part requirements.

Usually, in an application such as this, the datum points would be picked up by locating buttons. There is no implication that detents are to be applied to the part surface. If detents are desired, a note to this effect should be added to the drawing.

If positional dimensioning located by BASIC dimensions from an edge surface is employed and *no* datum points nor planes are specified, the high points or extremities of the implied datum surface will be used to establish datum planes and orient the position pattern of features to the edges. In this case, we would use conventional *flat* locating surfaces instead of buttons. Although the resulting control of the relationship of the position pattern of features to specific portions of the part edges would be less stringent, it may be adequate in many applications.

AS DRAWN

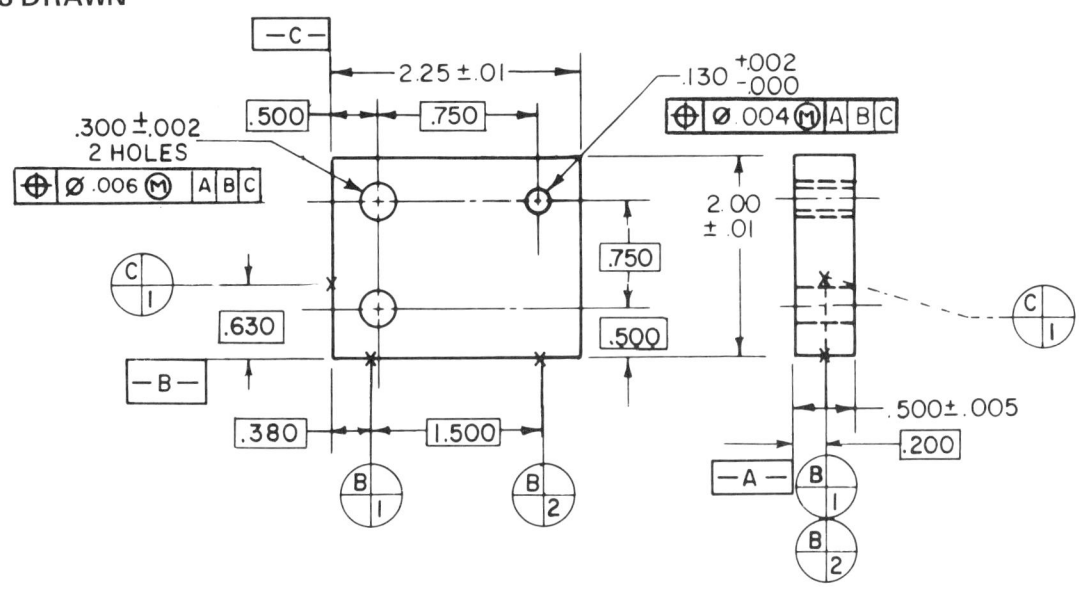

MEANING

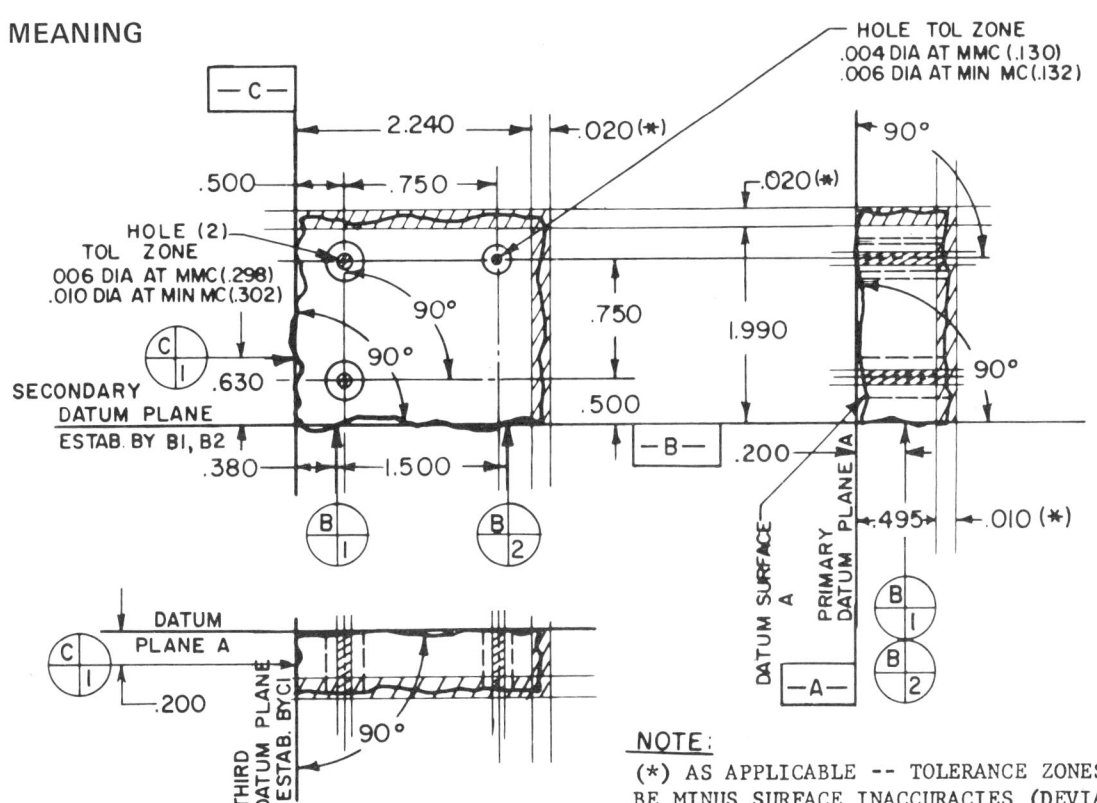

HOLE TOL ZONE
.004 DIA AT MMC (.130)
.006 DIA AT MIN MC (.132)

HOLE (2)
TOL ZONE
.006 DIA AT MMC (.298)
.010 DIA AT MIN MC (.302)

SECONDARY
DATUM PLANE
ESTAB. BY B1, B2

DATUM SURFACE A

PRIMARY DATUM PLANE A

DATUM PLANE A

THIRD DATUM PLANE ESTAB. BY C1

NOTE:

(*) AS APPLICABLE -- TOLERANCE ZONES TO
BE MINUS SURFACE INACCURACIES (DEVIATIONS
BETWEEN ACTUAL SURFACE & DATUM PLANE) AT
SURFACE COINCIDENTAL WITH ESTABLISHED
DATUM PLANE. EDGE TOLERANCE ZONE
INTERRELATIONSHIP TO EACH OTHER SUBJECT
TO CONSIDERATIONS OF GENERAL RULE #1.

In this example, we expand the use of datum points on surfaces. We see that selecting specific datum or reference points as the basis for establishing relationships to other features often has the advantage of providing effective and economical methods of achieving design intent or part function.

This illustration shows a rotor. It is not completely dimensioned—only enough to explain how the datums are applied.

Datum points A1 and A2 specify the location of two specific points on the rotor shaft which are to be the basis for the runout relationship with the rotor outside diameter.

Datum points A1 and A2, because they are points on a diameter, actually apply to an infinite number of points around the circumference of the shaft. As was previously explained with respect to the BASIC dimensions, exactness within tooling, gaging, or shop tolerances is implied. Where necessary, toleranced locations or sizes may be used with datum target symbols.

In actual applications, the gaging pickup of the datum points would, for example, be accomplished by two vee–blocks of very narrow width. The datum points essentially simulate the location of the bearing races and the accuracy required at these points. This eliminates the need for holding the critical runout requirement on a larger portion of the shaft length than is necessary to ensure proper function of the part. The remainder of the shaft is then controlled by shop tolerances or other specifications.

The runout notation would be interpreted to mean "the rotor diameter is to be within .001 FIM total runout with respect to datums A1 and A2 (and their common datum cylinder and axis) simultaneously."

AS DRAWN

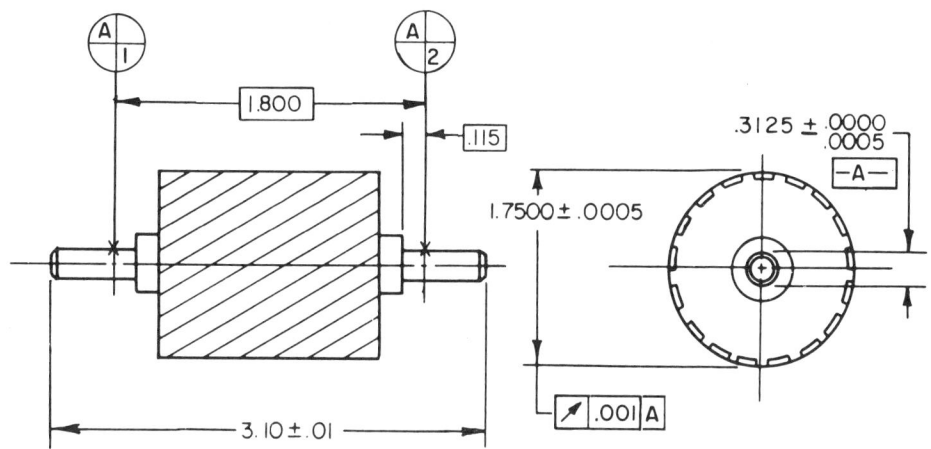

MEANING

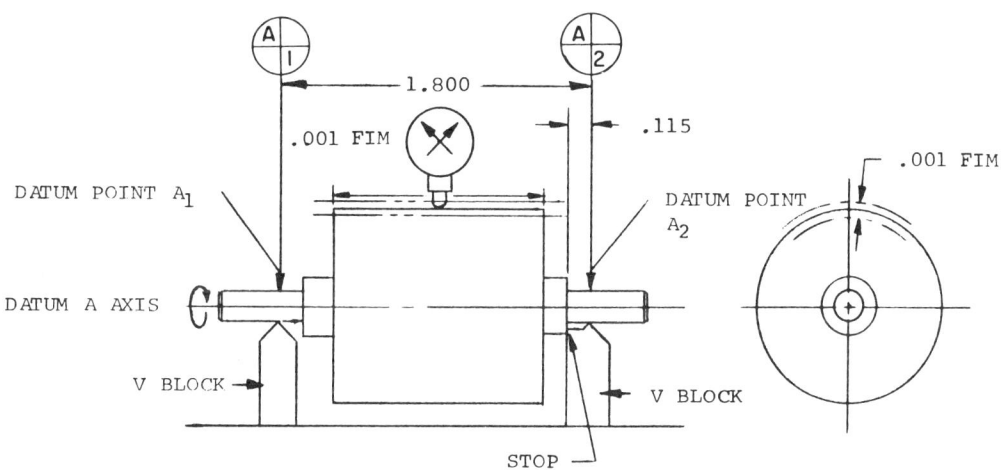

DATUM POINTS A_1 & A_2 CREATE DATUM AXIS A.

1.7500 DIA TO BE WITHIN TOTAL RUNOUT OF
.001 FIM WHEN MOUNTED ON POINTS A_1 & A_2.

237

DATUM APPLICATIONS
SPECIFIED DATUM POINTS ON SURFACES

Figure 1 shows a selection of two datum points, A1 and A2, used to specify a relationship of parallelism between two points on the part surface and the elongated slot.

FIGURE 1

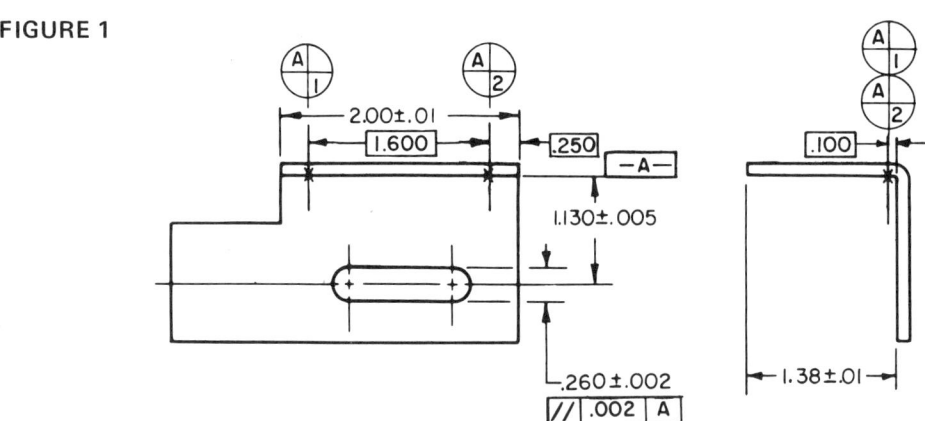

Figure 1 shows that these two datum points, A1 and A2, or more specifically, the theoretically exact datum line A extending between these two datum points (and not the entire surface), establish the datum reference from which the parallelism of the slot is to be established.

The surface upon which datum points A1 and A2 are established could have curvature or inaccuracies which would have no effect on the parallelism requirement. The accuracy of the entire surface is controlled by the vertical 1.130 ± .005 dimension. The parallelism requirement is specified to be from the common datum line A established by datum points A1 and A2 to the center line of the slot within .002 total.

Figure 2 shows datum point A1 (at the left) established on the angular surface of the part. Datum point A1 is used in conjunction with the center of datum feature B to establish a center plane which determines the reference from which the 30° BASIC angle to the slot is taken.

FIGURE 2

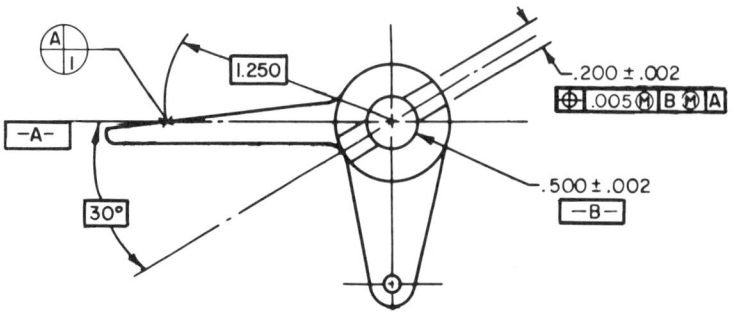

238

SPECIFIED DATUM AREAS ON SURFACES

FIGURE 3

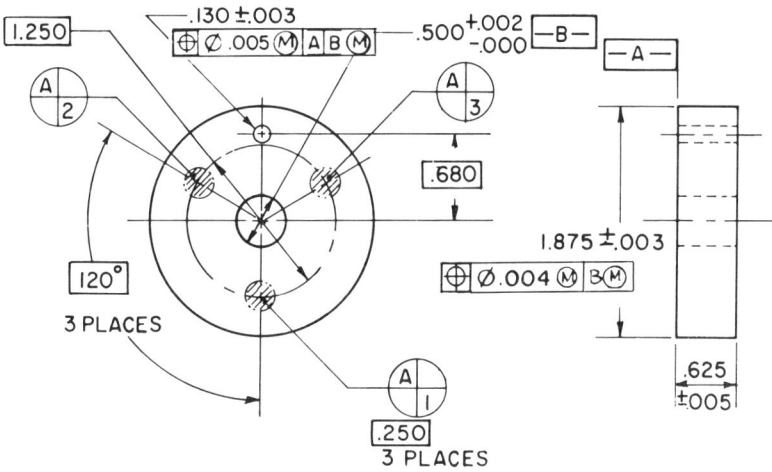

If the datum reference must be taken from given small areas on the primary datum surface (Fig. 3), the desired datum surface area may be shown as a diameter (or other shape), with location and size of the area controlled by BASIC dimensions. The datum surface area limits are enclosed (with phantom-line circles in this case) and the area is identified by diagonal slash lines.

The extremities of the three datum surface areas A1, A2, and A3 combine to form a single, or common, datum plane A.

Three .250 flat-ended buttons or locators contacting the surface at the prescribed locations will provide the actual location. The other requirements are then related only to these datum surface areas and their common resulting plane and not to the entire surface.

DATUM APPLICATIONS
SPECIFIED PARTIAL DATUM SURFACES

In some applications, it is convenient to select larger areas of surfaces as datums.

In Fig. 1, portions of the shaft diameter identified as datums A1 and A2 are the basis for the runout relationship to the shaft step diameter.

FIGURE 1

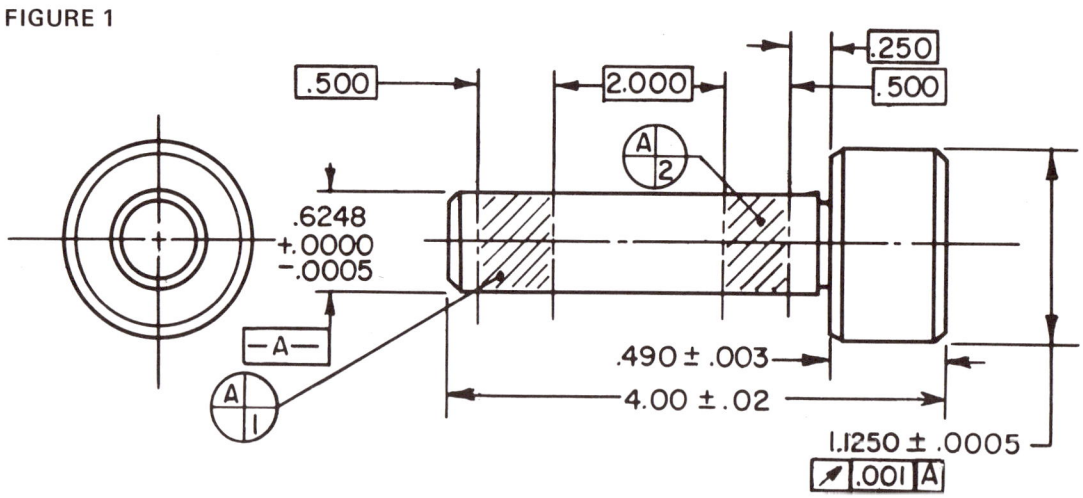

As in the previous example on datum points, the datum reference applies to the circumference around the shaft contained within the limits of the basic dimentions. The extent of the datum surfaces is the .500 BASIC wide reference bands or areas on the shaft diameter. These datum surface areas simulate the location of bearing races.

It can be seen that the shaft step diameter of 1.250 ± .0005 is required to be held within total runout of .001 with respect to the two datum surface diameters A1 and A2 and their established common datum reference axis A. The remainder of the shaft is permitted to deviate within more lenient standard or shop tolerances.

As in the previous examples, exactness within tooling, gaging, or shop tolerances with respect to the basic dimensions, is implied.

Figure 2 illustrates a metal blade. The 1.400 ± .010 dimension is to be held between the lower flat surface and the contact button. However, in addition, there is a .005 total parallelism requirement from that portion of the lower surface identified as datum A1 and the button.

The limitations of partial datum surface A and its established datum reference plane are specified by a phantom line and a BASIC dimension, with the area identified by diagonal slash lines drawn across the surface. This clearly shows that the surface so identified determines the datum reference plane.

FIGURE 2

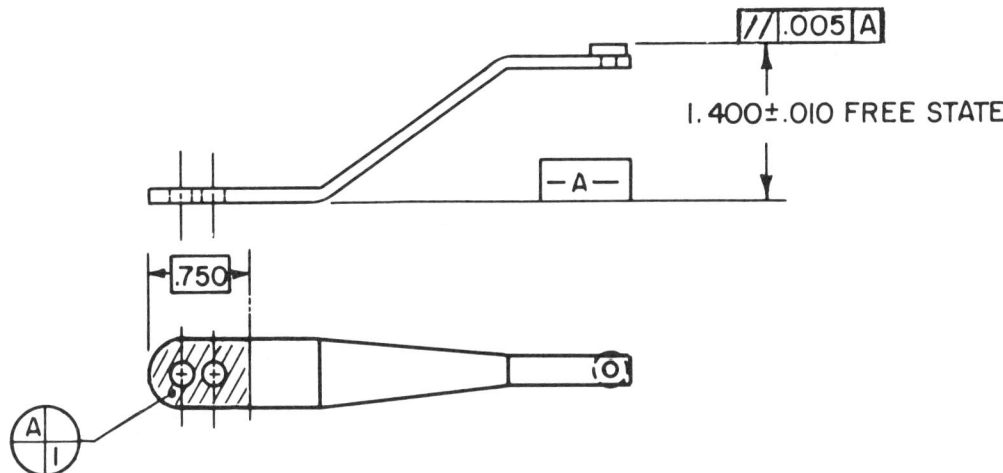

Figure 3 is another example in which a portion or area of a surface serves as a datum reference. This application was based on the part function and the anticipated difficulty of holding accuracy on any more of the formed part flat surface than necessary.

FIGURE 3

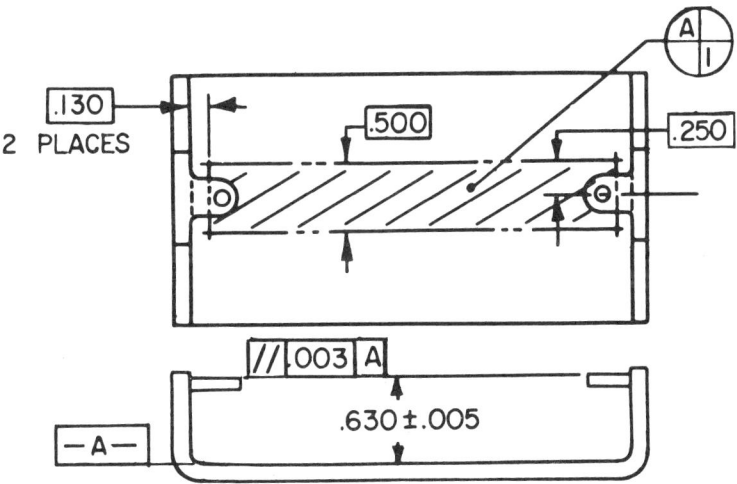

Note the manner in which the basic dimensions have been used to define the limits of the datum area A1. Area A1 establishes the datum reference plane A from which the measurement is taken.

The formed ears of the part are to be held parallel within .003 to the datum plane A.

Again, exactness within tooling, gaging, or shop tolerances is implied by the BASIC dimensions.

DATUM APPLICATIONS
SPECIFIED MULTIPLE DATUM SURFACES

In this example, three mounting bosses of a cast part are selected as datum target surfaces. The purpose is to clearly specify that the three bosses *only* are to establish datum plane A for the relationship of the parallelism requirement on the top surface. The extremities of the three individual boss surfaces would combine to provide the three-point (minimum) contact, establishing datum plane A.

The top surface of the part is to be parallel within .003 to the established datum plane A. This guarantees the assembled part relationship of the parallelism requirement to the mounting bosses.

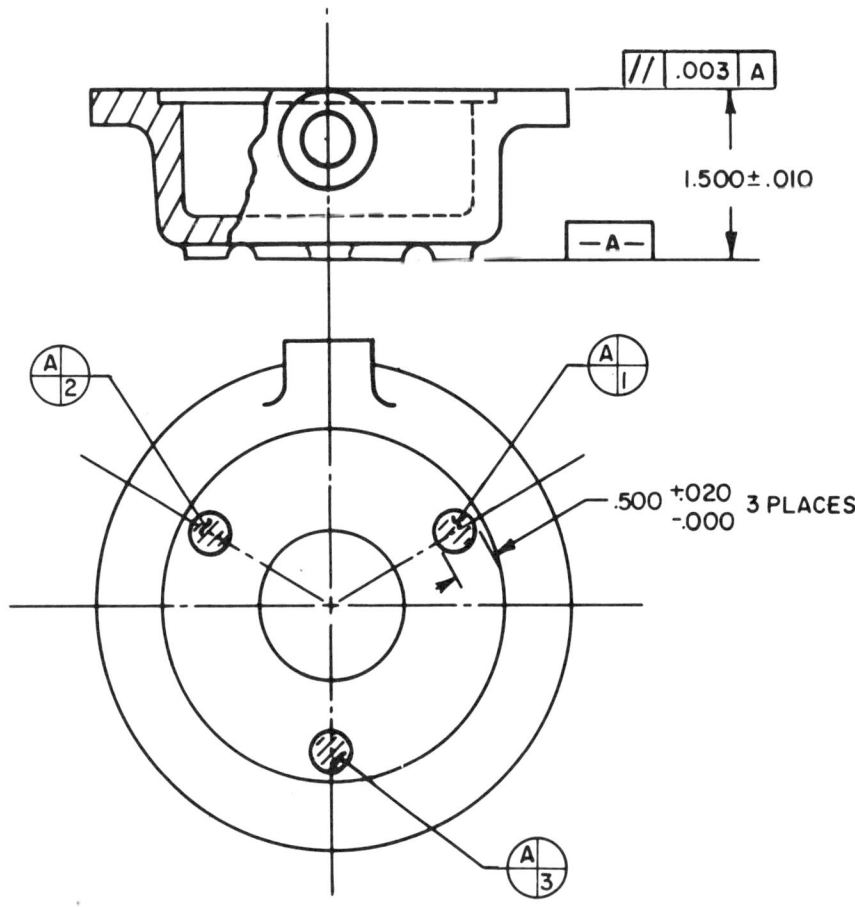

SPECIFIED "STEP" DATUMS

When we establish datum planes of orientation on parts of irregular shape such as a casting, the datum target points (and thus their datum pin locators) may involve surfaces which are at different levels and must therefore be *combined* to establish the desired plane of orientation.

Referring to the illustration on the next page, we see that datum points C1 and C2 are situated at different surfaces and at different levels (steps) in order to relate the entire bulk of the part to a common plane of orientation. This plane, identified as the common plane C, may then be related to the primary datum B and the tertiary datum D with the usual implied mutually perpendicular datum plane relationship. The step basic dimension is to be maintained within tooling or gaging tolerance accuracy.

Other features of the part, such as the position toleranced pattern, can then be related to the datum system as established from these "step" datums and the other interrelated datums to ensure compliance with design requirements and repeatability of manufacturing and inspection operations.

SPECIFIED "EQUALIZING" DATUMS

Page 245 illustrates a part on which the orientation is established from specified datum targets (lines and points) as a means of "equalizing" the bulk of the part to centralize the location of the three holes. Through the datum target method as specified, a guarantee of design requirement fulfillment is provided, and tooling and gaging repeatability is assured. The two planes of orientation necessary to this part are provided by the equalizing and centralizing placement of the datum targets. The specified position relationships controlling the hole locations are then established from the datum system A, B.

The primary datum plane A is established from points A1, A2, and A3 at the two levels of the part and relative to the step difference. These points and the step difference combine to give the common primary datum plane of orientation.

The secondary datum plane B is established by the equalizing action of the fixed and movable vee's as they contact tangent to the part at an angle of 45°. Datum center plane B is perpendicular to datum plane A. Some slight "cam-down" or hold-down of the part may be necessary in actual applications.

The effect of any "step" tolerance as it relates to the features located from this datum should be considered.

DATUM APPLICATIONS
SPECIFIED "STEP" DATUMS

AS DRAWN

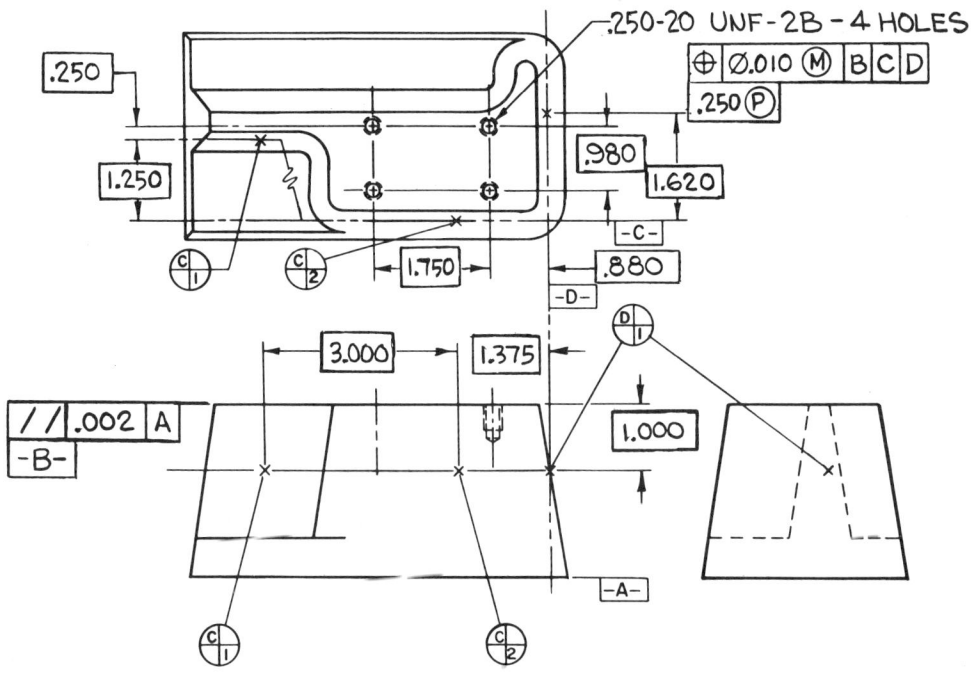

MEANING

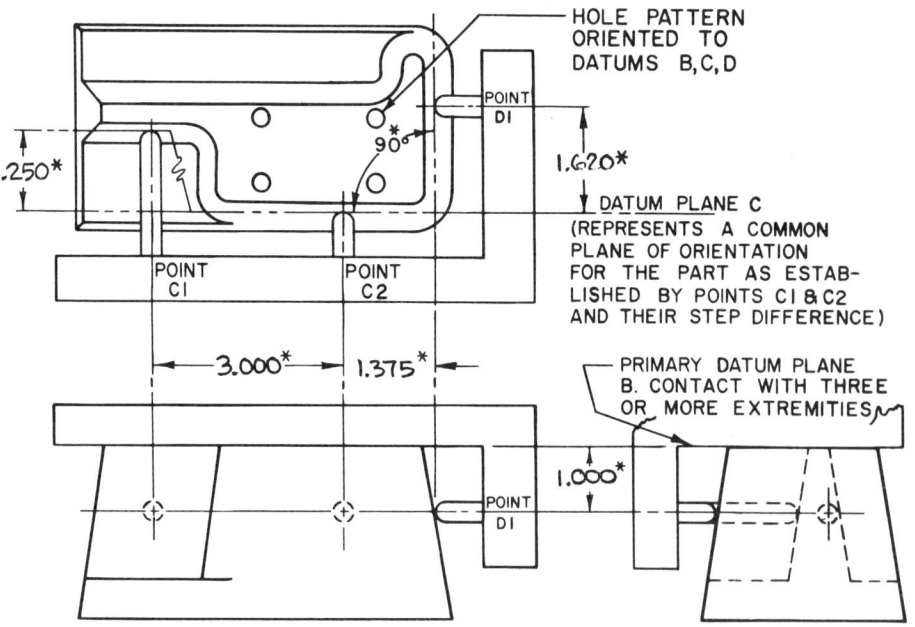

*DIMS MUTUALLY ⊥, ∥, & ON LOCATION WITHIN STANDARD OR ESTABLISHED TOOLING OR GAGING TOLERANCES UNLESS OTHERWISE SPECIFIED.

SPECIFIED "EQUALIZING" DATUMS

AS DRAWN

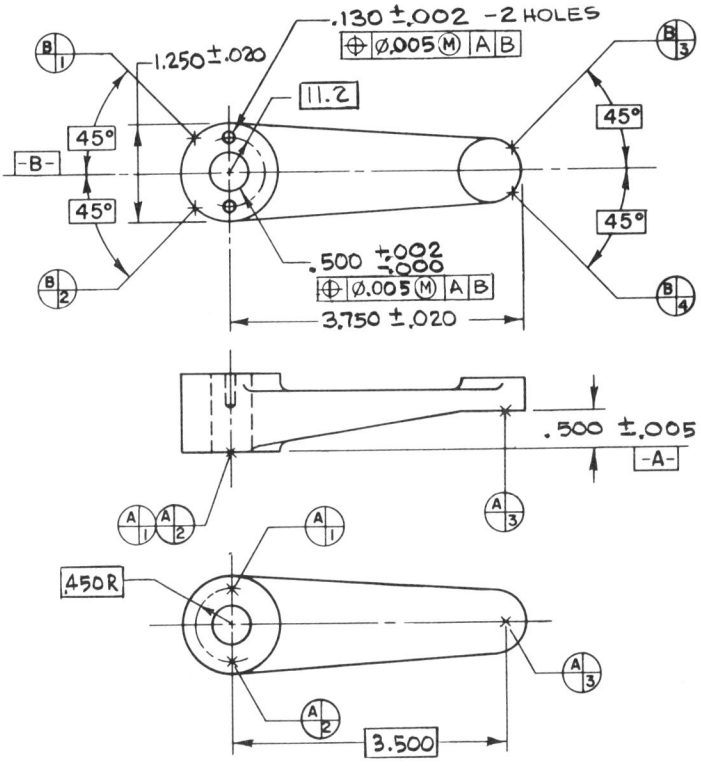

MEANING

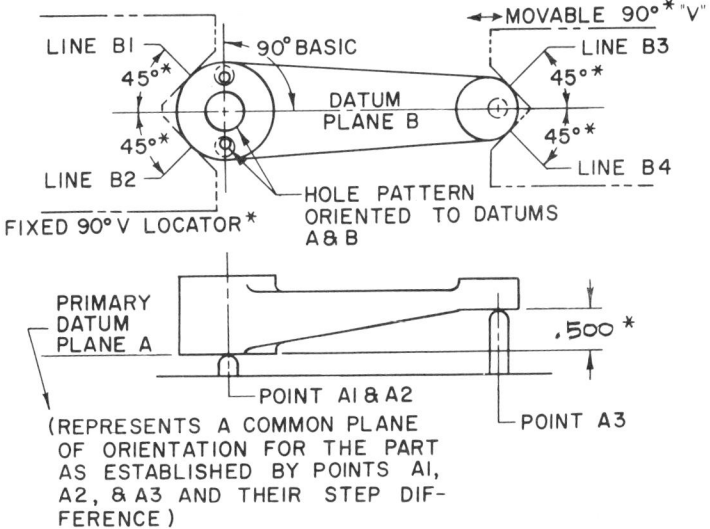

*DIMS. WITHIN STANDARD OR ESTABLISHED TOOLING OR GAGING TOLERANCES.

EXTENDED DATUM PRINCIPLES

MULTIPLE DATUM SYSTEMS WITH NONCOMMON DATUMS—
SEPARATE SYSTEMS

In multiple datum systems with noncommon datum references established on the same part to control specific part feature relationships, *no* specific relationship may exist between the datum systems. They are treated separately.

AS DRAWN

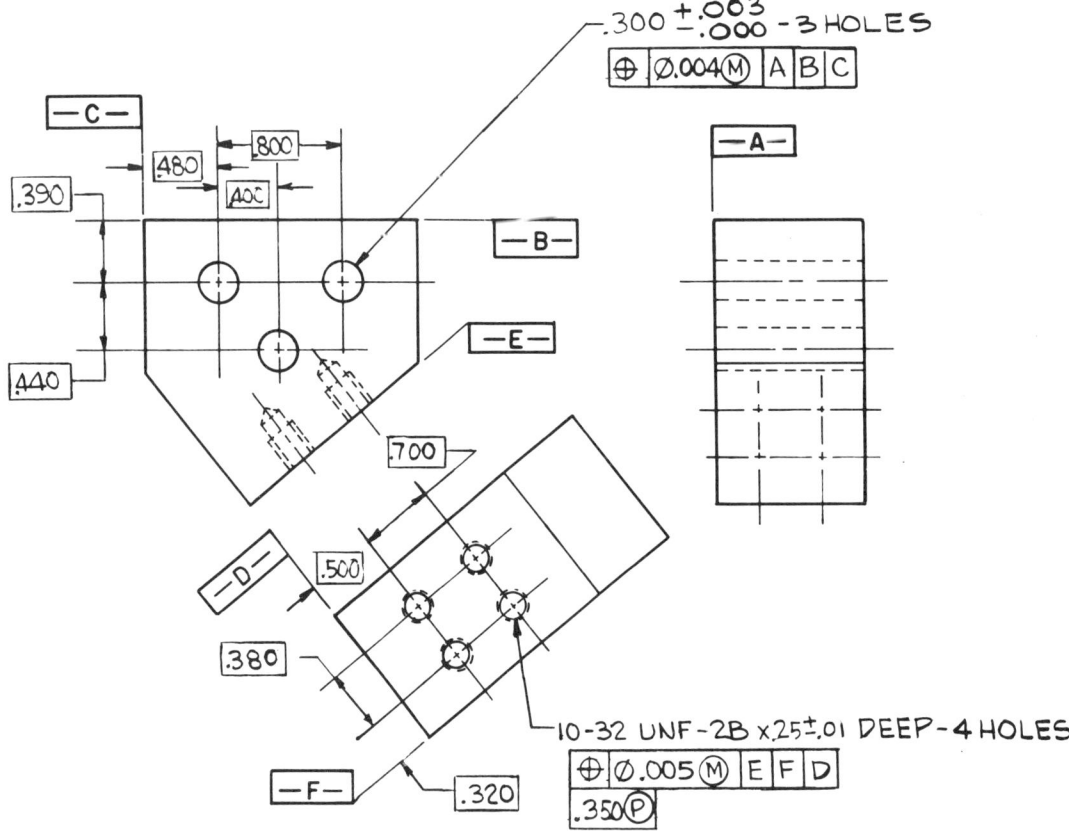

MEANING

ONE SYSTEM

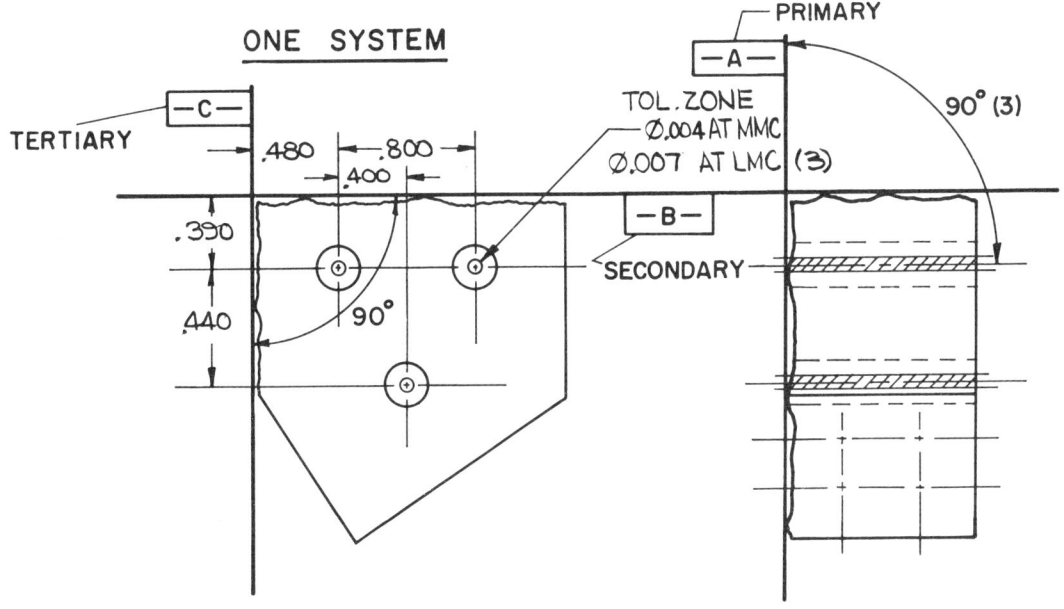

OTHER SYSTEM

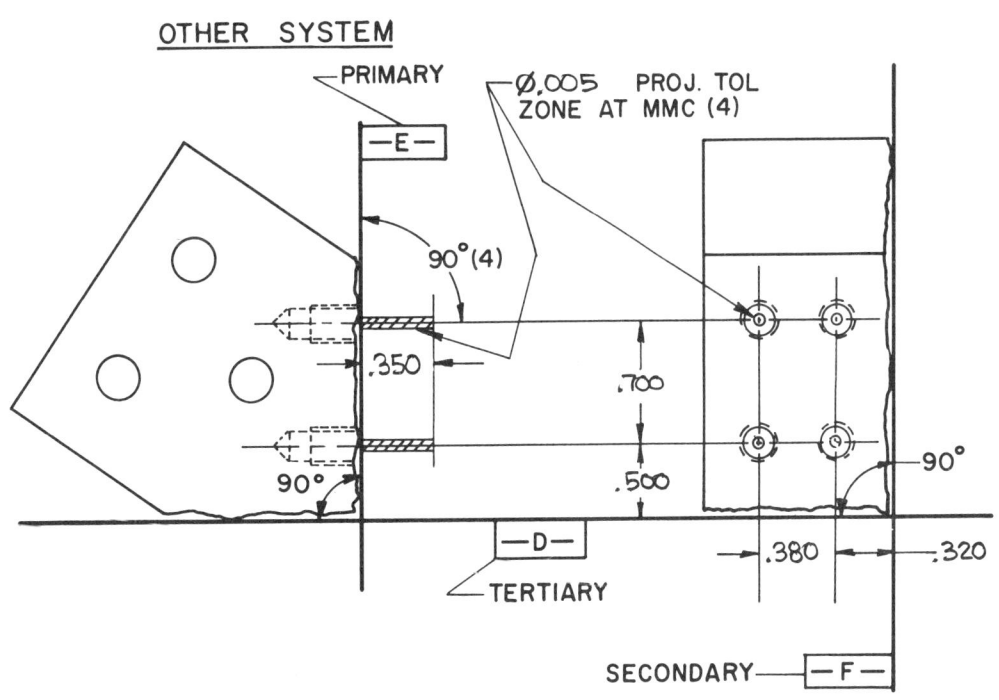

EXTENDED DATUM PRINCIPLES

INTERRELATED DATUM SYSTEMS—WITH COMMON DATUMS AND IMPLIED RELATIONSHIPS BETWEEN SYSTEMS

Datum references, either single or double datums, or complete three-plane datum reference systems, which are interrelated on the same part with other datum systems, retain a relationship with the other datum systems to the extent of common datum references.

In some cases, one datum system established within or related to another through common datum references involves refinement or extension of the other datum system, and, if necessary for orientation, may retain a relationship with the other datum system as implied by the drawing.

There are two position toleranced hole patterns on this part. The first one involves a single hole which is established relative to the datum system framework A, B, C. This hole then becomes the secondary datum for the interrelated second datum system framework A, D, B to which the second position pattern is related.

AS DRAWN

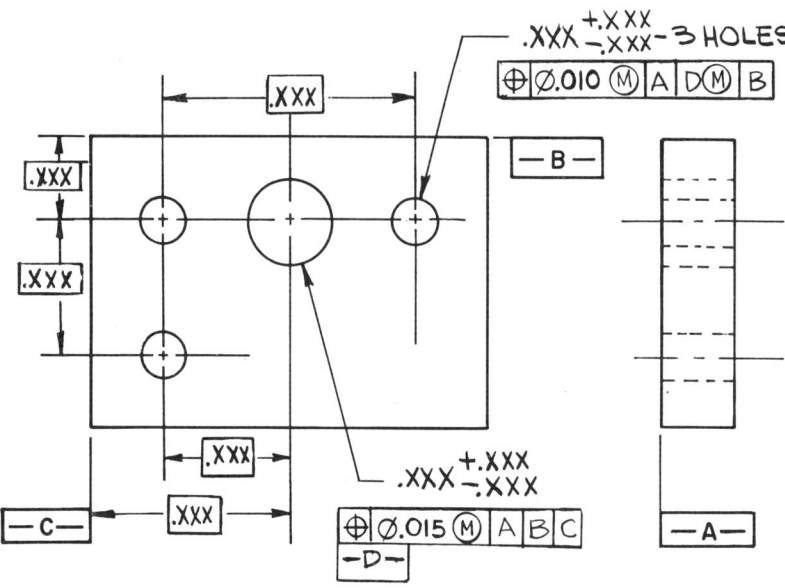

Datum D is directly related to the main datum system framework (as established by A–B–C) by specification. The three-hole pattern is then related to datums A, D, and B. Datum A is repeated for clarification to ensure correct selection of the primary datum.

Since the establishment of the interrelated datum system A–D–B within the main datum system A–B–C is to achieve a refinement of the relationship between the three small holes and the large hole (datum D), the positional relationship is to be between these features only as specified. Orientation (or rotation) is controlled by datum B as the tertiary datum.

MEANING

FIRST REQUIREMENT   (FIRST RELATIONSHIP-TO DATUM SYSTEM
A, B, C.)

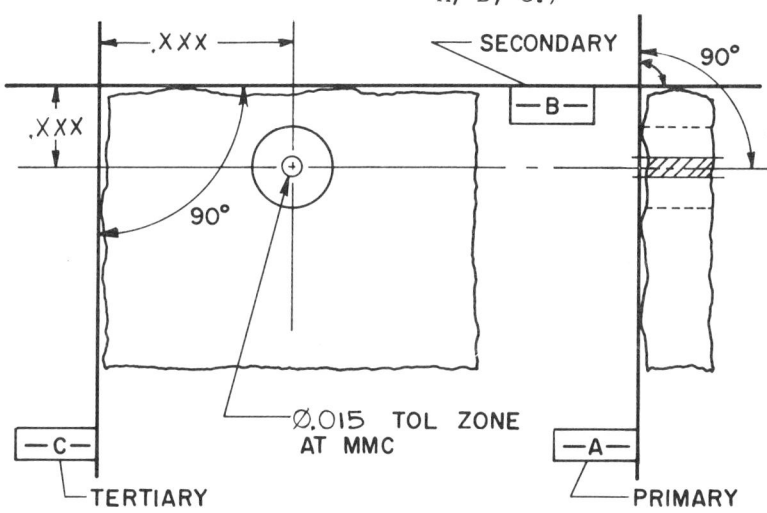

SECOND REQUIREMENT   (SECOND RELATIONSHIP-TO INTERRELATED DATUM SYSTEM A,D &C)

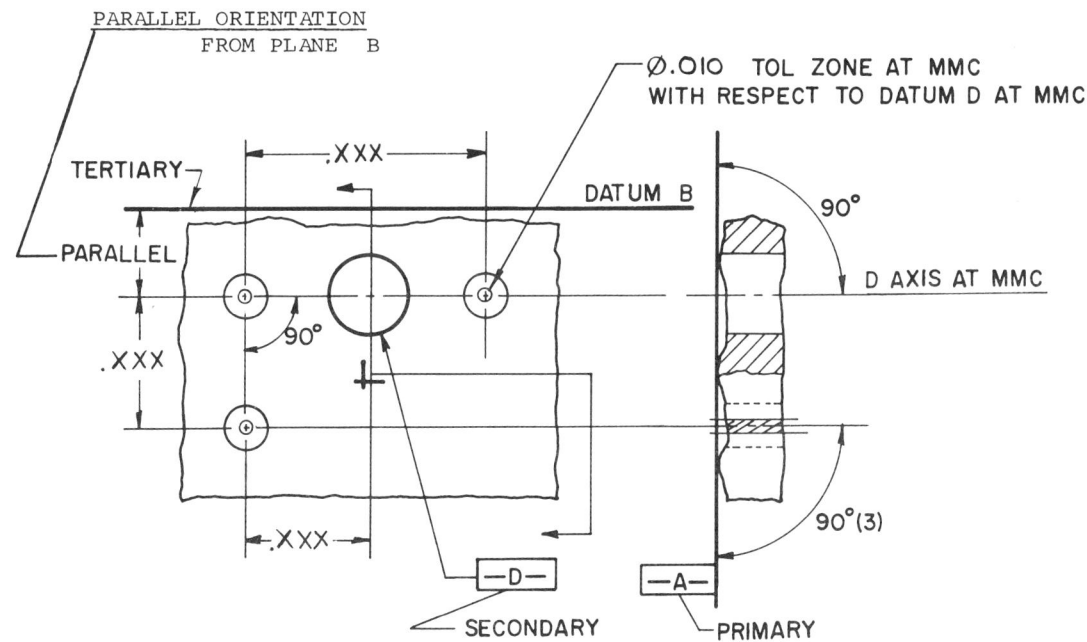

249

ESTABLISHING DATUMS

INTERRELATED DATUM SYSTEMS—WITH COMMON DATUMS, ONE SYSTEM BUILT UPON ANOTHER

There are two positionally toleranced hole patterns on this part. The first pattern is related to a complete datum system established by A, B, C. The two holes established with the first datum system are then used with the common datum A and become the secondary and tertiary datums for the second complete datum system A, D, E.

Note that this is an example illustrating the rather common requirement of pickup of *two* reference holes from which other features are to be related.

AS DRAWN

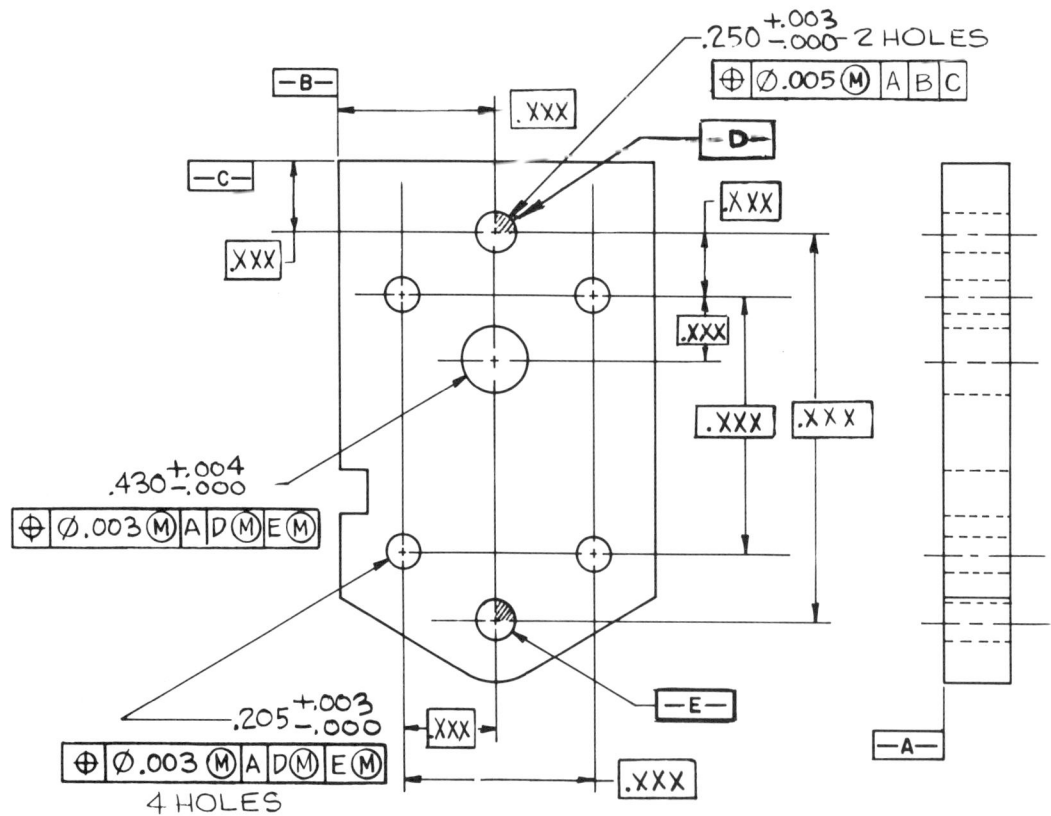

MEANING

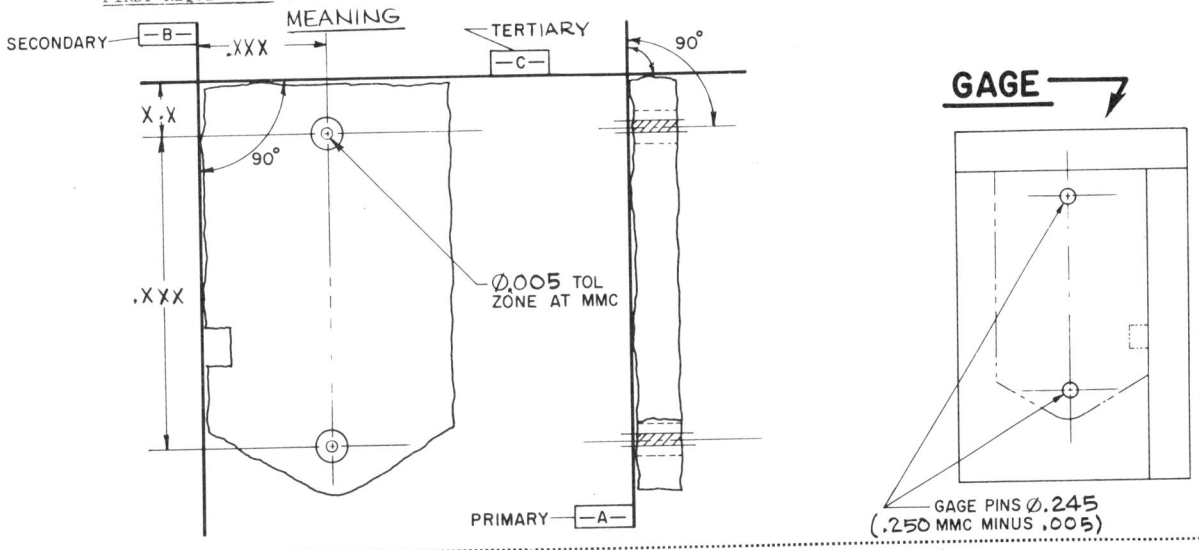

MEANING

SECONDARY —B—   .XXX   TERTIARY —C—   90°

X.X

90°

.XXX

Ø.005 TOL ZONE AT MMC

PRIMARY —A—

GAGE

GAGE PINS Ø.245
(.250 MMC MINUS .005)

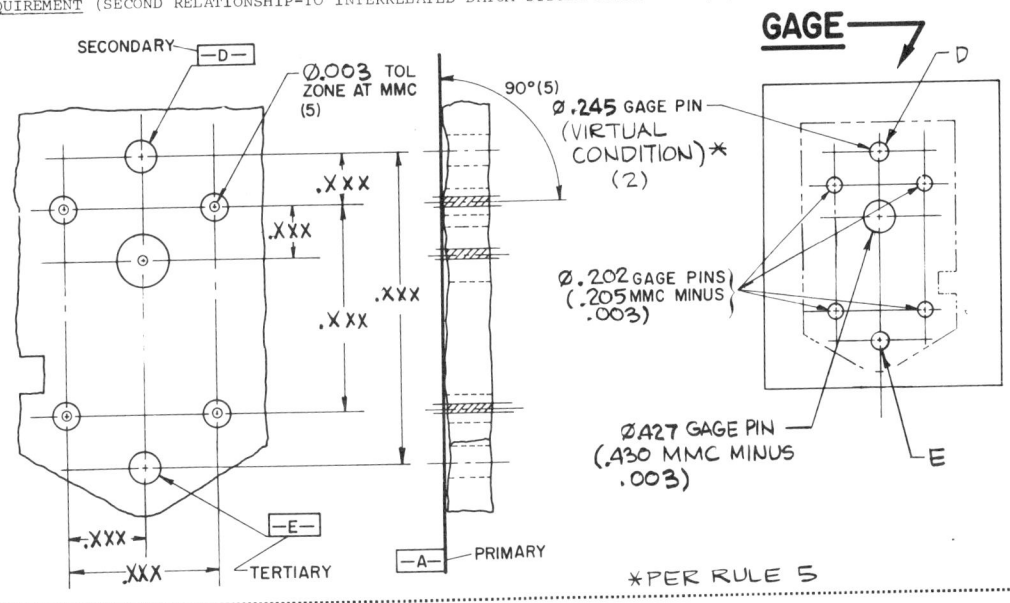

SECONDARY —D—   Ø.003 TOL ZONE AT MMC (5)

90°(5)   Ø.245 GAGE PIN (VIRTUAL CONDITION)* (2)

.XXX
.XXX
.XXX
.XXX

Ø.202 GAGE PINS
(.205 MMC MINUS .003)

ØA27 GAGE PIN
(.430 MMC MINUS .003)

GAGE   D

E

—A— PRIMARY

.XXX
.XXX   —E— TERTIARY

*PER RULE 5

The similarity of the above situation to the simpler one shown here on a cylindrical part may serve to clarify this principle of orientation through the secondary and tertiary datums.

Note that the secondary datum cylinder D on this cylindrical part could be compared to the secondary datum hole (cylinder) D on the above flat part and that datum hole E on both parts give orientation (rotation control) as the tertiary datum. A complete datum system framework is established in both parts to which other features on the parts can be related.

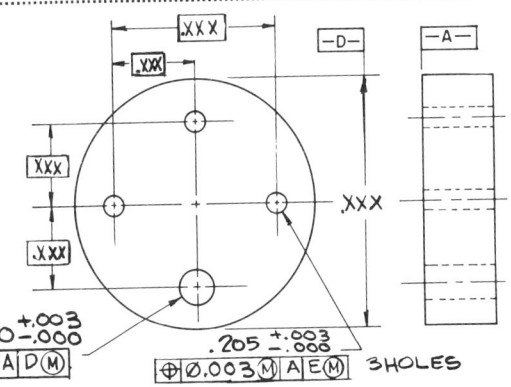

.XXX
.XXX   —D—   —A—

XXX
XXX
XXX
XXX

.250 +.003 -.000
⊕ Ø.003 Ⓜ A D Ⓜ
—E—

.205 +.003 -.000
⊕ Ø.003 Ⓜ A E Ⓜ   3 HOLES

251

EXTENDED DATUM PRINCIPLES

INTERRELATED DATUM SYSTEMS—WITH COMMON DATUMS BUT ESTABLISHING SEPARATE SYSTEMS

In some cases a relationship between interrelated datums is established only to the extent of the common datum references. Hence separate datum systems may be established on the same part which are related only to the extent of their common datum references, with no further orientation implied or necessary.

MEANING

1. Several *separate* datum systems are established using common datums.

2. Datum C (slot) is a common tertiary datum for position toleranced hole patterns on each end of the part.

3. Datum D (hole) is a primary datum for runout of the step diameter, and is a secondary datum for runout of the counterbore and position of the slot.

AS DRAWN

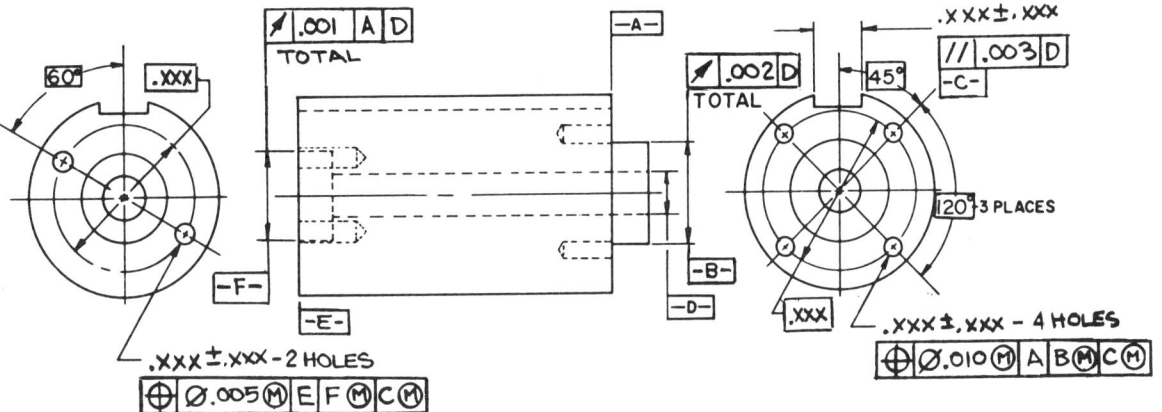

EXTENDED DATUM PRINCIPLES

INTERRELATED DATUM SYSTEMS—FUNCTIONAL TOLERANCE TO DATUM*

Where particular datum references are necessary such as for orientation, as in this example, but the restriction of Rule 5 (perfect form at MMC interrelationship or virtual condition) will not satisfy intent, a functional tolerance to the datum may be required.

Referring to the .140 holes, we see that the position tolerance pattern orientation is with respect to datums A, B, and C. However, since the hole pattern requires less restrictive orientation with the datums B and C, tolerances of the relationship to the datums are specified. In this manner, location and orientation reference to the datums is controlled more leniently, yet the desired amount of control is stated.

The .190 slot is controlled by separate, more restrictive size and position tolerances to meet its requirement of relationship with datums A and B (at MMC). The principle involved is shown in the representative functional gage (Gage 1) shown on page 255. Gage 1 is conventional with pickup of datum A for attitude, datum B at .630 MMC, and the .190 part slot gage pin at .184. Gage 2 introduces the principle of selection of datum pick-up sizes which can represent unique design intent or relax implied restrictions of usual datum relationships.

Note in Gage 2 that the gage member for the center hole (datum B) is undersize (.620), and the gage member for the slot (datum C) is undersize (.170) to the extent of the functional location and orientation tolerances established by the unique design requirement.

---

*Advisory only. Not per ANSI Y14.5-1973.

INTERRELATED DATUM SYSTEMS—FUNCTIONAL TOLERANCE TO DATUM*

EXAMPLE

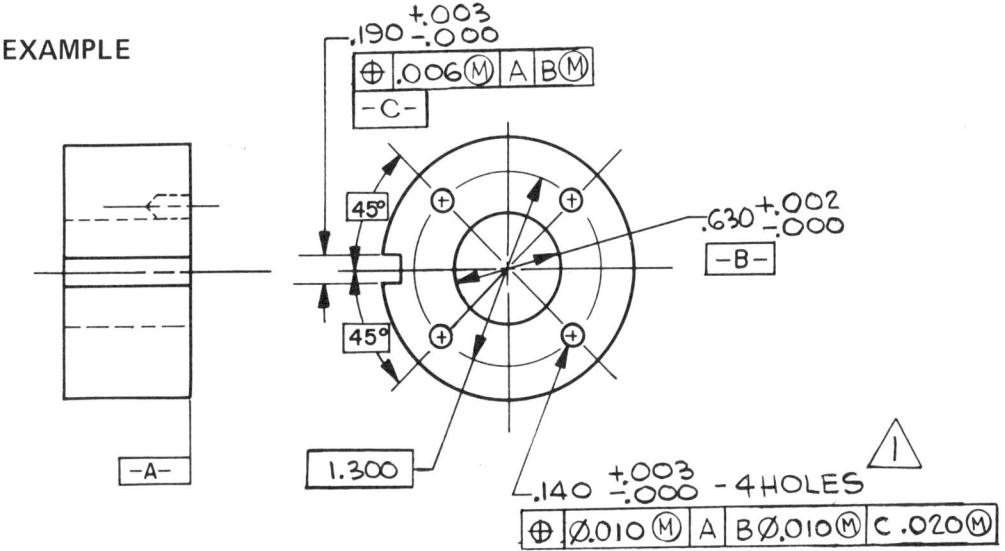

$.190 {}^{+.003}_{-.000}$

| ⊕ | .006 Ⓜ | A | B Ⓜ |

-C-

$.630 {}^{+.002}_{-.000}$

-B-

45°

45°

-A-

1.300

$.140 {}^{+.003}_{-.000}$ - 4 HOLES

| ⊕ | Ø.010 Ⓜ | A | B Ø.010 Ⓜ | C .020 Ⓜ |

⚠ RULE #5 ANSI Y 14.5–1973 DOES NOT APPLY

GAGE 1

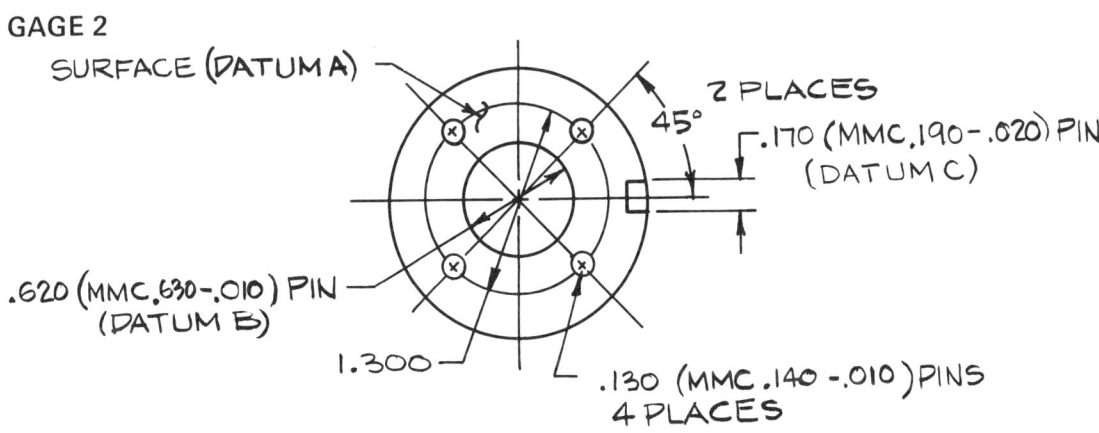

SURFACE (DATUM A)

.184 (MMC .190 –.006) PIN

.630 (MMC) PIN
(DATUM B)

GAGE 2

SURFACE (DATUM A)

2 PLACES

45°

.170 (MMC .190 –.020) PIN
(DATUM C)

.620 (MMC .630 –.010) PIN
(DATUM B)

1.300

.130 (MMC .140 –.010) PINS
4 PLACES

---

* Advisory only. Not per ANSI Y14.5–1973

255

CONCENTRICITY

CONCENTRICITY

Definition. Concentricity is the condition where the axes of all cross-sectional elements of a feature's surface of revolution are common to the axis of a datum feature.

Concentricity tolerance. Concentricity tolerance is the diameter of the cylindrical tolerance zone within which the axis of the feature(s) must lie; the axis of the tolerance zone must coincide with the axis of the datum feature(s).

EXAMPLE

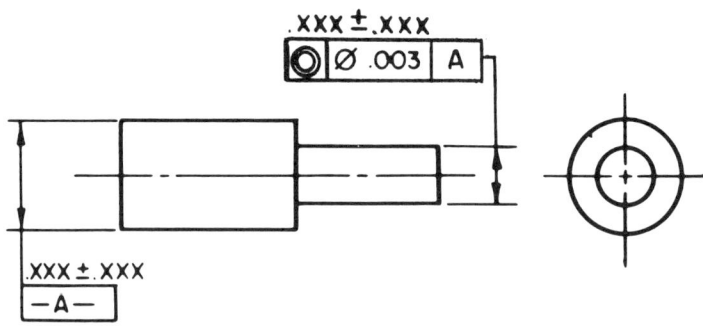

MEANING

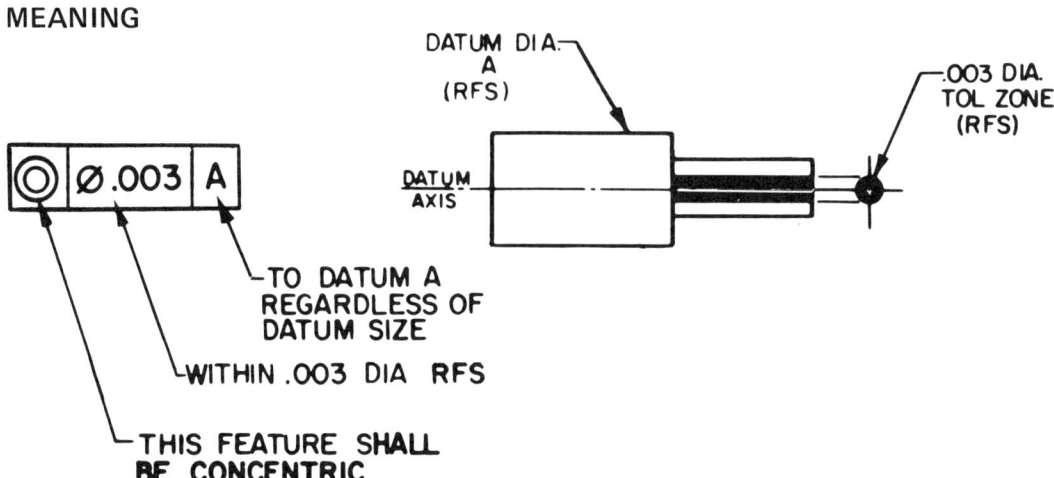

Concentricity is a type of location tolerancing. It always involves two or more basically coaxial features of size and controls the amount by which the *axes* of the features may fail to coincide. Concentricity tolerance, due to its unique characteristic, is always used on an RFS basis.

Where interrelated features are basically coaxial, we must first consider the possibility of using the more economical position or runout controls before considering concentricity. (See also "Coaxial Features—Selection of Proper Control" page 264.)

Concentricity relates to a condition in which two or more features (cylinders, cones, spheres, hexagons, etc.) in any combination have a common axis. Concentricity tolerance is the diameter of the cylindrical tolerance zone within which the axis of the feature (or features) must lie, the axis of the tolerance zone must coincide with the axis of the datum feature or features.

The surface of a feature must be used to establish its axis. Therefore, all the irregularities or errors of form of a feature surface must be considered in establishing the axis. For instance, the surface may be bowed, out of round, etc., in addition to being offset from its datum feature. This usually involves a complex inspection analysis of the entire surface, and therefore requires a more time-consuming and costly procedure.

Concentricity requirements are required less frequently than position or runout requirements. However, where concentricity is required, it provides effective control over the more unique applications of coaxial relationships. For example, concentricity might be applied to the coaxiality requirements of a tape drive pulley or of a capstan on a computer mechanism or a motor generator rotor. Often where balance is required, the out-of-roundness or lobing effect (or possible other form errors) may be permissible although it may exceed the conventional FIM requirement. Hence, any basically symmetrical form of revolution (hexagons, cones, etc.), or consistently symmetrical variation of such a form, could satisfy a concentricity tolerance where a runout requirement may not.

As stated before, the axis of a feature is determined by the *surface* of that feature. We compare that feature axis with the datum feature axis; the result of this comparison will determine whether or not the feature axis satisfies the concentricity requirement.

A common method of determining the coaxiality of a feature with its datum (such as in runout) uses dial indicator (FIM) readings, with the part mounted and rotated on its datum feature. This method is also often adequate as a means of determining whether concentricity requirements have been satisfied. However, in concentricity, by the very nature of its characteristics, a FIM reading in excess of the stated permissible tolerance may not necessarily mean that the feature *axis* lies outside the concentricity tolerance zone. That is, if a feature has a surface error such as out-of-roundness, its surface will register a FIM dial reading relative to its datum diameter but its axis may not have actually exceeded the concentricity tolerance zone.

CONCENTRICITY

Figures 1 through 5 illustrate this comparison with FIM and the principles involved in concentricity tolerancing. Using the sample requirement shown. Figure 1 shows a part that is theoretically perfect; it also shows the location of the cylindrical concentricity tolerance zone about the datum axis.

Figure 2 illustrates the part feature at its maximum displacement (or combination of various form errors). Note that the displacement registered by .003 FIM at the *surface* is equivalent to the *diametral* displacement of the *axis* within the concentricity tolerance zone. Figure 3 illustrates an out-of-roundness condition for which the .003 FIM reading is also *within* the specified concentricity zone. Under the conditions illustrated in Figs. 2 and 3, since the FIM is within the stated concentricity tolerance, the parts are acceptable with no further considerations. In this case, the results are essentially identical to a requirement which might have been stated using total runout on the basis of the FIM only. Therefore, to some extent, concentricity and runout can be considered identical except for interpretation, that is, surface variation (FIM direct) vs. axial displacement (as derived from FIM or some other comparable method).

Also, the above provides the uniform interpretation of any drawings for which concentricity (RFS) as previously defined (now generally equivalent to a total runout requirement) is used. However, where the stated concentricity tolerance is exceeded by the FIM, the characteristics of concentricity as now interpreted may possibly authorize acceptance of the part as being concentric within the specified limits.

Figure 4 illustrates the part with a .005 FIM reading which appears to exceed the stated concentricity tolerance. Yet closer observation will show that the part is actually *perfectly concentric* (as a hypothetical example) because the out-of-roundness error (possibly including other form errors) is symmetrically uniform. The FIM reading as conventionally applied is no longer conclusive and, in fact, could reject good parts. Closer scrutiny of Fig. 3 also shows that the FIM reading is again inconclusive with respect to the part's actual concentricity. The part is perfectly concentric although a FIM reading was registered.

Therefore, whenever a part is being evaluated for concentricity with conventional FIM techniques and is found to be within the concentricity tolerance stated, the part has met the concentricity tolerance requirement. However, when the part *exceeds* the FIM as conventionally checked or when actual concentricity is to be checked, more extensive inspection techniques are required; i.e., further consideration must be given to the question of whether additional form errors, such as out-of-roundness, are permissible within the concentricity tolerance. The form errors may vary anywhere within the tolerances of size so long as the concentricity tolerance is met.

Figure 5 illustrates the principles involved in checking actual concentricity. A precision collet and indicator set-up method is illustrated for purposes of explanation. However, other methods can be used, such as precision spindle inspection machines which are capable of picking up the true axes of rotation.

In the method shown, a measurement (M1) is taken of the part feature at a given point (X). The part is rotated 180°, and another measurement (M2) is taken.

COMPARISON WITH FIM METHODS
AS DRAWN

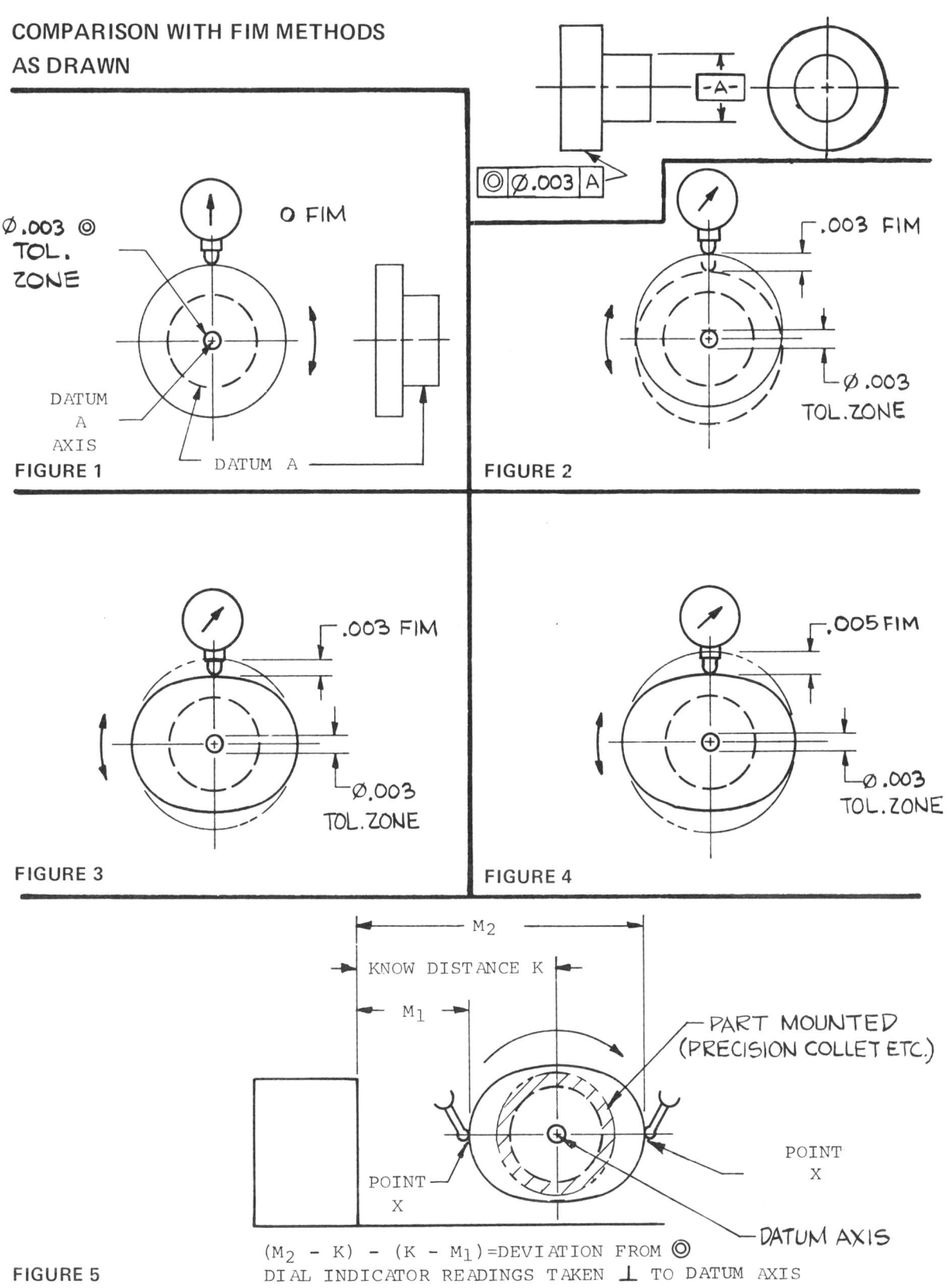

FIGURE 1

FIGURE 2

FIGURE 3

FIGURE 4

FIGURE 5

$(M_2 - K) - (K - M_1) =$ DEVIATION FROM ◎
DIAL INDICATOR READINGS TAKEN ⊥ TO DATUM AXIS

CONCENTRICITY

The difference between these readings (see formula) is the deviation from concentricity (as a diameter) at that cross-sectional plane which is perpendicular to the datum axis.

Measurements are made at as many diametrical cross sections on the entire surface as necessary to adequately satisfy the requirement. For part acceptance, the maximum deviation from the concentricity calculated should be within the permissible concentricity tolerance. Since the deviation from the concentricity calculated is expressed as a diameter, it can be compared directly with the stated concentricity tolerance zone.

As previously described, when checking concentricity, we consider all the irregularities or errors of form of a feature surface in composite to determine its axis, and thus the displacement of this axis relative to the datum axis.

Since the dial indicator (or other measuring method) sweeps the surface, the type of errors involved are, for all practical purposes, indistinguishable from one another.

To illustrate the various basic types of error involved, the example shown on the next page explains the general effect of feature eccentricity, error of parallelism of axis, out-of-straightness of axis, out-of-roundness, and out-of-cylindricity on concentricity tolerance requirements using FIM methods. Note that the sample part has a concentricity tolerance requirement of $\varnothing$.001.

Eccentricity is seen as the allowable parallel displacement of the feature axis with respect to the datum axis. The .0005 eccentricity is a displacement to one side but must be considered as a $\varnothing$.001 total tolerance cylinder.

Out-of-straightness and error of parallelism of axis is seen as a tip up, down, sideways, or bow of the feature with respect to the datum axis. The $\varnothing$.001 cylindrical tolerance zone represents the limits of the allowable error of this type.

Out-of-roundness and out-of-cylindricity (as it affects concentricity) is the difference in the diameters of two concentric circles and cylinders (in planes normal to the axis) between which the actual surface of the feature must fall.

As is seen, concentricity tolerance is a composite control of all elements of the feature concerned as they create an axis. It is this *axis* which serves as the design requirement and thus its basis for part evaluation or inspection.

For all practical purposes, when concentricity is the requirement, these errors are viewed as indistinguishable from one another; as a result they are considered in composite when we determine whether the part complies with the concentricity tolerance.

EXAMPLE

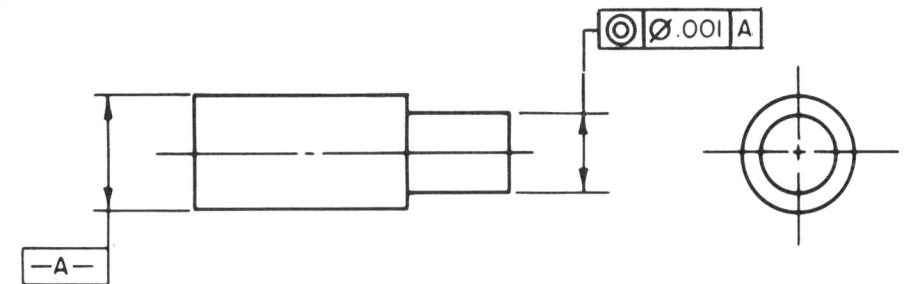

⌖ ⌀.001 A

—A—

MEANING

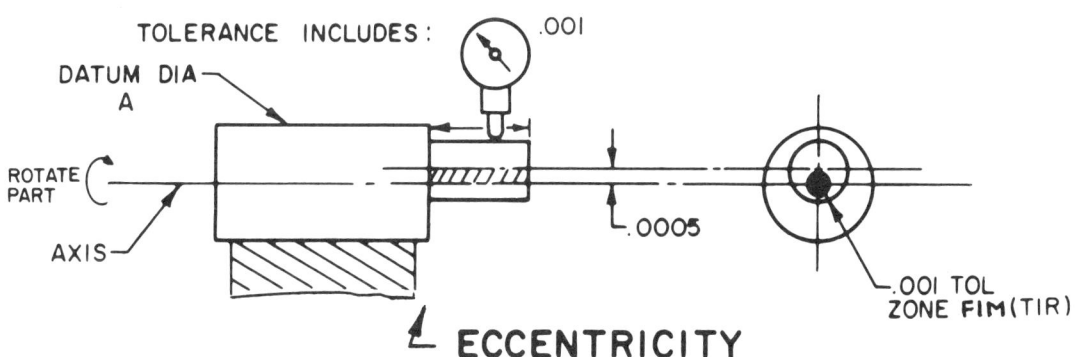

TOLERANCE INCLUDES:

DATUM DIA A

ROTATE PART

AXIS

.001

.0005

.001 TOL ZONE FIM(TIR)

ECCENTRICITY

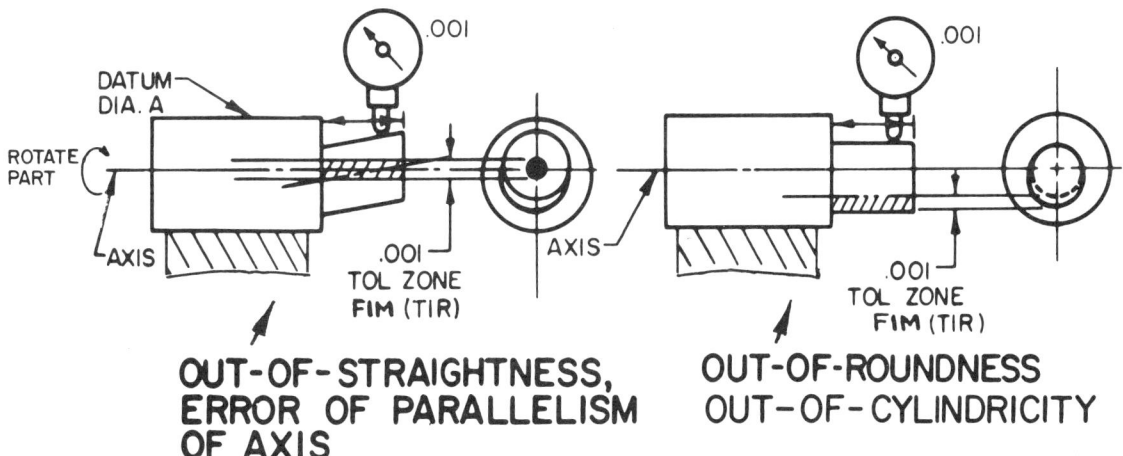

DATUM DIA. A

ROTATE PART

AXIS

.001

.001 TOL ZONE FIM (TIR)

OUT-OF-STRAIGHTNESS,
ERROR OF PARALLELISM
OF AXIS

AXIS

.001

.001 TOL ZONE FIM (TIR)

OUT-OF-ROUNDNESS
OUT-OF-CYLINDRICITY

263

COAXIAL FEATURES — SELECTION OF PROPER CONTROL

There are three methods of controlling interrelated coaxial features:

1. RUNOUT TOLERANCE (circular or total) (RFS)
2. POSITION TOLERANCE   (MMC)
3. CONCENTRICITY TOLERANCE   (RFS)

Any of the above methods provide effective control. However it is important to select the *most appropriate* one to both meet the design requirements and provide the most economical manufacturing conditions. (See also details of above sections).

Below are recommendations to assist in selecting the proper control:

If the need is to control only CIRCULAR cross sectional elements in a composite relationship to the datum axis, RFS, e.g., multi-diameters on a shaft, use

CIRCULAR RUNOUT   EXAMPLE: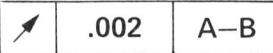

(This method controls any composite error effect of roundness and circular cross-sectional profile variations)

If the need is to control the TOTAL cylindrical or profile surface in composite relative to the datum axis' RFS, e.g., multi-diameters on a shaft, bearing mounting diameters, etc., use

    EXAMPLE: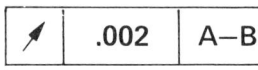

TOTAL RUNOUT   TOTAL   TOTAL

(This method controls any composite error effect of roundness, cylindricity, straightness, angularity, and parallelism.)

Note: Runout is always implied as an RFS application. It cannot be applied on an MMC basis, since an MMC situation involves functional interchangeability or assemblability (probably of mating parts), in which case POSITION tolerance would be used. See below.

If the need is to control the total cylindrical or profile surface and its axis in composite location relative to the datum axis on an MMC basis, e.g., on mating parts to assure interchangeability or assemblability, use

POSITION   EXAMPLE:

If the need is to control the *axis* of one or more features in composite relative to a *datum axis*, RFS, e.g., to control balance of a rotating part, use

CONCENTRICITY   EXAMPLE: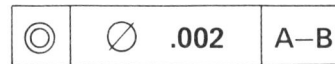

Note: Concentricity is always implied as an RFS application. Variations in size (departure from MMC size, out-of-roundness, out-of-cylindricity, etc.) do not in themselves conclude *axis* error.

EXAMPLE

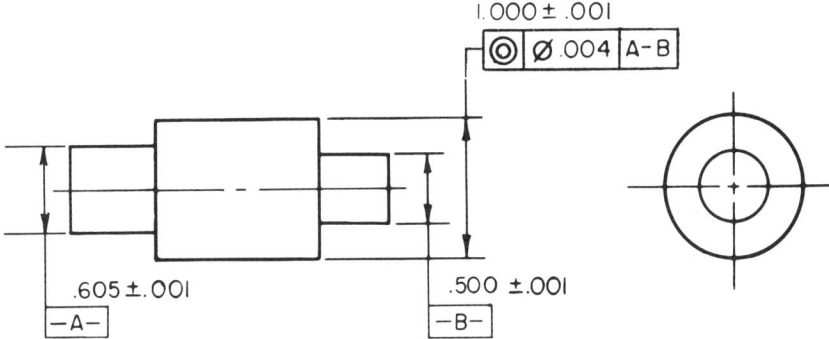

MEANING

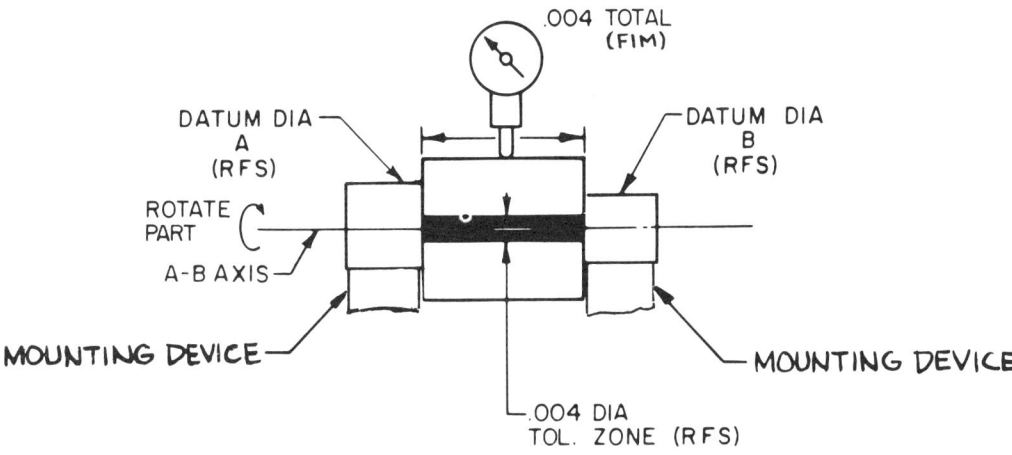

The example above illustrates a concentricity tolerance, using two datums to establish a common datum axis.

The 1.000 ± .001 outside diameter is to be concentric to the common datum axis established by datum diameters A and B, within ∅.004 regardless of the sizes to which any of the diameters are produced. The diameter size tolerances must be evaluated separately.

The lower portion of the illustration shows the part mounted on both datum diameters A and B, so that the concentricity of the 1.000 diameter with respect to both diameters A and B and thus to their common axis can be checked in one operation. The FIR method is not conclusive when the reading exceeds the stated concentricity tolerance. In this case, other methods must be used (see previous pages) to determine whether the feature does exceed the concentricity tolerance in an actual concentricity check.

SYMMETRY

SYMMETRY $=$

The third type of locational tolerancing is SYMMETRY:

Like the two previously discussed types of locational tolerance, position and concentricity, symmetry deals with the location of actual features with respect to established center planes or axes. As its name implies, the purpose of any symmetry tolerance is to specify a symmetrical relationship for the toleranced feature—usually with the outside limits of the part—for reasons of appearance, clearance, or fit to related or mating parts.

The example shows a part using symmetry tolerancing on the slot; the part width is established as the datum. The requirement is to relate the slot location to the outside width of the part. To simplify explanation, size dimensions and tolerances of the slot and the datum width have been omitted.

In order to establish some basis for the relationship of the slot feature to the datum feature, we must have a common reference plane. This is the center, or median plane, of the datum feature. We also determine the center, or median plane, of the symmetry-toleranced feature and then compare these two median planes. The result of this comparison shows whether or not the symmetry tolerance is met.

The interpretation shows the relationship and establishment of the symmetry tolerance zone with respect to the datum center plane. Note that the center plane, or median plane, of datum A establishes the midpoint of the .005 tolerance zone.

The median plane of the produced slot, regardless of the slot size, must lie between these two planes which are .005 apart and equidistant (or .0025) from the median plane of the datum, regardless of datum size.

Note that the median plane of the slot, as determined by the actual slot, may fall at an angle or anywhere within the tolerance zone.

Symmetry is a type of positional tolerancing. Where part features of a symmetrical shape are to be geometrically toleranced, it is recommended that the positional characteristic be used instead of symmetry. Symmetry is, however, a valid characteristic and may be applied, if desired, on an RFS basis only. In MMC applications of symmetrically shaped features, position tolerancing should be used.

FOR MMC APPLICATION (USE POSITION)

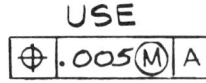

FOR RFS APPLICATION (USE SYMMETRY OR POSITION)

RECOMMENDED   PERMISSIBLE

| ⊕ | .005 Ⓢ | A Ⓢ | | = | .005 | A |

268

Definitions. Symmetry is a condition in which a feature(s) is symmetrically disposed about the center plane of a datum feature.

Symmetry Tolerance. The tolerance governing the symmetry of a feature with respect to a datum is the distance between two parallel planes between which the median plane of the feature must lie, the parallel planes being parallel to, and equally disposed about, the median plane, or axis, of the datum feature.

EXAMPLE

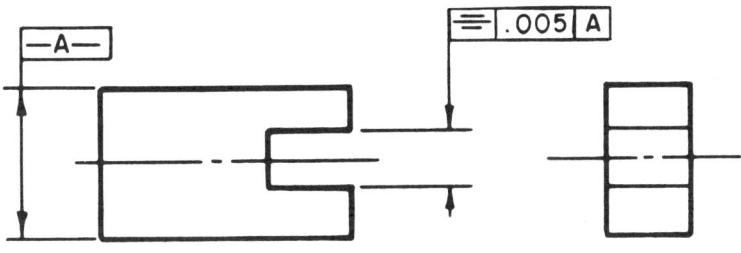

MEANING

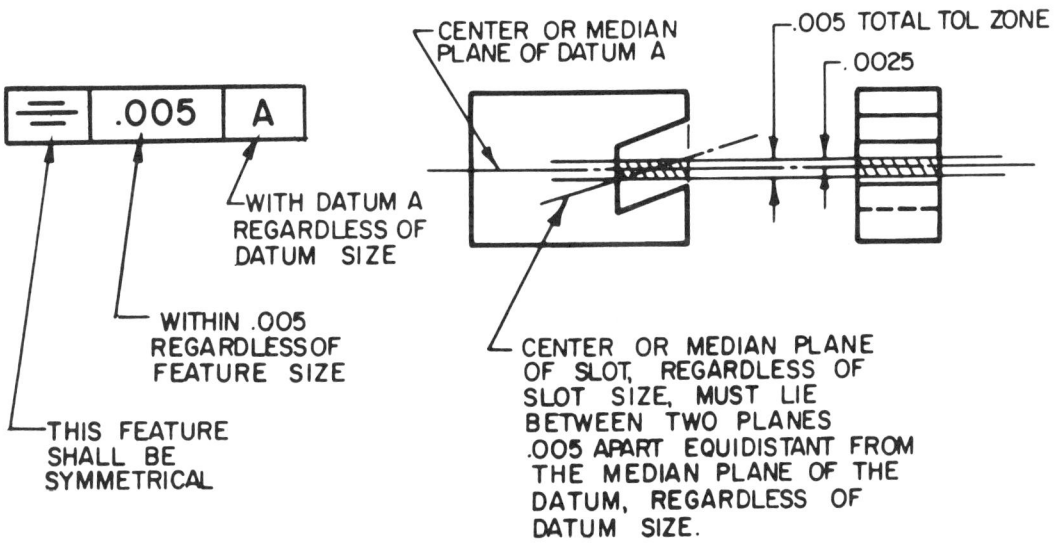

In this example, a part similar to that in the previous illustration has been used. However, position tolerance control has been used instead of symmetry. Size dimensions and tolerances have also been added to the slot and datum features. The tolerance is the same as before except that positional tolerancing, has been substituted as the preferred method, and an added datum has been specified. Since symmetry and position tolerance derive exactly the same result in an application of this kind, position tolerance is preferred, as it is more consistent with general position tolerance practice. This example, as the previous one, is also an RFS application.

The meaning below the example illustrates the establishment of the .005 tolerance zone about the center plane of the datum feature. Note that the tolerance zone remains the same, .005, when the datum feature size and the slot size vary anywhere in their size tolerance range.

In either case, and also at any place between these sizes, the center plane is established by the midpoint of the datum size but the .005 tolerance zone remains the maximum permissible tolerance.

The total tolerance zone for the slot is therefore .005 regardless of the size of the datum feature or the slot.

In 2(a) and 2(b) we show the manner in which the center plane of the slot, regardless of its size, may vary within the symmetry tolerance zones. In 2(a) the slot is shown at .500 and in 2(b) at .504. To be acceptable, the center planes of either of these extremes of the slot size must fall within the .005 tolerance zone. Note that the slot center planes in either example fall within the tolerance zone so these parts are acceptable.

Figure 3 shows that the slot center plane may fall anywhere within this tolerance zone.

In Figure 4 we show a method of gaging the position tolerance on this part in order to clarify how theoretical center planes presented on drawings are converted to actual measurements. The slot has been offset to the limit of the .005 tolerance zone. A .500 wide slot and a 1.000 wide datum feature have been assumed for this example. The part is rested on one side of the 1.000 datum feature, or its width. A maximum measurement is taken from the gaging surface to the top underside of the slot.

The part is then rotated 180°, or turned over, to the opposite side of the 1.000 datum feature and a similar maximum measurement is taken. The difference between these two measurements, as may be seen in the illustration, actually determines the total variation of the slot center plane with respect to the datum center plane and the tolerance zone.

Comparison between the resulting value and the tolerance requirement determines the acceptance of the parts' position (symmetry).

As seen in this example, the resulting .005 total variation of the measurements is equal to the .005 required tolerance. Therefore, the slot is positional, or central, to datum A within the required tolerance and is also square (perpendicular) to the primary datum B.

EXAMPLE

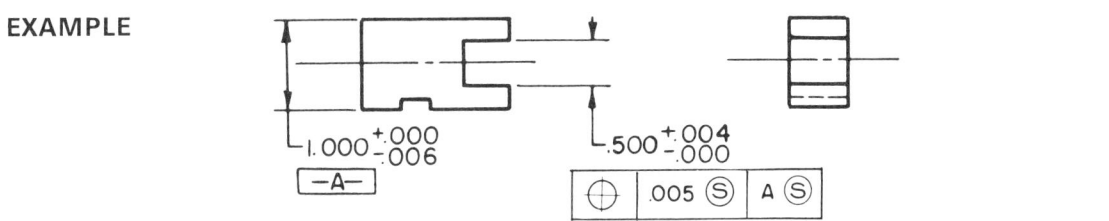

$1.000^{+.000}_{-.006}$

⎯A⎯

$.500^{+.004}_{-.000}$

⊕ | .005 Ⓢ | A Ⓢ

MEANING

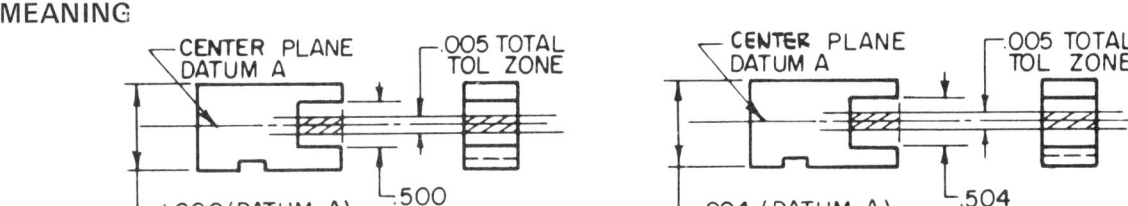

CENTER PLANE DATUM A

.005 TOTAL TOL ZONE

1.000 (DATUM A)

.500

CENTER PLANE DATUM A

.005 TOTAL TOL ZONE

.994 (DATUM A)

.504

TOTAL TOL ZONE FOR SLOT IS .005 REGARDLESS OF SIZE OF DATUM OR SLOT

FIGURE 1(a)   FIGURE 1(b)

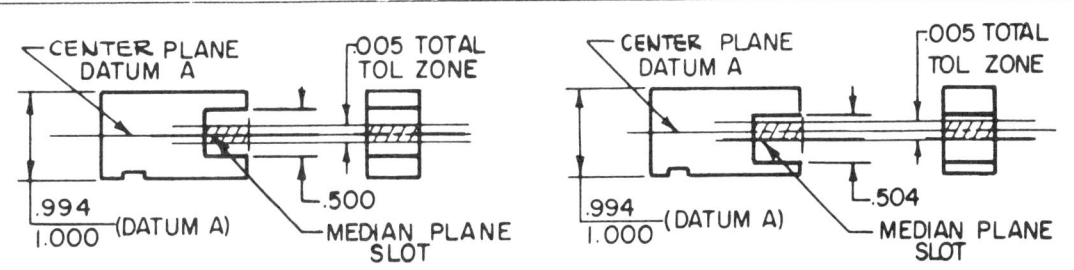

CENTER PLANE DATUM A

.005 TOTAL TOL ZONE

.994 / 1.000 (DATUM A)

.500

MEDIAN PLANE SLOT

CENTER PLANE DATUM A

.005 TOTAL TOL ZONE

.994 / 1.000 (DATUM A)

.504

MEDIAN PLANE SLOT

CENTER PLANE OF SLOT, REGARDLESS OF SLOT SIZE, MUST FALL WITHIN .005 TOTAL TOL ZONE

FIGURE 2(a)   FIGURE 2(b)

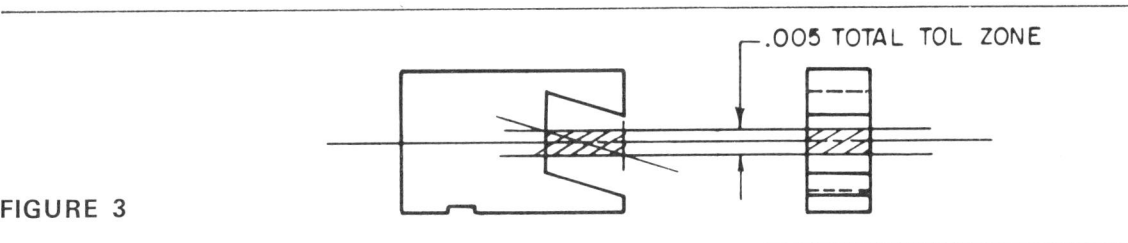

.005 TOTAL TOL ZONE

FIGURE 3

GAGING

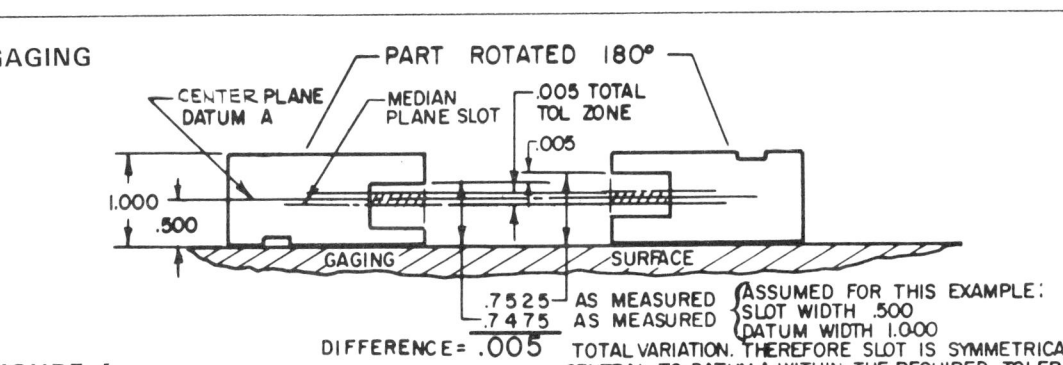

PART ROTATED 180°

CENTER PLANE DATUM A

MEDIAN PLANE SLOT

.005 TOTAL TOL ZONE

.005

1.000

.500

GAGING   SURFACE

.7525 AS MEASURED
.7475 AS MEASURED

ASSUMED FOR THIS EXAMPLE:
SLOT WIDTH .500
DATUM WIDTH 1.000

DIFFERENCE= .005   TOTAL VARIATION. THEREFORE SLOT IS SYMMETRICAL OR CENTRAL TO DATUM A WITHIN THE REQUIRED TOLERANCE.

FIGURE 4

271

CONCLUSION

The following APPENDIX presents additional material helpful to a more complete understanding of geometric dimensioning and tolerancing. It discusses some of the fundamentals underlying the principles previously introduced, and covers some advanced techniques of interest to those who work in areas of advanced technologies. The material presented suggests new avenues of approach to more unusual applications.

However, with the information you now have from the foregoing text it should be possible for you to effectively apply the techniques of geometric dimensioning and tolerancing. Remember: the best teacher is experience. Use these techniques and, as you gain experience, you will find many advantageous solutions to a wide variety of problems.

Geometric dimensioning and tolerancing properly and adequately applied play a major role in building a better product at an increased profit. It can turn disadvantages into advantages; it can make order out of chaos; it can open up new doors of engineering opportunity.

APPENDIX

Conversion formula–position tolerance to/from coordinate (formula) 274

Conversion charts 275

USA practices and International practices 278

ISO R1101 practices 279

Comparison between ANSI Y14.5–1966 and ISO 281

Pickup points; functional and gaging references 282

Position; least material condition principle 283

Free state variation 286

Average diameter 287

Clarification of Rule 1 288

Clarification of "feature" .291

Position tolerance conversion to tool tolerance 292

APPENDIX

CONVERSION OF ⊕ POSITION (CYLINDRICAL) TOLERANCE ZONES TO/FROM COORDINATE TOLERANCE ZONES

⊕ TO ±

TOTAL ⊕ TOL ZONE X .70711 = TOTAL COORDINATE TOLERANCE ZONE

EXAMPLE: .007 ⊕ TOL X .70711 = .00495 $\left.\begin{array}{l} \text{.005 TOTAL COORDINATE TOL} \\ \text{OR} \\ \text{.0025 BILATERAL TOL} \end{array}\right\}$

RULE OF THUMB:

USE .7 (OR 70%) OF TOTAL ⊕ TOL ZONE TO CONVERT QUICKLY IN NON-CRITICAL APPLICATIONS, e.g..7 X .007 = .0049 OR .005 (±0025)

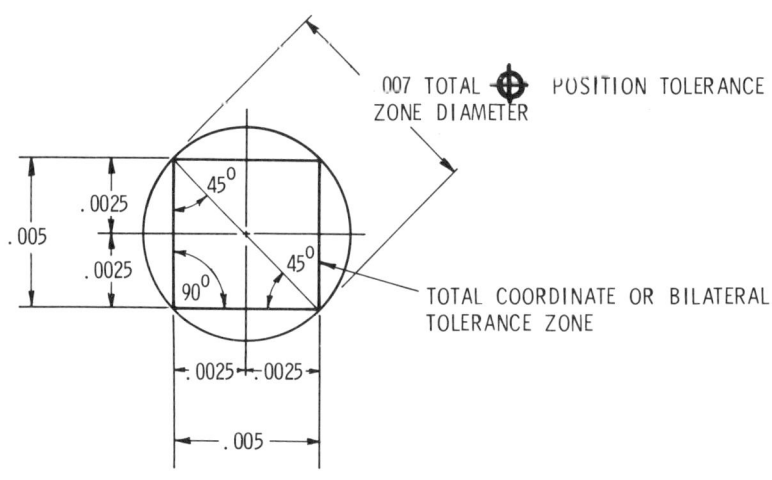

.007 TOTAL ⊕ POSITION TOLERANCE ZONE DIAMETER

TOTAL COORDINATE OR BILATERAL TOLERANCE ZONE

± TO ⊕

TOTAL COORDINATE TOL ZONE X 1.4142 = TOTAL ⊕ TOL ZONE

EXAMPLE: $\left.\begin{array}{l} \text{.005 TOTAL COORDINATE TOL} \\ \text{OR} \\ \text{.0025 BILATERAL TOL X 2} \end{array}\right\}$ X 1.4142 = .007 TOTAL ⊕ TOL

RULE OF THUMB:

USE 1.4 TIMES TOTAL COORD TOL ZONE TO CONVERT QUICKLY IN NON-CRITICAL APPLICATIONS, e.g. 1.4 X .005 = .007 ⊕ TOL

CONVERSION CHART

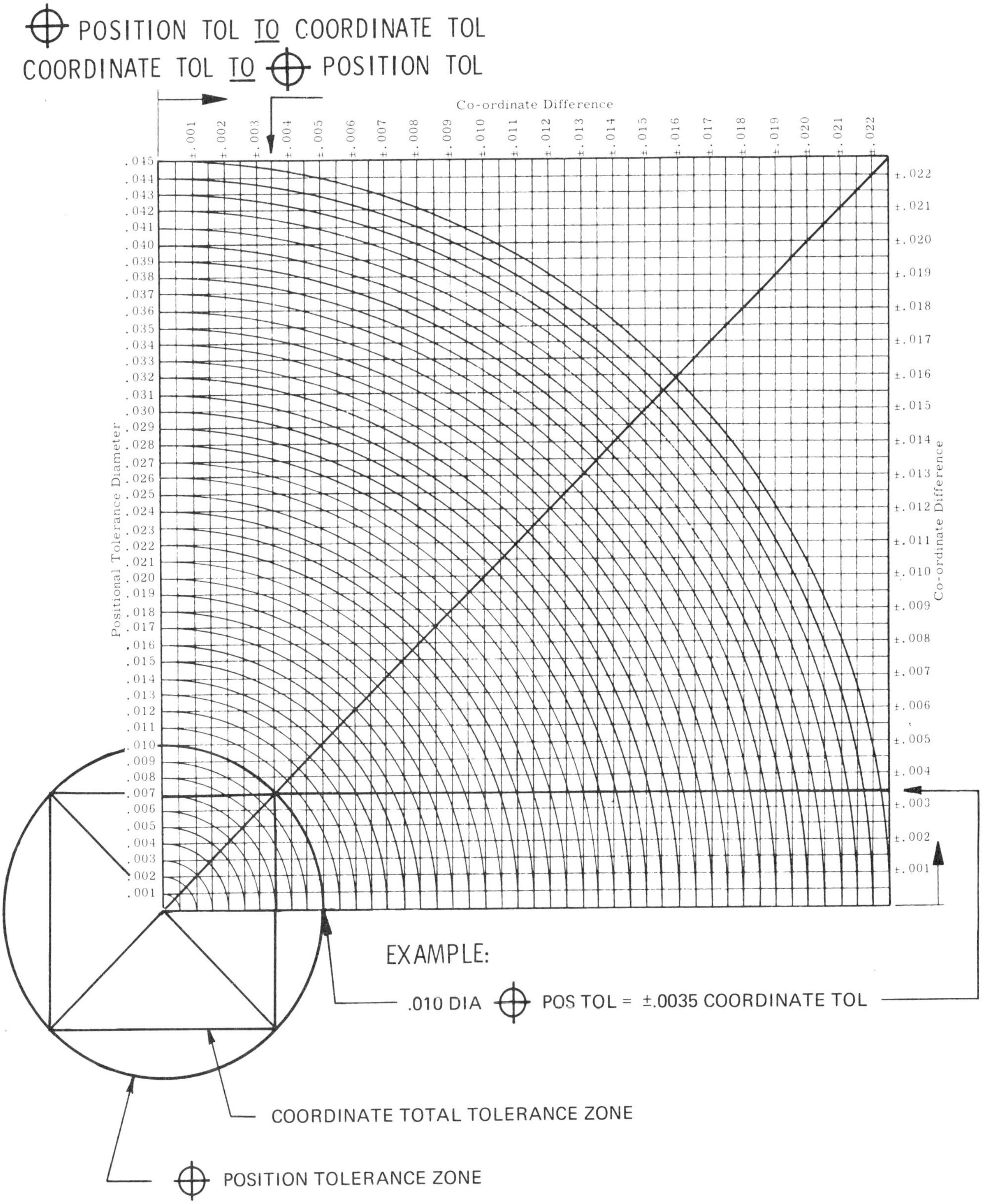

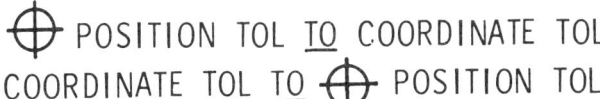

POSITION TOL <u>TO</u> COORDINATE TOL
COORDINATE TOL <u>TO</u> POSITION TOL

Co-ordinate Difference

±.001 ±.002 ±.003 ±.004 ±.005 ±.006 ±.007 ±.008 ±.009 ±.010 ±.011 ±.012 ±.013 ±.014 ±.015 ±.016 ±.017 ±.018 ±.019 ±.020 ±.021 ±.022

Positional Tolerance Diameter

.045 .044 .043 .042 .041 .040 .039 .038 .037 .036 .035 .034 .033 .032 .031 .030 .029 .028 .027 .026 .025 .024 .023 .022 .021 .020 .019 .018 .017 .016 .015 .014 .013 .012 .011 .010 .009 .008 .007 .006 .005 .004 .003 .002 .001

Co-ordinate Difference

±.022 ±.021 ±.020 ±.019 ±.018 ±.017 ±.016 ±.015 ±.014 ±.013 ±.012 ±.011 ±.010 ±.009 ±.008 ±.007 ±.006 ±.005 ±.004 ±.003 ±.002 ±.001

EXAMPLE:

.010 DIA ⊕ POS TOL = ±.0035 COORDINATE TOL

COORDINATE TOTAL TOLERANCE ZONE

⊕ POSITION TOLERANCE ZONE

APPENDIX

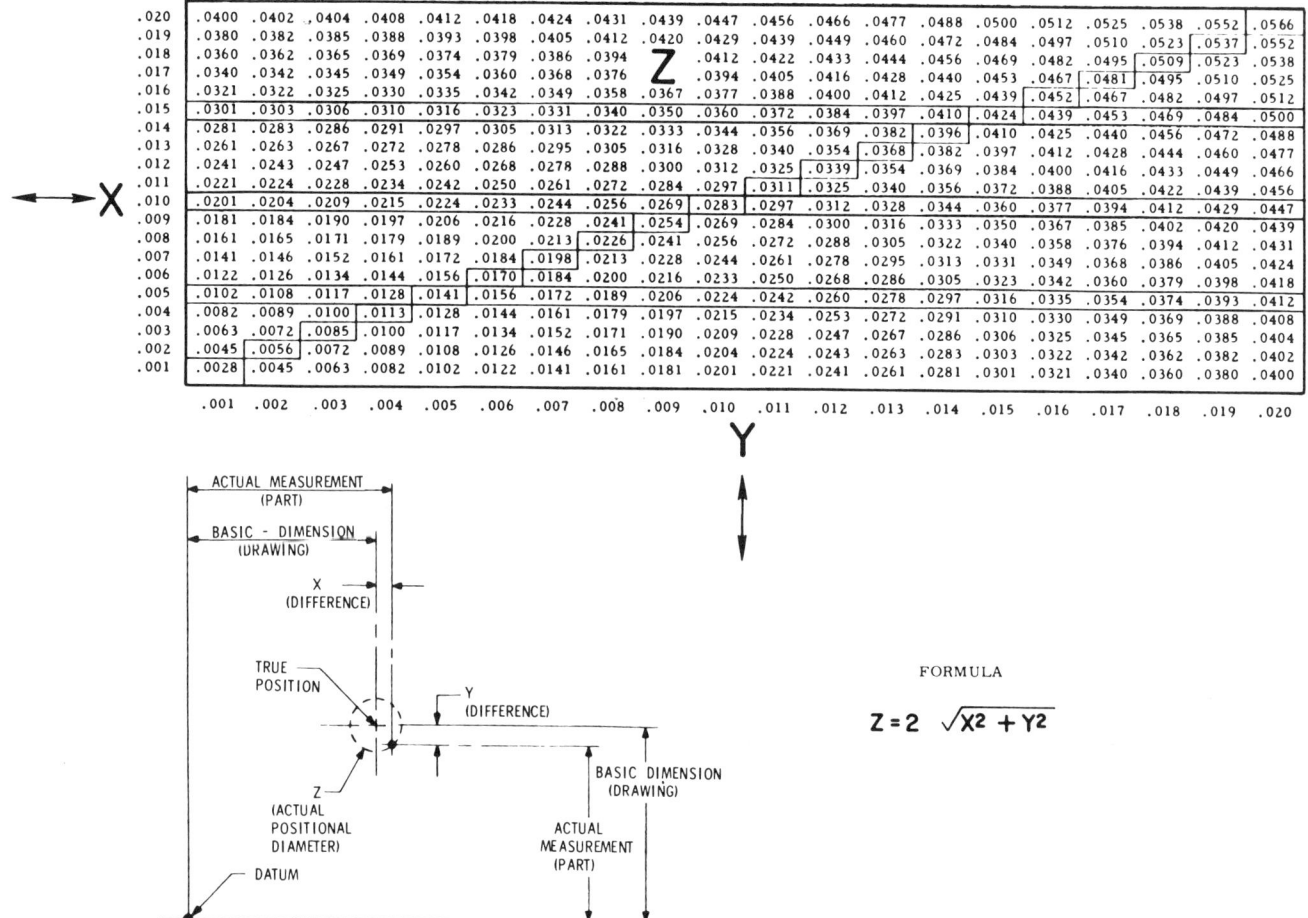

CONVERSION OF COORDINATE MEASUREMENTS TO ⊕ POSITION LOCATION

X	.001	.002	.003	.004	.005	.006	.007	.008	.009	.010	.011	.012	.013	.014	.015	.016	.017	.018	.019	.020
.020	.0400	.0402	.0404	.0408	.0412	.0418	.0424	.0431	.0439	.0447	.0456	.0466	.0477	.0488	.0500	.0512	.0525	.0538	.0552	.0566
.019	.0380	.0382	.0385	.0388	.0393	.0398	.0405	.0412	.0420	.0429	.0439	.0449	.0460	.0472	.0484	.0497	.0510	.0523	.0537	.0552
.018	.0360	.0362	.0365	.0369	.0374	.0379	.0386	.0394	Z	.0412	.0422	.0433	.0444	.0456	.0469	.0482	.0495	.0509	.0523	.0538
.017	.0340	.0342	.0345	.0349	.0354	.0360	.0368	.0376		.0394	.0405	.0416	.0428	.0440	.0453	.0467	.0481	.0495	.0510	.0525
.016	.0321	.0322	.0325	.0330	.0335	.0342	.0349	.0358	.0367	.0377	.0388	.0400	.0412	.0425	.0439	.0452	.0467	.0482	.0497	.0512
.015	.0301	.0303	.0306	.0310	.0316	.0323	.0331	.0340	.0350	.0360	.0372	.0384	.0397	.0410	.0424	.0439	.0453	.0469	.0484	.0500
.014	.0281	.0283	.0286	.0291	.0297	.0305	.0313	.0322	.0333	.0344	.0356	.0369	.0382	.0396	.0410	.0425	.0440	.0456	.0472	.0488
.013	.0261	.0263	.0267	.0272	.0278	.0286	.0295	.0305	.0316	.0328	.0340	.0354	.0368	.0382	.0397	.0412	.0428	.0444	.0460	.0477
.012	.0241	.0243	.0247	.0253	.0260	.0268	.0278	.0288	.0300	.0312	.0325	.0339	.0354	.0369	.0384	.0400	.0416	.0433	.0449	.0466
.011	.0221	.0224	.0228	.0234	.0242	.0250	.0261	.0272	.0284	.0297	.0311	.0325	.0340	.0356	.0372	.0388	.0405	.0422	.0439	.0456
.010	.0201	.0204	.0209	.0215	.0224	.0233	.0244	.0256	.0269	.0283	.0297	.0312	.0328	.0344	.0360	.0377	.0394	.0412	.0429	.0447
.009	.0181	.0184	.0190	.0197	.0206	.0216	.0228	.0241	.0254	.0269	.0284	.0300	.0316	.0333	.0350	.0367	.0385	.0402	.0420	.0439
.008	.0161	.0165	.0171	.0179	.0189	.0200	.0213	.0226	.0241	.0256	.0272	.0288	.0305	.0322	.0340	.0358	.0376	.0394	.0412	.0431
.007	.0141	.0146	.0152	.0161	.0172	.0184	.0198	.0213	.0228	.0244	.0261	.0278	.0295	.0313	.0331	.0349	.0368	.0386	.0405	.0424
.006	.0122	.0126	.0134	.0144	.0156	.0170	.0184	.0200	.0216	.0233	.0250	.0268	.0286	.0305	.0323	.0342	.0360	.0379	.0398	.0418
.005	.0102	.0108	.0117	.0128	.0141	.0156	.0172	.0189	.0206	.0224	.0242	.0260	.0278	.0297	.0316	.0335	.0354	.0374	.0393	.0412
.004	.0082	.0089	.0100	.0113	.0128	.0144	.0161	.0179	.0197	.0215	.0234	.0253	.0272	.0291	.0310	.0330	.0349	.0369	.0388	.0408
.003	.0063	.0072	.0085	.0100	.0117	.0134	.0152	.0171	.0190	.0209	.0228	.0247	.0267	.0286	.0306	.0325	.0345	.0365	.0385	.0404
.002	.0045	.0056	.0072	.0089	.0108	.0126	.0146	.0165	.0184	.0204	.0224	.0243	.0263	.0283	.0303	.0322	.0342	.0362	.0382	.0402
.001	.0028	.0045	.0063	.0082	.0102	.0122	.0141	.0161	.0181	.0201	.0221	.0241	.0261	.0281	.0301	.0321	.0340	.0360	.0380	.0400

Y

ACTUAL MEASUREMENT (PART)

BASIC - DIMENSION (DRAWING)

X (DIFFERENCE)

TRUE POSITION

Y (DIFFERENCE)

Z (ACTUAL POSITIONAL DIAMETER)

DATUM

BASIC DIMENSION (DRAWING)

ACTUAL MEASUREMENT (PART)

FORMULA

$$Z = 2\sqrt{X^2 + Y^2}$$

EXAMPLE

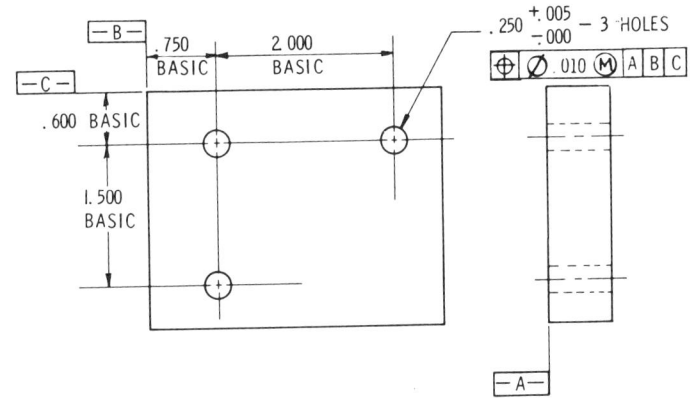

CONVERSION

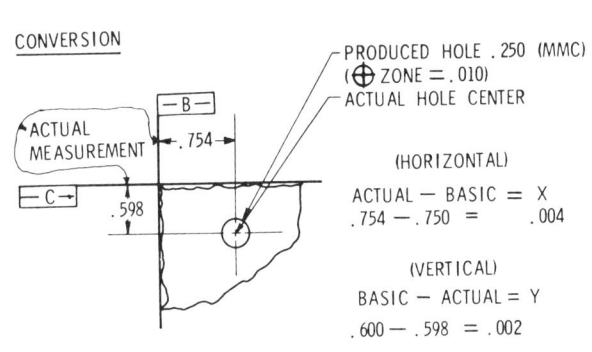

PRODUCED HOLE .250 (MMC)
(⊕ ZONE = .010)
ACTUAL HOLE CENTER

(HORIZONTAL)

ACTUAL — BASIC = X
.754 — .750 =   .004

(VERTICAL)

BASIC — ACTUAL = Y
.600 — .598 = .002

FROM CHART ABOVE .004 (X)
AND .002 (Y) GIVE .0089 DIA.
HOLE LOCATION IS WITHIN
SPECIFIED .010 DIA. HOLE LOCATION
ACCEPTABLE

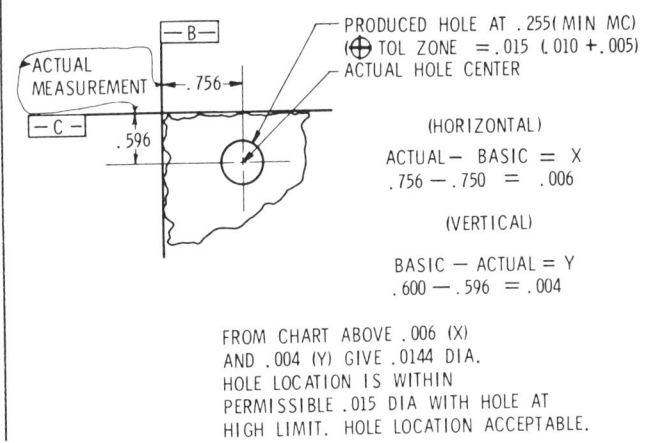

PRODUCED HOLE AT .255(MIN MC)
(⊕ TOL ZONE = .015 (.010 +.005)
ACTUAL HOLE CENTER

(HORIZONTAL)

ACTUAL— BASIC = X
.756 —.750 =   .006

(VERTICAL)

BASIC — ACTUAL = Y
.600 — .596 = .004

FROM CHART ABOVE .006 (X)
AND .004 (Y) GIVE .0144 DIA.
HOLE LOCATION IS WITHIN
PERMISSIBLE .015 DIA WITH HOLE AT
HIGH LIMIT. HOLE LOCATION ACCEPTABLE.

APPENDIX

USA PRACTICES AND INTERNATIONAL PRACTICES

With the advent of ANSI Y14.5–1973, an improved base of compatibility exists between USA and ISO (International) practices. Further, with the current U.S. involvement and input to revisions and expansion of international standards, even greater improvement will evolve in the future.

Where there is concern that U.S. practices be implemented for greatest possible compatibility with international practices, the following points of discussion may be helpful.

MMC AND RFS CONDITIONS

Referring to RULE 2, it is noted that the condition desired (MMC or RFS) must be specified on all applications of POSITION tolerance and appropriate related datums as preferred practice. This achieves a degree of agreement with international practices. ISO practices require MMC be specified when desired; no indication of condition implies regardless-of-feature-size application. Incorporating other new refinements and options of Y14.5–1973, the below comparison and similarities are found.

USA (ANSI Y14.5–1973)   ISO (R1101)

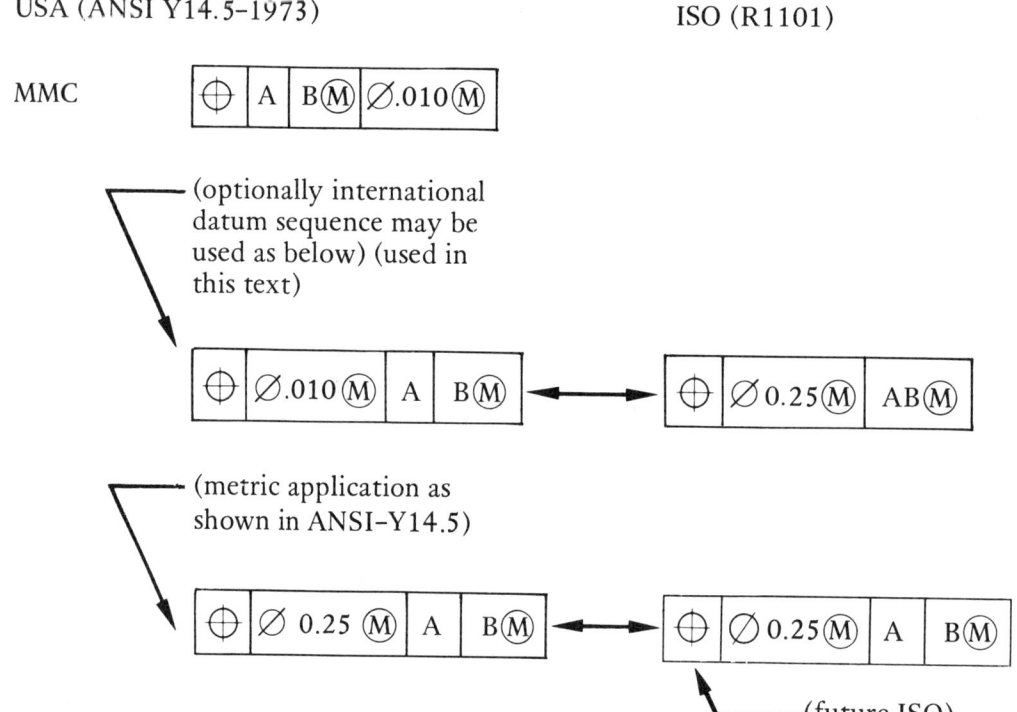

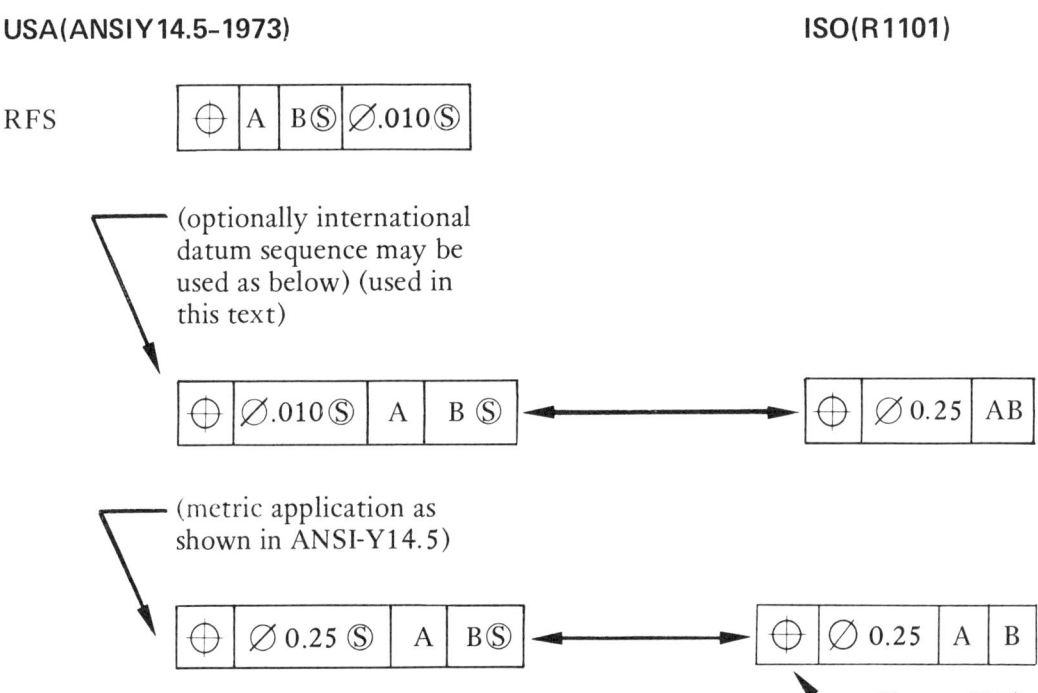

USA(ANSI Y 14.5–1973)   **ISO(R 1101)**

RFS

(optionally international datum sequence may be used as below) (used in this text)

(metric application as shown in ANSI-Y14.5)

(future ISO)

The ⓈS. condition symbol is required in U.S. practices as retained from the previous standard (ANSI-Y14.5-1966) to assure uniform interpretation. It is anticipated that future expansion of U.S. standards will drop the need for Ⓢ. Datum precedence and the vertical line separation in the ISO symbol method is now being studied in international standards development as proposed by the U.S.

ISO R110I PRACTICES

The decimal comma (,) is used in some metric countries as shown on the following page. This principle could be used to no prejudice of the illustration in this text. These examples also illustrate R1101 datum letter sequence and datum identification triangle symbol (⊥) use.

APPENDIX

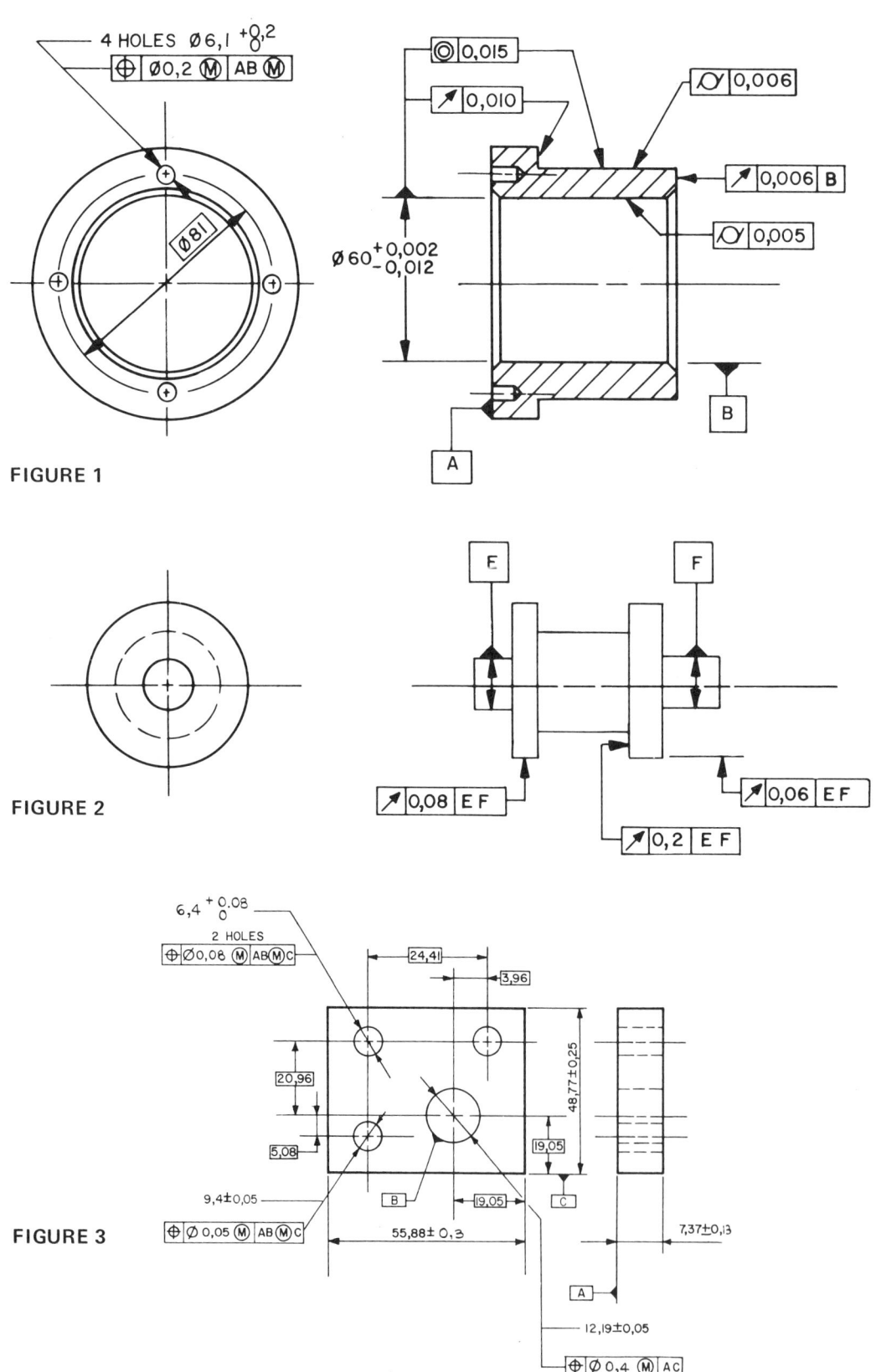

FIGURE 1

FIGURE 2

FIGURE 3

COMPARISON ANSI Y14.5, ISO SYMBOLS

CHARACTERISTIC	ANSI – Y14.5	ISO-R1101
STRAIGHTNESS	—	—
FLATNESS	▱	▱
ANGULARITY	∠	∠
PERPENDICULARITY (SQUARENESS)	⊥	⊥
PARALLELISM	//	//
CONCENTRICITY	◎	◎
POSITION	⊕	⊕
ROUNDNESS (CIRCULARITY)	○	○
SYMMETRY	≡	≡
PROFILE OF ANY LINE	⌒	⌒
PROFILE OF ANY SURFACE	⌓	⌓
RUNOUT (CIRCULAR)	↗	↗
RUNOUT (TOTAL)	TOTAL ↗ *	(PROPOSED) *
CYLINDRICITY	⌭	⌭
DATUM FEATURE	–A–	A
MAXIMUM MATERIAL CONDITION (MMC)	Ⓜ	Ⓜ
REGARDLESS OF FEATURE SIZE (RFS)	Ⓢ	NONE (ASSUMED UNLESS SPECIFIED MMC)

* USA & ISO PROPOSED

* USA and ISO proposed.

APPENDIX

PICKUP POINTS (FUNCTIONAL AND GAGING REFERENCES)

AS DRAWN

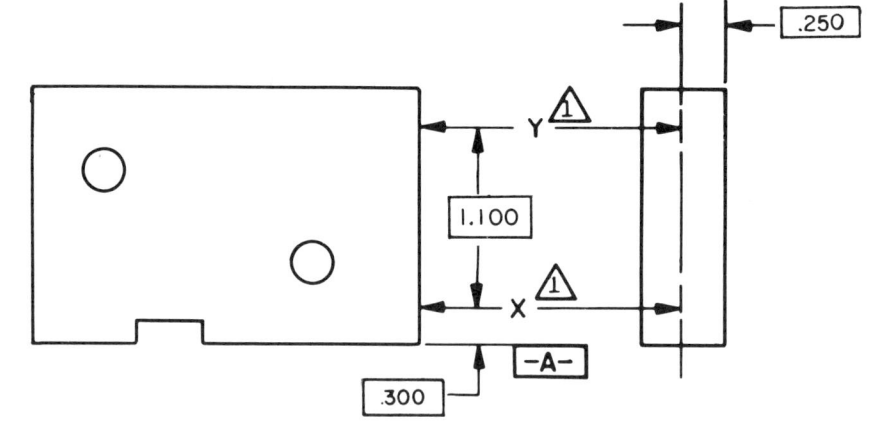

$\underset{\boxed{1}}{\triangle}$ LINE ESTABLISHED BY PICKUP POINTS
X & Y TO BE $\perp$ TO DATUM A WITHIN .020
TOTAL.

MEANING

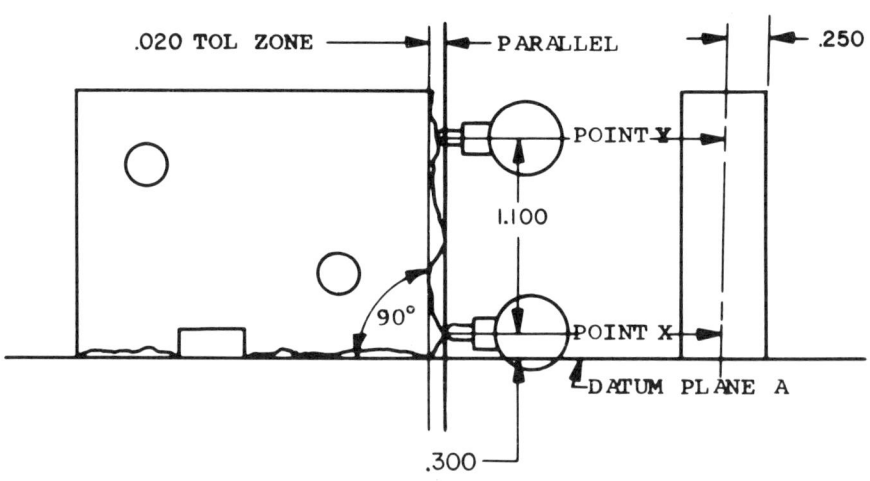

Where specific functional and gaging pickup points are necessary, they should be specified approximately as above. It should be noted that pickup points which are *"Measured to"* on a part are *not* to be considered *datum* points; datums are measured *from* as pickup references for orientation of the measurement. This distinction must be kept in mind.

POSITION TOLERANCE LEAST MATERIAL CONDITION (LMC)*

Occasionally we need a method to control a situation which is essentially the reverse of the usual position relationship; that is, the stated position tolerance applies at the *least material condition*, LMC, of the feature or datum, instead of at MMC, and increases as the feature or datum *departs from* the least material condition.

Definition. Least material condition is the condition opposite to MMC. For example, a shaft is at least material condition when it is at its LOW limit of size and a hole is at least material condition when it is at its HIGH limit of size.

This method is applicable to special design requirements that will not permit MMC or that do not warrant the exacting requirements of RFS. It can be used to maintain critical wall thicknesses or critical center locations of features for which accuracy of location can be relaxed (position tolerance increased) when the feature leaves least material condition and approaches MMC. The amount of increase of positional tolerance permissible is equal to the feature size departure from least material condition.

The term "least material condition" and the abbreviation LMC have been used instead of "minimum material condition" (which is synonymous) to avoid confusion, since the abbreviation would be the same as that for maximum material condition. The symbol modifier Ⓛ is used to indicate the LMC requirement applicable to feature or datum.

Although the use of LMC does impose exacting requirements on both manufacturing and inspection, it does permit additional tolerances.

Whenever (LMC) or Ⓛ is specified on a drawing, the position tolerance applies only when the feature is produced at its LMC size. See Fig. 1.

Additional positional tolerance is permissible but is dependent on, and equal to, the difference between the actually produced feature size (within its size tolerance) and LMC. See Fig. 2.

---

* Not per ANSI Y14.5–1973.

APPENDIX

POSITION TOLERANCE LEAST MATERIAL CONDITION (LMC)

EXAMPLE

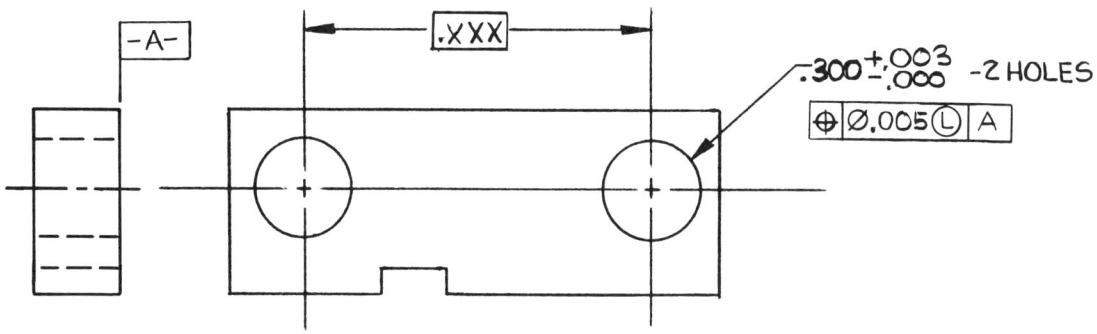

MEANING

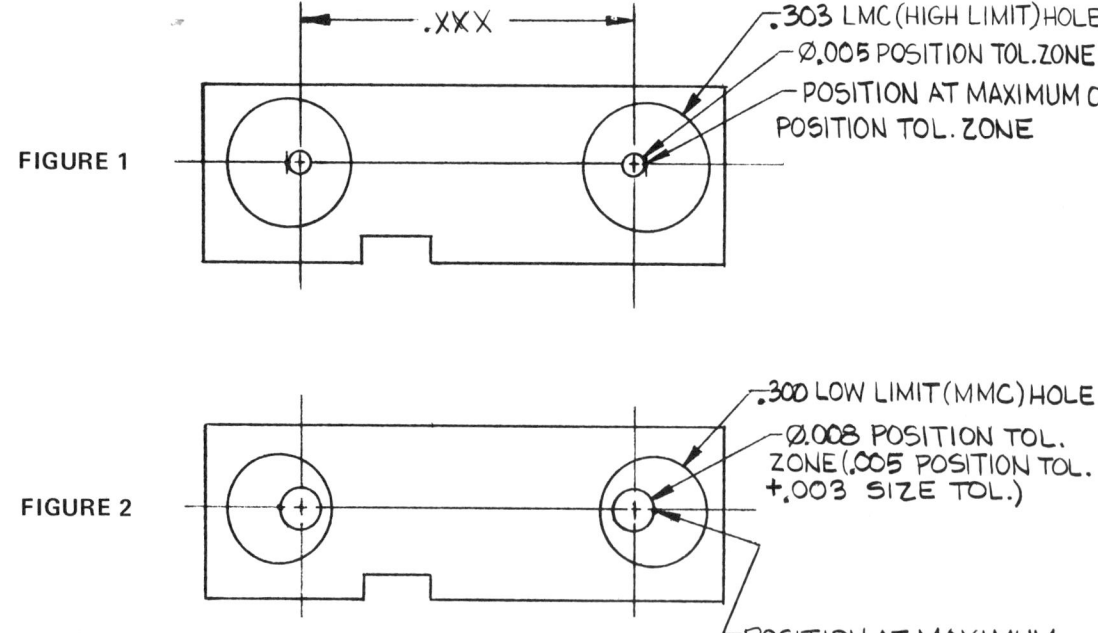

EXAMPLE

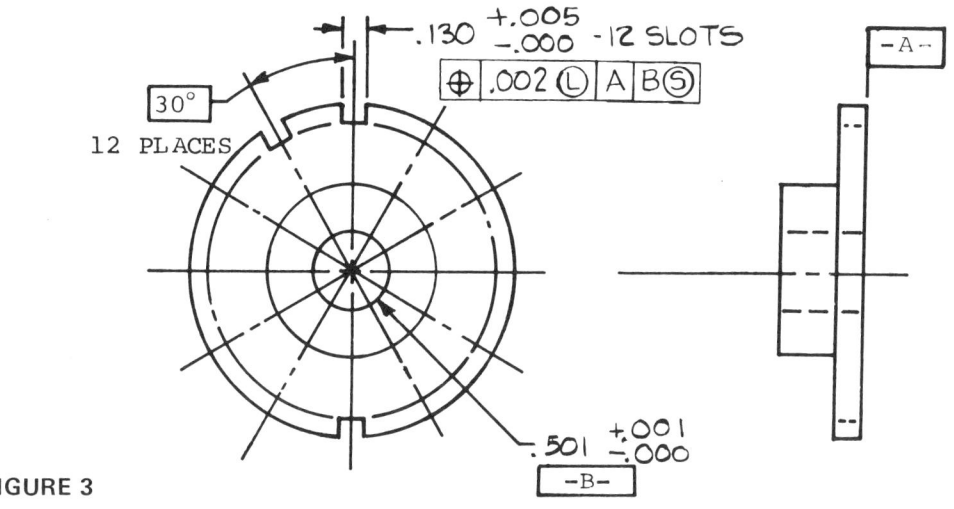

FIGURE 3

MEANING

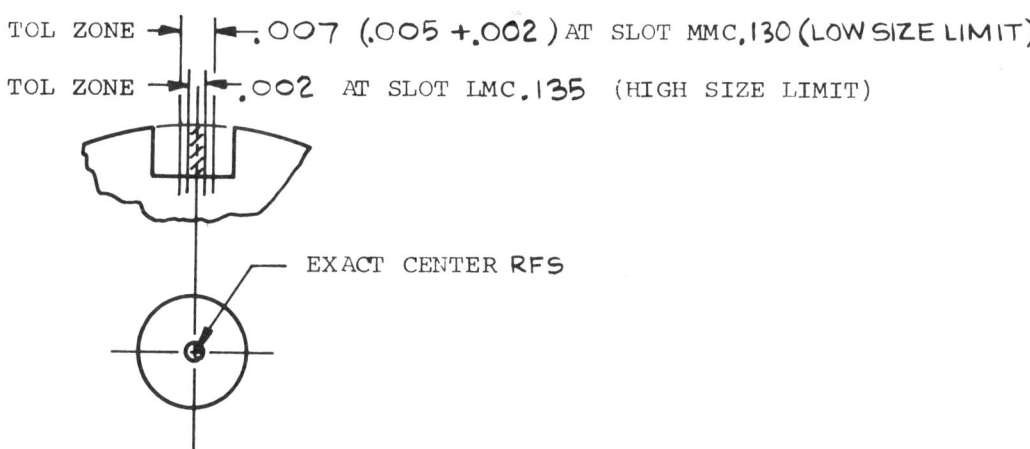

TOL ZONE → .007 (.005 +.002) AT SLOT MMC .130 (LOW SIZE LIMIT)

TOL ZONE → .002 AT SLOT LMC .135 (HIGH SIZE LIMIT)

EXACT CENTER RFS

APPENDIX

FREE-STATE VARIATION: "RESTRAINED" APPLICATION

Free-state variation is the amount a part distorts after removal of external forces applied during manufacture, for instance, parts consisting essentially of shells or tubes with a thin wall thickness in proportion to the diameter. Geometric tolerances (such as roundness, cylindricity, and concentricity) cannot be properly applied without controlling free-state variation on parts of this type.

Variations in the free state can exist in two ways: (1) distortion due to the weight or flexibility of the part, or (2) distortion due to internal stresses set up in fabrication.

Where free-state variation control is necessary, any datums and the features in control may require specification of their allowable free-state variation or the maximum force necessary to restrain each of them to drawing tolerance; so that the desired assembly relationship can be stated and represented for evaluation of the requirement, thus achieving compliance.

EXAMPLE

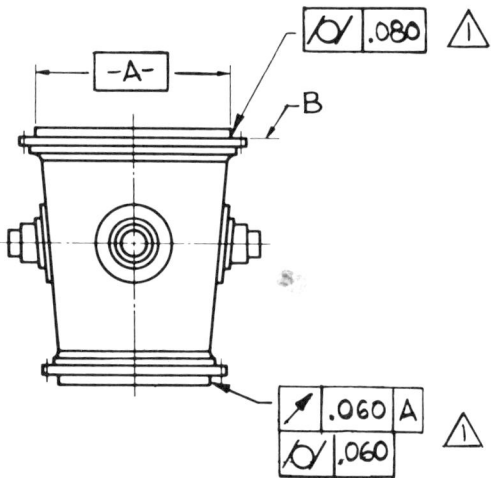

⚠ APPLIES IN FREE STATE WHEN DATUM ∅ A IS RESTRAINED TO DRAWING TOLERANCE AND SURFACE B IS HELD FLAT WITH THE FORCE OF 150/175 POUNDS WITH PART IN HORIZ. POSITION.

MEANING

Compliance with the drawing note as stated.

AVERAGE DIAMETER

An average diameter is the mean of several diameters (not less than four) across a circular or spherical part used to determine conformance to *diameter* tolerance only. If practicable, the average diameter may be determined by using a periphery tape. Only when a diameter is allowed a maximum roundness tolerance in the free state should it be specified as AVG. $\emptyset$.

A part of this kind is normally expected to flex to proper form or shape at assembly. The reason for control in its free state is to facilitate hand or automated assembly and handling while the part is in its free-state configuration, to restrict distortion to within safe eleastic limits of the material, etc.

EXAMPLE

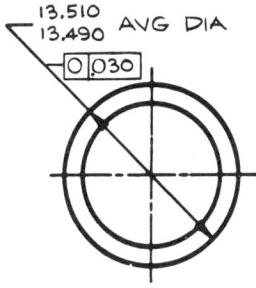

MEANING

1. The high and low diameter size measurements (four minimum) must differ by no more than .060 (because tolerance is .030 round on R). The calculated *average* (mean) of these diameters must be within the specified size tolerance 13.490–13.510 diameters.

2. The entire surface at the cross section measured must lie within the .030 wide tolerance zone.

If the part in free state is of the general shape shown in the figure below, the 13.540 could be the highest diameter measurement, and 13.480 could be the lowest diameter measurement.

For example, four measurements are:

13.540	(high)
13.525	
13.495	
13.480	(low)
54.040	

13.510 average diameter (within *size* tolerance)

13.540 – 13.480 = .060
AVERAGE DIA = 13.510

APPENDIX

High and low measurements do not differ by more than .060; therefore, part is within free-state *roundness* tolerance.

If the part in free state is of the general shape shown in the figure below, the 13.520 could be the highest diameter measurement, and 13.460 could be the lowest diameter measurement.

For example, four measurements taken are:

13.520	(high)
13.500	
13.480	
13.460	(low)
53.960	

13.490 average diameter (within *size* tolerance)

13.520 − 13.460 = .060
AVERAGE DIA = 13.490

High and low measurements do not differ by more than .060; therefore, part is within free-state *roundness* tolerance.

CLARIFICATION OF RULE 1 OF ANSI Y14.5

That Rule 1 applies to "individual" features *only* and *not* to an "interrelationship" of individual features, and that form variation of the individual feature is contained within the PERFECT FORM AT MMC ENVELOPE, will be explained in detail in the following text and illustrations.

Rule 1 applies to "individual" features whether the form variation is through permissible SIZE variation (form tolerance *not* specified) or whether the tolerance is specified.

The illustrations on the following pages show that the individual features involved in the form tolerances of FLATNESS, STRAIGHTNESS, PARALLELISM, ROUNDNESS and CYLINDRICITY readily comply with the definition of "individual" feature (see definition in section of "Clarification of Feature") and are to be within the perfect form at MMC envelope. If no form tolerance relationship had been specified, we would nonetheless assume that these features were also confined to the perfect form at MMC envelope.

ANGULARITY is a unique characteristic involving an *interrelationship* of individual features. Rule 1 does *not* apply to such interrelationship, so angularity control must be specified. However, an extremity of the angular surface which is adjacent to, or part of, a surface which is size-dimensioned must be contained within the perfect form at MMC envelope, such as, for example, the lower left surface extremity near the angle vertex in the figure illustrating angularity.

PERPENDICULARITY defines an interrelationship of individual features; thus Rule 1 does *not* apply. Where perpendicularity is required, it must be specified. However, the individual surface features must be contained within the perfect form at MMC envelope as established by the size dimensions (such as the X and Y dimensions shown).

A PROFILE OF ANY SURFACE tolerance is usually applied as a combined form *and* size control (i.e. a BASIC overall dimension with a profile of surface tolerance). In this case, it is *not* subject to Rule 1. Where a PROFILE OF ANY LINE control is used as a refinement of size control, Rule 1 *would* apply.

RUNOUT, being a unique variety of rotating surface tolerance, involves an interrelationship of individual features; thus Rule 1 does *not* apply. However, the individual features must be confined to their size dimensions and perfect form at MMC envelope.

From the above we can state in summary that unless otherwise specified or controlled, the "size" limits of an individual feature control the applicable "form" variation within these size limits and the envelope of perfect form when no form tolerance is stated; furthermore, when a form tolerance *is* specified, the form variation must also be contained within these size limits and the envelope of perfect form at MMC (exception: straightness specified on $\emptyset$ basis and related to size dimension). Rule 1 does not control and is not intended to control "interrelationships" of individual features. When an "interrelationship" is required, it must be specified.

PERFECT FORM AT MMC AS APPLIED TO FORM TOLERANCE

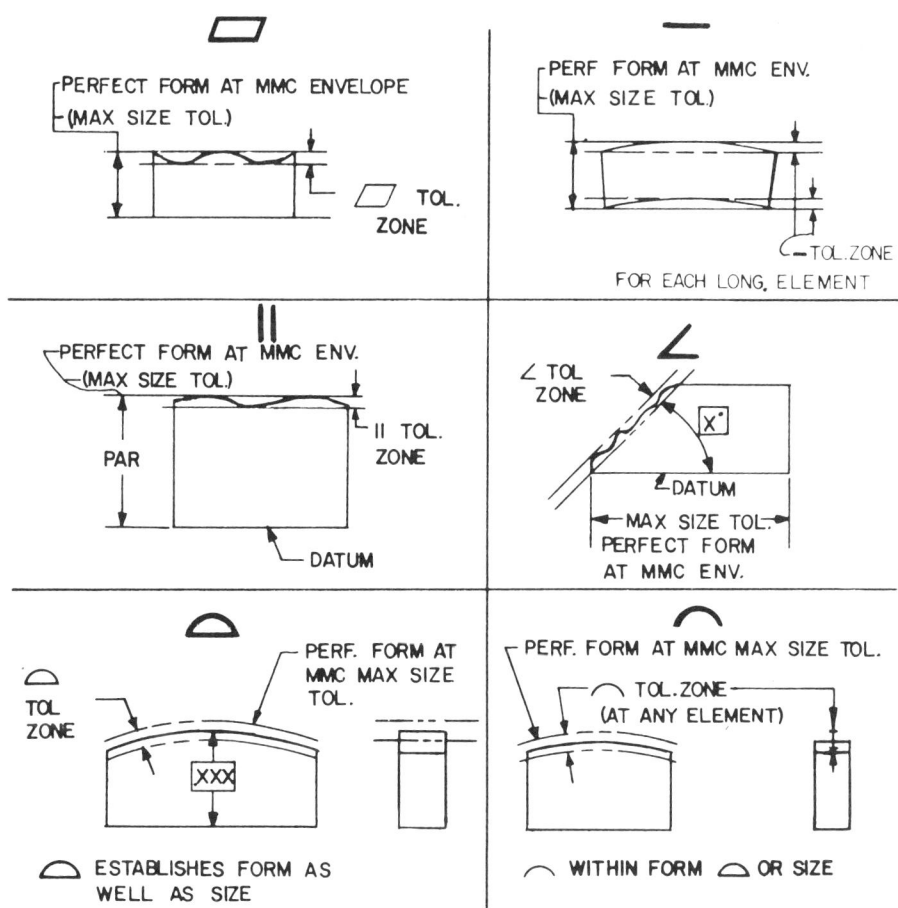

APPENDIX

PERFECT FORM AT MMC AS APPLIED TO FORM TOLERANCE (cont.)

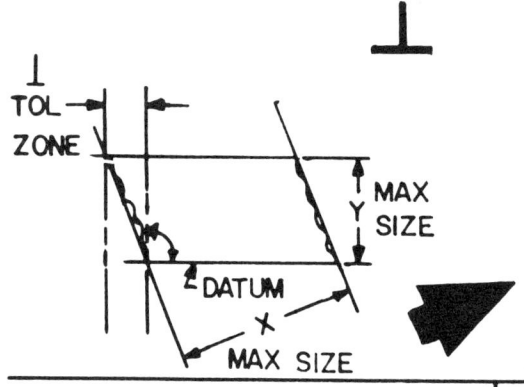

PERF. FORM AT MMC
ENVELOPE APPLIES TO
X & Y SEPARATELY.

INTERRELATIONSHIP
CONTROLLED BY FORM
TOL. ⊥ AS STATED

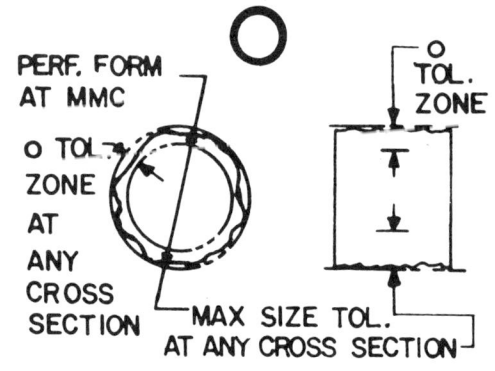

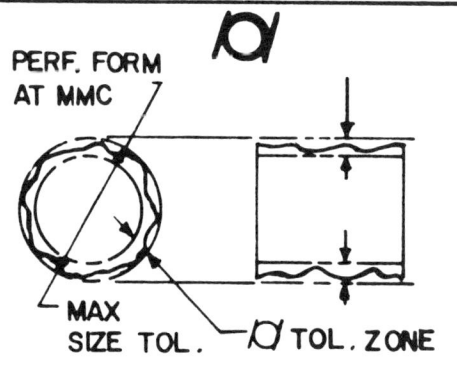

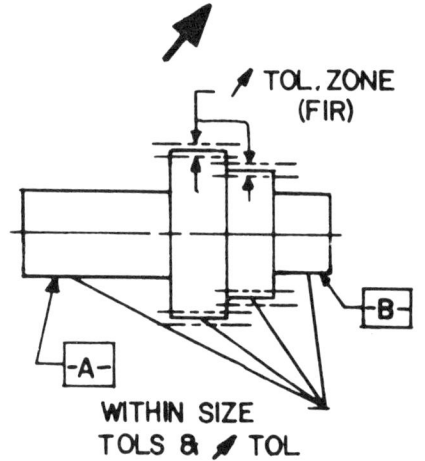

PERFECT FORM AT MMC
ENV. APPLIES TO EACH
FEATURE SEPARATELY.

INTERRELATIONSHIP
CONTROLLED BY FORM /
POSITION TOL.

CLARIFICATION OF FEATURE

The meaning of the term "feature" is defined below. The definitions clearly show the flexibility of this term.

Definitions.

Features. Features are specific characteristics or component portions of a part and may include one or more surfaces such as cylinders, holes, screw threads, profiles, faces, slots, etc.

Individual Feature. The "individual" feature, as implied by Rule 1, is a feature related to the part size. There are two types of individual features: "single surface" and "opposed surface."

A "single surface" individual feature, for example, is the flat surface extremity of a size dimension, or a surface given a "flatness" tolerance, etc.

An "opposed surface" individual feature, for example, is the cylindrical feature shown under the RULE 1 explanation (page 22), or a specified "roundness" or "cylindricity" tolerance, etc. Specified "parallelism," which involves opposed surfaces, becomes a variety of interrelationship but is considered to retain the connotation of an individual feature and is subject to the confinement of form error within the perfect form at MMC envelope.

Interrelationship Feature. An "interrelationship" feature involves a relationship of two or more individual features and always requires specified geometric controls and a datum or datum system. "Perpendicularity" is the most predominant example of an interrelationship type feature. One exception to the need for a datum would be the use of "straightness' RFS or MMC applied on a $\emptyset$ basis to the part diameter size dimension. In this instance, an application of straightness (all longitudinal elements assumed collectively to establish a *size* feature consideration), would change from the category of "individual" feature to that of "interrelationship" feature because of the interrelationship of size and form which results in a "virtual condition" size exceeding the MMC perfect form envelope.

Due to the variety of applications in which a "feature" can be used, we cannot clearly define it without relating it to a specific drawing requirement. Just as it would be difficult to describe a form or location tolerance application without a drawing, we need pictorial representation and specification of the involved requirements to clearly "see" the "feature."

A drawing illustrates the "feature" relationship within the specification and clarifies the functional intent of the requirement.

See the drawing that follows as an example.

APPENDIX

EXAMPLES

Question   How many features are on this part?

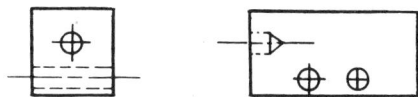

Answer   At least thirteen. However, they need not and cannot be identified until the drawing specification is completed.

Question   What are the features on the part shown below in its completed version?

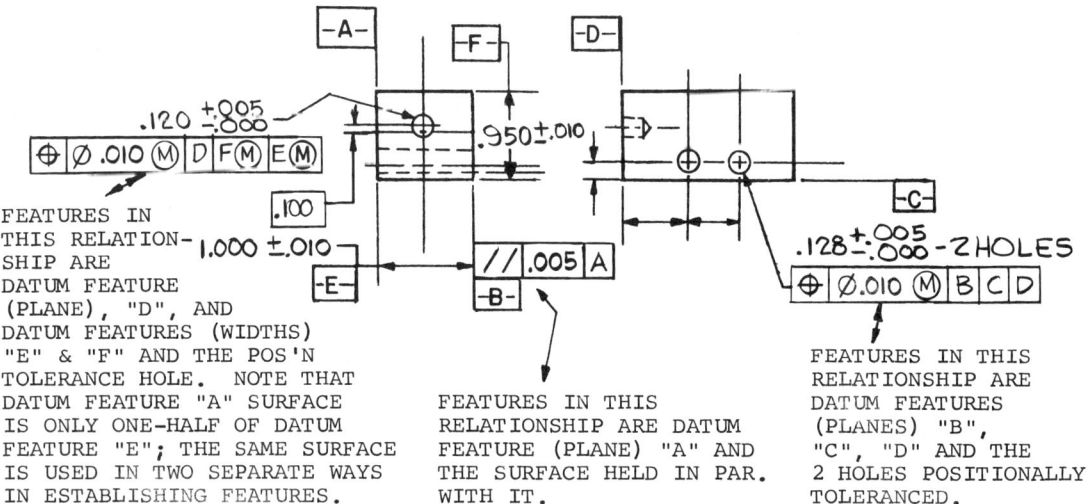

FEATURES IN THIS RELATIONSHIP ARE DATUM FEATURE (PLANE), "D", AND DATUM FEATURES (WIDTHS) "E" & "F" AND THE POS'N TOLERANCE HOLE. NOTE THAT DATUM FEATURE "A" SURFACE IS ONLY ONE-HALF OF DATUM FEATURE "E"; THE SAME SURFACE IS USED IN TWO SEPARATE WAYS IN ESTABLISHING FEATURES.

FEATURES IN THIS RELATIONSHIP ARE DATUM FEATURE (PLANE) "A" AND THE SURFACE HELD IN PAR. WITH IT.

FEATURES IN THIS RELATIONSHIP ARE DATUM FEATURES (PLANES) "B", "C", "D" AND THE 2 HOLES POSITIONALLY TOLERANCED.

POSITION TOLERANCE CONVERSION TO TOOL TOLERANCE

The question is often raised concerning relationship between a part dimensioned and toleranced with positional tolerance and the tooling necessary to produce the part. Positional tolerancing of cylindrical features normally derives cylindrical tolerance zones, whereas the tooling for the part can be based upon X and Y movement in its design or construction. The two geometrically opposed concepts are at times misunderstood as being incompatible with one another.

This point is often also used as an argument of resistance against the use of positional tolerancing using cylindrical tolerance zones. The uninformed fails to make the important distinction between the necessity (and normal practice) of defining the part *end-product* (reflecting its function) on the engineering drawing as the objective, rather than *methods* of manufacture of the part or its tooling.

Since, however, most hard tooling is produced in a shop provided with machinery based upon conventional X and Y axis movement, the methods of translating requirements between the product drawing and the tooling mode is important.

In the example on the following page, a sample part is shown using position tolerancing of three holes. Below the part representation is a typical box type drill jig design, which for purposes of our explanation, is the method used in producing the holes in this part during its manufacture.

Note that the X and Y nominal coordinates are readily transferable to the jig design from the product design. The nest surface and rails in the jig provide the datum location and origination of the jig coordinate dimensions to the drill bushing holes. The part clamping device in the jig must hold the part down against its primary datum A (for attitude), against the datum B rail (for orientation and location), and in contact with the third datum (implied—for location).

The matter of translating (converting) the product positional tolerance ($\varnothing$.007) to an X and Y value is illustrated below the jig drawing.

Imagine that in the evolutionary process from part drawing to part build, a tool order is written from the processing engineer to the tool engineer. This tool order requires the tool engineer to design a drill jig capable of producing 5,000 parts. Suppose as a hypothetical situation, that this tool order specifies, within a company standard or a contractual requirement, an "X" class tool to set the life, quality, and thus the percentage of product tolerance which can be used for this part.

The tool engineer simply converts as shown the total product design positional tolerance ($\varnothing$.007) to the equivalent total coordinate ± tolerance (.005) based upon standard conversion factors (.7 X .007) and derives .005. Reducing the .005 to 20% usable total tolerance for the "X" class jig derives .001 or ± .0005 coordinate tooling tolerance for each dimension in X and Y.

The lower illustration recaps the dimension and tolerance allocation in one plane for clarification.

This reasoning of conversion from positional tolerancing on the product drawing to equivalent coordinate ±tolerancing may be applied to other situations in the process from design to building of parts; e.g. building prototype parts without hard tooling, building temporary tooling without tool design, and inspecting prototype parts.

Note that the transition from positional tolerance to coordinate tolerance is simply done and illustrates the flexibility inherent in the use of this system.

APPENDIX

POSITION TOLERANCE CONVERSION TO TOOL TOLERANCE

AS DRAWN

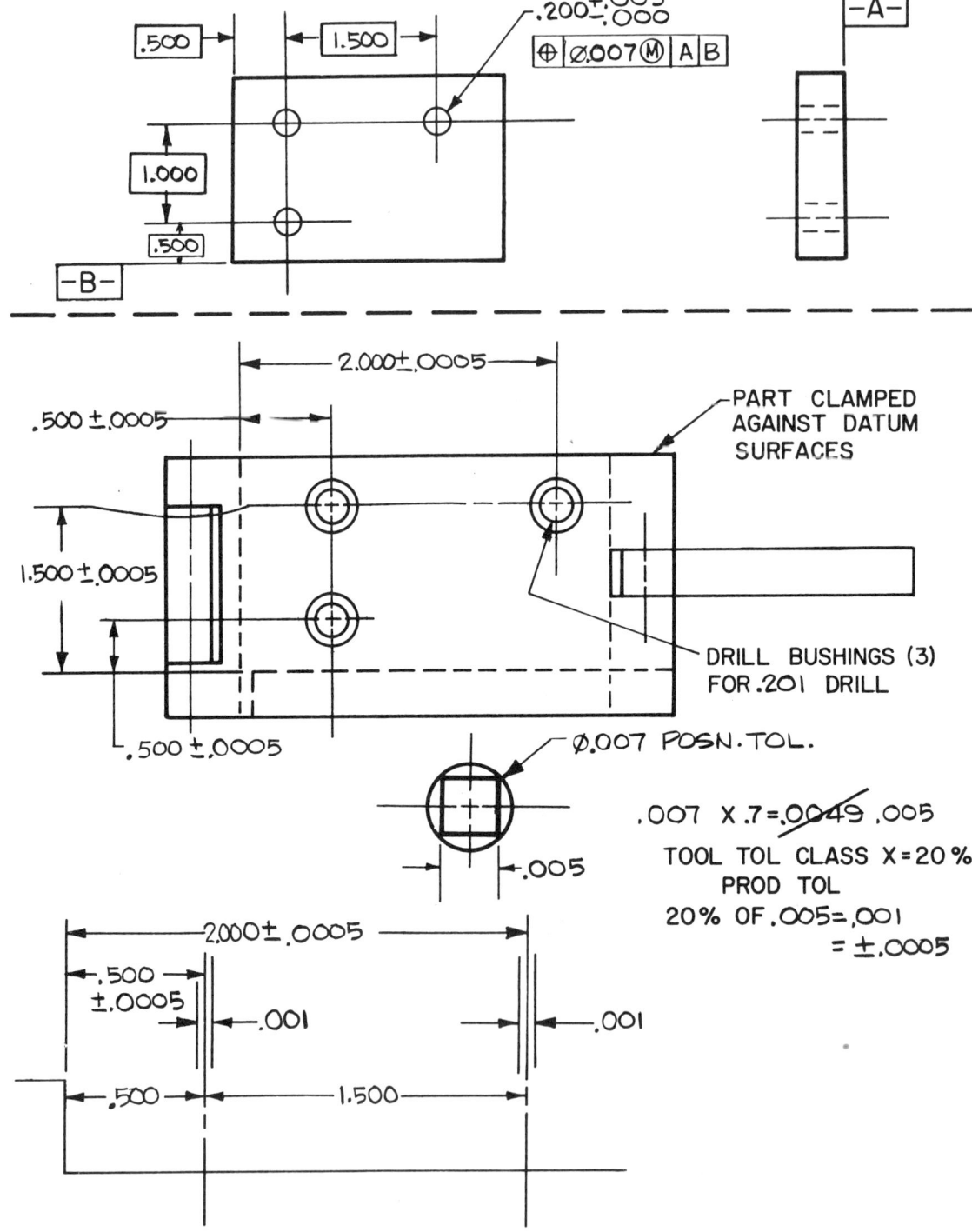

INDEX

Angularity, 58–61
   definition of and examples, 58–61
   features of at RFS and MMC, 60, 61
   of plane surfaces, 59
Appendix, 273
Average diameters, 287, 288

Basic (BSC), 13, 14
   definition of, 13

Circularity, see Roundness
Coaxial features, 79, 157, 264
Comparison between ANSI Y14.5, and ISO
   standards, 281
Combined feature control and datum symbol, 17
Concentricity, 257–265
   definition of and example, 258
   theory and checking, 259–262
Conclusion, 272
Conversion charts and formulas, 274–277
Cylindricity, 43–47
   definition of and example, 43
   checking, 45, 47
Coplanar surfaces, 74, 75

Datum, 14, 15, 48, 201–255
   accuracy, 221, 222
   application of, 15, 16, 220–244
   extended principles of 246–255
   identification symbol for, 14, 15
   implied 110, 111, 222, 223
   index, 202, 203
   reference to, 18, 19
   selection, 220, 221
   step and equalizing, 243–245
   specified 112, 113
   three-plane concept, 211
   types of 204–209
Datum targets, 205, 206, 212,213, 232-242
Diameter symbol, 18

Element control, 34, 35, 40, 54, 55, 80

Feature, 16, 291, 292
Feature control symbol and datum identifying
   symbol, 16, 17, 19
Flatness, 32, 33
   definition of and example, 32
Form tolerances, 21, 29
   angularity, 58-61

cylindricity, 43–47
individual features, 31-47
flatness, 32-33
parallelism, 62-65
perpendicularity, 50–57
   related features, 49
   profile, 66-74
   roundness, 40-42, 45–47
   runout, 76-95
   straightness, 33-39
Free-state variation, 286

Gaging; see each section
General rules, 22-25, 283-290
   exceptions to (rule 1), 23, 289
   optional use of, 24
   use of modifiers, 24
Geometric characteristics, 9, 21
Geometric dimensioning and tolerancing, 8
   function and use of, 8, 9
   reasons for using symbols, 10, 21
   symbols for (geometric characteristics)
      9, 21
   types of, 20
Glossary, 3–7

Introduction, 1
ISO practices, 2, 278–281

Least material condition (LMC), 283-285
Location tolerances 21, 97-200, 257-271
   position, 97-200
   concentricity, 257-265
   symmetry, 267-271

Maximum material condition (MMC), 11, 12, 26
   definition, 12
   applicability, 26
Metric system and ISO application, 2, 278-281

Normality, see Perpendicularity

Parallelism, 62–65
   definition of and example, 62
   with feature at MMC, datum A plane, 64
   with feature at MMC, datum at RFS, 65
   with feature at RFS, datum A plane, 64
   with feature at RFS, datum at RFS, 65
   with surface to datum plane, 63
Perfect form at MMC, 22, 23, 25, 34,

INDEX

Perpendicularity, 50-57
  with cylindrical feature at MMC,
    datum A plane, 53, 54
  with cylindrical feature at RFS,
    datum A cylinder RFS, 53
  definition of and example, 50
  with noncylindrical feature at MMC,
    datum A plane, 52
  with noncylindrical feature at RFS,
    datum A plane, 52
  radial, 54
  with surface to datum plane, 51, 55
  zero tolerance at MMC, 56, 57
Pickup points, 282
Position tolerances, 97-200
  coaxial features and, 155-167
  combination and unique applications of, 172-200
  composite position tolerancing, 116-121
  conversion formulas, charts for, 274-277
  conversion to tool tolerance, 292-294
  extended principles of, 171-200
  of mating parts, fixed fastener, 106-109, 185-191
  of mating parts, floating fastener, 103-105, 190
  where MMC is related to MMC datum
    feature, 122-125, 132-137
  where MMC is related to RFS datum
    feature, 126-129
  noncylindrical features and, 139-154
  projected tolerance zone and, 132-134, 175
  in relation to datum surfaces with
    larger tolerances between datum and
    pattern of features, 114, 115
  in relation to datum surfaces—radial
    hole pattern, 168, 169
  in relation to implied datum surfaces, 110, 111
  in relation to specified datum
    surfaces, 112, 113
  where RFS is related to RFS datum, 130-131
  separate patterns of holes, 173
  theory, 98-102
  zero tolerancing and, 176-184
Profile of a line, 66, 68, 72, 73
  definition of and example, 73
Profile of a surface, 66, 68, 69-72
  definition of and examples, 66, 69-71

Regardless of feature size (RFS), definition of,
  12, 26
  applicability, 26
Roundness, 40-42
  checking, 45, 46, 47
  of a cone, 42
  of a cylinder, 41
  definition of and example, 40, 41
  and fallacies of vee block and "miking," 45-47
  free-state variation in, 286
  of a sphere, 42
Runout, 76
  analysis, 92-95
  application, 76
  circular, definition of and examples, 77-83, 90
  definition of and example, 76
  and related surfaces, axis, 80-87
  surface, axis, 89-91
  types, circular and total, 77
  total, definition and examples 77, 84-95

Straightness, 33-39
  definition of and examples, 33, 34-39
Symbols, 9, 21
Symmetry, 267-271
  definition of, 269
  examples of, 269-271

Theoretical exact value (basic) 13
Tooling tolerance, 292-294
Tolerance zone shape, 27

Virtual condition, 28

Zero tolerancing
  perpendicularity, 56-57
  position and, 176-184